Advances in Applied Mechanics

Volume 19

ADVANCES IN
APPLIED MECHANICS

Edited by Chia-Shun Yih

DEPARTMENT OF APPLIED MECHANICS AND ENGINEERING SCIENCE
THE UNIVERSITY OF MICHIGAN
ANN ARBOR, MICHIGAN

VOLUME 19

1979

ACADEMIC PRESS
A Subsidiary of Harcourt Brace Jovanovich, Publishers

New York   London   Toronto   Sydney   San Francisco

ACADEMIC PRESS, INC.
111 Fifth Avenue, New York, New York 10003

United Kingdom Edition published by
ACADEMIC PRESS, INC. (LONDON) LTD.
24/28 Oval Road, London NW1 7DX

LIBRARY OF CONGRESS CATALOG CARD NUMBER: 48–8503

ISBN 0–12–002019–X

PRINTED IN THE UNITED STATES OF AMERICA

79 80 81 82   9 8 7 6 5 4 3 2 1

Contents

Stress Analysis for Fiber-Reinforced Materials

Allen C. Pipkin

Theory of Water-Wave Refraction

R. E. Meyer

Polymer Fluid Mechanics

J. D. Goddard

Contents

Relaminarization of Fluid Flows

R. Narasimha and K. R. Sreenivasan

List of Contributors

Numbers in parentheses indicate the pages on which the authors' contributions begin.

J. D. GODDARD, Department of Chemical Engineering, University of Southern California, University Park, Los Angeles, California 90007 (143)

R. E. MEYER, Department of Mathematics, University of Wisconsin, Madison, Wisconsin 53706 (53)

R. NARASIMHA, Indian Institute of Science, Bangalore 560 012, India (221)

ALLEN C. PIPKIN, Division of Applied Mathematics, Brown University, Providence, Rhode Island 02912 (1)

K. R. SREENIVASAN,* Indian Institute of Science, Bangalore 560 012, India (221)

* Present address: Department of Engineering and Applied Science, Yale University, New Haven, Connecticut 06520.

ADVANCES IN APPLIED MECHANICS, VOLUME 19

Stress Analysis for Fiber-Reinforced Materials

ALLEN C. PIPKIN

Division of Applied Mathematics
Brown University
Providence, Rhode Island

1

I. **Introduction**

Stress analysis for highly anisotropic materials is a matter of some current interest because of the growing practical importance of fiber-reinforced composites. Some of the most important problems concerning composites deal with the prediction of their properties from those of their constituents, or stress analysis on the microscopic scale, in which individual fibers are recognized. However, there is also a need for practical methods of stress analysis at the macroscopic level. For this purpose, several closely related continuum models of the mechanical behavior of fibrous materials have been developed in recent years, and this is the subject of the present review. These models treat the material as a homogeneous continuum with known properties.

Since fibrous materials include wood and fabrics, the subject is not entirely a new one. In fact, the theories that we discuss here originated in connection with a composite material that has been of practical importance for a long time, rubber reinforced with strong cords. However, the recent interest in the mechanics of fibrous materials has arisen from the development of high-strength composites such as glass or graphite fibers bonded together by a resin, or strong metallic fibers in a matrix of ductile metal. All of these materials have the property that the tensile modulus in the fiber direction is large in comparison to the shear moduli of the material. The theories that we discuss here take advantage of this property by treating the fibers as *inextensible*, either exactly or approximately.

The assumption of fiber inextensibility produces some important simplifications, but the theory is simpler still if the material is *incompressible* in bulk. The assumption of bulk incompressibility can be used if the bulk modulus of the material is large in comparison to its shear moduli, as in the case of rubber. Although this assumption is not so widely applicable as that of fiber inextensibility, we put some emphasis on the theory of *ideal* (inextensible, incompressible) composites, because this idealized theory leads toward an understanding of the more realistic *inextensible* theory, in which the assumption of bulk incompressibility is not used.

Much of the theory of ideal composites is independent of the nature of the shearing stress response of the material. Thus, although the first theory of the sort that we consider here concerned finite elastic shearing response (Green and Adkins, 1960), the main development of the theory came in the context of rigid-plastic response (Mulhern *et al.*, 1967). In the present review we assume elastic response when some such assumption is necessary, but many of the papers that we cite deal with plasticity theory.

With the two kinematic constraints of inextensibility and incompressibility, a further restriction to two-dimensional problems reduces much of the analysis to kinematics. For this reason, most published work concerns either plane or axisymmetric deformations. Here we explain only the plane-deformation theory, which is particularly simple. Papers concerning axisymmetric deformations are cited, but we do not explain the axisymmetric theory in detail.

Although the idealized theory was originally introduced in order to facilitate the solution of finite deformation problems, it was soon realized that there was much to learn from the even simpler small deformation theory (Rogers and Pipkin, 1971a), and in fact this is the appropriate theory for most problems involving high-strength composites. In the present review we begin, in Section II, with the theory of infinitesimal plane strain of ideal composites. The theory is also applicable to plane stress of fabrics. Displacement boundary value problems are trivial in this theory, but traction problems lead to a system of differential-difference equations of unusual form (England, 1972).

In the context of a more exact theory, the constraint conditions used in the idealized theory can be regarded as approximations. As a more exact theory within which the nature of these approximations can be examined, we can use the classical theory of transversely isotropic elastic materials, with the fiber direction as the preferred direction. In Section III we set up the basic equations of the plane theory for such materials, and examine the solutions of some simple problems.

Incompressibility produces no simplification when plane stress is considered. (By *plane stress*, we always mean the theory usually called *generalized plane stress*.) Consequently, the plane stress theory formulated by England *et al.* (1973) and by Morland (1973) concerns inextensible but compressible materials. In Section IV we derive the inextensible theory as an asymptotic approximation to elasticity theory, valid in the limit as a small extensibility parameter ε approaches zero. Boundary value problems in the inextensible theory are not so simple as in the idealized theory, since they involve the determination of one harmonic function, but they are much simpler than problems in the exact theory of elasticity.

Solutions in the ideal or inextensible theories often involve *singular fibers* in which the tensile stress is infinite (Pipkin and Rogers, 1971a). A single fiber can carry a finite force because there is no accompanying extension. In order to show how singular fibers are to be interpreted when it is admitted that real fibers are extensible, Everstine and Pipkin (1971) examined some solutions in classical anisotropic elasticity, and showed that singular fibers appear there as thin regions of high tensile stress (Section III,D). The same phenomenon had been pointed out much earlier by Green and Taylor (1939). Everstine and Pipkin (1973) also formulated a singular perturbation theory of stress concentration layers, in the context of a particular problem, and

Spencer (1974a) has discussed stress concentration layers more generally. In Section IV we derive the inextensible and stress concentration layer theories as outer and inner asymptotic approximations to the equations of elasticity theory.

In Section V we return to the original form of the idealized theory, the theory of finite plane deformations. The discussion we give here is relatively brief because an extended review of this subject has been given previously (Pipkin, 1973).

In every section of this review there is some discussion of the problem of fracture. In the idealized theory (Section II,E), the fiber and normal line passing through the tip of a crack carry finite forces, and the values of these forces at the crack tip are analogous to the stress intensity factors used in elastic fracture analysis. The characteristic elastic crack-tip stress singularity (Section III,E) is not seen in the idealized theory because it is buried inside the stress concentration layers represented by the singular lines through the crack tip (Section IV,G). Nevertheless, the elastic stress intensity factors can be expressed directly in terms of the tip forces obtained from the idealized theory (Section III,E).

Several topics of importance are not discussed in this survey. First, there is no discussion of thermal effects. Papers on this subject include those by Trapp (1971), Gurtin and Podio Guidugli (1973), and Alts (1976). Second, we consider only static equilibrium. Wave propagation in inextensible materials has been discussed by Weitsman (1972), Chen and Gurtin (1974), Chen and Nunziato (1975), and Scott and Hayes (1976). Beam impact problems involving plastic response have been studied by Spencer (1974b, 1976), Jones (1976), and Shaw and Spencer (1977). Parker (1975) has considered the dynamics of elastic beams, and Aifantis and Beskos (1976) have given some exact dynamical solutions for finite elastic deformations.

The book by Spencer (1972a) is a general introduction to the theories that we consider here. In addition to topics that we discuss, this book includes general three-dimensional constitutive equations for elastic and plastic materials reinforced with one or two families of fibers, and discusses a theory of materials reinforced with inextensible surfaces. In view of the existence of this book, in the present review we emphasize work performed since 1972, particularly work on methods of solution of boundary value problems.

II. Small, Plane Deformations of Ideal Composites

We consider materials composed of strong fibers, which may be embedded in a matrix of more compliant material. Although the fibers may in general be curved, for simplicity we restrict attention to cases in which the fibers are initially straight and parallel, lying along the x-direction of a system of Cartesian coordinates. Furthermore, we consider only plane deformations in the x, y plane.

The fibrous material is treated as a continuum with no distinction between particles lying in a physical fiber and particles in the matrix material or coating. Every line parallel to the x-direction is called a *fiber*.

Since we are considering plane strain, a line $y = $ constant actually represents a sheet of fibers of some length in the z-direction, but it is convenient to use the two-dimensional terminology. Similarly, when we mention a force we mean a force per unit of length in the z-direction.

In the present section we outline the theory of infinitesimal plane strain of incompressible materials reinforced with inextensible fibers. This theory turns out to have the same form as the theory of plane stress or plane strain of a material composed of two orthogonal families of inextensible fibers. Although the theory is easier to understand in the latter interpretation, we prefer to speak of only one family of fibers and to give a different name, *normal lines*, to the orthogonal family of lines. This will facilitate comparison with the finite deformation theory outlined in Section V.

A. KINEMATIC CONSTRAINTS

We suppose that the tensile compliance of the fibers is very small in comparison to the shear compliances of the material, and as an idealization of this property we postulate that the fibers are *inextensible*. This means that the displacement component u parallel to the fiber direction must be constant along each fiber, and thus a function of y alone.

Throughout Section II we restrict attention to materials that can be treated as *incompressible* in bulk. Since the volume dilation is $u_{,x} + v_{,y}$ in plane strain, and since $u_{,x} = 0$ according to the inextensibility postulate, incompressibility implies that $v_{,y} = 0$. Thus, it is not possible to squeeze two fibers closer together than they were originally. Then the displacement field has the form

$$u = u(y), \qquad v = v(x). \qquad (2.1)$$

Lines parallel to the y-direction are called *normal lines*. The conclusion that v is constant along normal lines means that normal lines act as if they, too, were inextensible fibers. Because of the complete duality between fibers and normal lines in the infinitesimal theory, the theory applies equally well to materials with fibers in the y-direction, or to materials with fibers in both directions. In the latter case, the theory also applies to plane stress.

For materials with only one family of fibers, plane stress is not as simple as the plane strain theory considered here, because the result $v = v(x)$ is not valid. We consider plane stress in Section IV. The plane stress theory has the same form as the theory of plane strain for a compressible material reinforced with inextensible fibers or with inextensible sheets (Westergaard, 1938; Spencer, 1972a).

The theory also applies to materials with inextensible fibers in the z-direction as well as the x-direction; the fibers in the z-direction merely enforce the assumed plane strain conditions. Spencer (1972b) has also applied the theory to cases in which the fibers lie in planes $y = $ constant but are at different angles to the x-axis in different layers $y = $ constant.

B. STRESS—REACTIONS—SINGULAR FIBERS

Because of results such as (2.1), much of the theory of ideal composites is independent of the nature of the shearing stress response of the material. However, for concreteness and simplicity we restrict attention to elastic response. Then for infinitesimal deformations, in view of (2.1) the shearing stress is

$$\sigma_{xy} = G[u'(y) + v'(x)], \tag{2.2}$$

where G is the shear modulus.

The stress components σ_{xx} and σ_{yy} are reactions to the constraints of fiber inextensibility and constant fiber spacing. These reactions are determined from the equilibrium equations, which with (2.2), are

$$\sigma_{xx,x} = -\sigma_{xy,y} = -Gu''(y) \tag{2.3}$$

and

$$\sigma_{yy,y} = -\sigma_{yx,x} = -Gv''(x). \tag{2.4}$$

Integration yields

$$\sigma_{xx}/G = -xu''(y) + f(y) \tag{2.5}$$

and

$$\sigma_{yy}/G = -yv''(x) + g(x), \tag{2.6}$$

where f and g are to be determined from traction boundary conditions.

Solutions of boundary value problems often involve jump discontinuities in σ_{xy} across a fiber or normal line. At a discontinuity in $u'(y)$, $u''(y)$ is a Dirac delta, and (2.5) shows that the tensile stress is infinite in the discontinuity line. This singularity represents a finite force carried by a geometric fiber of zero thickness (Pipkin and Rogers, 1971a). If the line $y = y_0$ is a singular fiber, the tensile stress in it has the form

$$\sigma_{xx} = F(x)\delta(y - y_0), \tag{2.7}$$

where $F(x)$ is the force carried by the singular fiber. The variation of $F(x)$ along the fiber is related to the stress discontinuity across it by

$$F'(x) = -\sigma_{xy}(x, y_0+) + \sigma_{xy}(x, y_0-). \tag{2.8}$$

When the singular fiber lies on the interior of the body, the shearing stress discontinuity is independent of x and then F varies linearly along the fiber. If the fiber ends at a boundary point where finite surface tractions are prescribed, the force F must vanish there; this is the condition most often used to determine the constant of integration when (2.8) is integrated.

When the fiber lies along an external boundary of the body, the prescribed shearing traction on the outside of the boundary fiber need not agree with the interior stress given by the stress–strain relation (2.2), and this is the most common reason for the appearance of singular fibers. For example, if the body is the region $y \geq 0$ and no traction is applied to the boundary $y = 0$, then (2.8) yields

$$F(x)/G = -xu'(0) - v(x) + v(0). \tag{2.9}$$

Evidently the shearing stress could be continuous across the boundary only if $u'(0) = 0$ and $v(x) = v(0)$, conditions that will usually be contradicted by other results.

When the present theory is regarded as an approximation to elasticity theory, it is found that singular fibers represent thin layers in which the tensile stress is unusually high, with a finite resultant equal to the force $F(x)$ (Section III).

Normal lines can also be singular in the present theory; we denote the force in a singular normal line by $N(y)$. However, in the inextensible theory (Section IV), normal lines cannot be singular, and in the finite deformation theory discussed in Section V, they are unusual. In that theory they usually dissolve into regions of finite width (Section V,G).

C. Boundary Value Problems

The basic equations of the idealized theory are hyperbolic, rather than elliptic, with the fibers and normal lines as characteristics. Since problems in solid mechanics are most naturally posed as boundary value problems, the hyperbolic character of the equations might cause some concern. Boundary values will be specified at more than one point on each characteristic, and often data will be prescribed along characteristics.

Nevertheless, problems that are well set in elasticity theory are also well set in the idealized theory, if some obvious restrictions are observed: u should not be prescribed at more than one point on each fiber, nor v at more than one point on each normal line.

The simplest case is that in which u is prescribed at one end of each fiber and the x-component of traction, T_x, is prescribed at the other end, while v and T_y are prescribed at opposite ends of each normal line. The displacement field (2.1) is then determined directly by the boundary values, and the stress field is given by (2.2), (2.5), and (2.6), with f and g in the latter expressions determined from the traction boundary conditions.

Problems are less trivial when the complete displacement field is not determined immediately by the boundary data. In such problems it turns out that the simplest cases are those in which data are prescribed along characteristics. As an example of this, and to illustrate a general method of solution, let us consider the deformation of a cantilever (Fig. 1) bounded by the lines $x = 0, L$ and $y = 0, H$ (Rogers and Pipkin, 1971a). The end $x = 0$ is bonded to a rigid support, so $u = 0$ everywhere and $v(0) = 0$. The sides $y = 0$ and $y = H$ are free from traction, and no normal traction is applied to the end $x = L$, but there is tangential traction on the end, with resultant F_y.

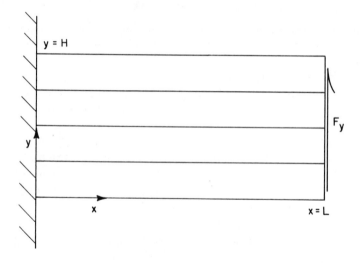

y = H

F_y

y

x

x = L

FIG. 1.   Cantilever under end load.

The resultant shearing stress on a line $x = $ constant is $GHv'(x)$, and this must be equal to the force F_y, so $v(x) = xF_y/GH$. Thus, the displacement field is determined without taking into account the details of the distribution of surface traction. This method based on stress resultants is of frequent use.

The displacement field has been found without any attention to the reactions σ_{xx} and σ_{yy}. These are now determined by direct integration. Beginning with the value zero on $y = 0$, σ_{yy} is found by integrating (2.4) along the normal line $x = $ constant, and is found to be zero everywhere on that line, including the point where the normal line crosses the boundary $y = H$. The fact that the correct value $\sigma_{yy} = 0$ is attained at this second point on the same characteristic is not accidental; the form of $v(x)$ determined by the method of stress resultants ensures compatibility of the two boundary values.

Beginning with $\sigma_{xx} = 0$ at the end $x = L$, σ_{xx} is found by integrating (2.3) or (2.8) along each fiber:

$$\sigma_{xx} = (F_y/H)(L - x)[\delta(y) - \delta(y - H)]. \tag{2.10}$$

There is no stress except in the two boundary fibers $y = 0$ and $y = H$, each of which carries a finite force. The condition of zero applied shearing traction on these lines was not used at all in determining the displacement field. The tangential traction could be ignored because these boundaries are characteristics; specifying data along characteristics leads to simpler problems, not more difficult ones.

The present example shows that Saint-Venant's principle is not valid for highly anisotropic materials, or at least that it must be used with great caution. The end condition $u = 0$ affects the displacement field arbitrarily far from the end. In Section III,D we show that in elasticity theory this propagation of disturbances along fibers to an infinite distance is replaced by a very slow decay.

D. Traction Boundary Value Problems

Problems with no prescribed boundary displacements are relatively difficult because it is not possible to deduce the displacement field directly from the kinematic constraints. However, England (1972) has developed a method based on stress resultants, which makes it possible to find the displacement field without detailed consideration of the partial differential equations of equilibrium.

For the present explanation, we restrict the boundary of the body to have no more than two points in common with any fiber or normal line, unless the boundary lies along a characteristic. Then the equation of the boundary can be written as

$$y = y_+(x) \text{ and } y = y_-(x), \text{ or } x = x_+(y) \text{ and } x = x_-(y), \tag{2.11}$$

with $y_+ \geq y_-$ and $x_+ \geq x_-$. The technique can be used for other boundary shapes, but the notation becomes more complicated.

The difference between the values of any function $f(x, y)$ at the two ends of a fiber or normal line are denoted by

$$\Delta f(y) = f(x_+(y), y) - f(x_-(y), y),$$
$$\Delta f(x) = f(x, y_+(x)) - f(x, y_-(x)). \tag{2.12}$$

In particular, $\Delta x(y)$ is the width of the body along the fiber y, and $\Delta y(x)$ is the length of the normal line x.

Let $F_x(y)$ be the x-component of the resultant surface traction on the part of the boundary above the fiber y, and let $F_y(x)$ be the y-component of the resultant surface traction on the part of the boundary to the right of the normal line x. These are equal to the resultant shearing forces on the lines $y = $ constant and $x = $ constant, respectively. Then on evaluating these shearing forces by integrating (2.2), we find that

$$u'(y)\Delta x(y) + \Delta v(y) = F_x(y)/G = f_x(y) \qquad \text{(say)}, \qquad (2.13a)$$

$$v'(x)\Delta y(x) + \Delta u(x) = F_y(x)/G = f_y(x). \qquad (2.13b)$$

The compact notation hides the complexity of this system of differential-difference equations. The degree of complexity in a given case can be determined graphically by using a *ricochet diagram*. The ricochet diagram is a path in the region occupied by the body, composed of straight segments in the x and y directions, and changing from one direction to the other when the path strikes the boundary of the body. An exception occurs when the path strikes a part of the boundary that lies along a fiber or normal line, so that the path is perpendicular to the boundary. In this case the path terminates.

The values of u on the horizontal parts of the path and the values of v on the verticals are coupled to one another by the difference-equation aspect of the system (2.13). If the path eventually closes upon itself or terminates at fiber or normal boundaries, then the system can be treated as a finite system of first-order differential equations for the simultaneous determination of the u and v values encountered on the path. For example, for a rectangular body like the cantilever considered in Section II,C, every ricochet path consists of a single horizontal or vertical segment, and every stress-resultant equation has only one unknown variable (plus two unknown constants, the boundary displacements); in this case each equation can be integrated immediately.

If the body has reflectional symmetry about a fiber or normal line, every ricochet path is a rectangle, and the system reduces to four differential equations involving the values of u and v at two places each. Splitting the solution into odd and even parts then reduces the system to second order, and the problem can be reduced to quadratures (England, 1972). This technique has been extended to crack problems by England and Rogers (1973) and to inclusion problems by England (1975).

The system can also be reduced to second order, and integrated, in some cases in which the ricochet path terminates after only one or two bounces, with both of its ends on fiber or normal boundaries. Thomas and England

(1974) consider a rectangular plate with a hole of arbitrary shape, Thomas (1974) considers a notched rectangular plate, and Sanchez-Moya and Pipkin (1977) consider bodies with one boundary along a fiber and a straight crack along a normal line.

When the ricochet path has an excessive number of segments or does not terminate at all, approximate solutions of (2.13) can be obtained by iteration. By integrating (2.13a) over the length of the normal line x and (2.13b) over the length of the fiber y, we obtain a system involving only the displacement differences Δu and Δv:

$$\Delta u(x) = \int U(x, y)[f_x(y) - \Delta v(y)](\Delta x(y))^{-1} \, dy,$$
$$\Delta v(y) = \int U(x, y)[f_y(x) - \Delta u(x)](\Delta y(x))^{-1} \, dx. \tag{2.14}$$

Here $U(x, y)$ is unity at points inside the body and zero outside it, and the limits of integration can now be taken as any fixed limits that include the whole body; introduction of the function U is a notational trick to avoid complicated limits of integration.

Pipkin and Sanchez-Moya (1974) have proved that a solution of the system (2.14) exists and iteration converges to a solution, provided that the prescribed tractions are in global translational and rotational equilibrium. The solution is nonunique to the extent of a rigid rotation, which is represented in terms of the displacement differences by $\Delta u = C \, \Delta y$ and $\Delta v = -C \, \Delta x$, where C is an arbitrary constant. Once the differences have been found, (2.13) can be integrated to find $u(y)$ and $v(x)$, and the stress field is then obtained in the usual way.

As a general rule, one should seek to determine the displacement differences $\Delta u(x)$ and $\Delta v(y)$ first, although not necessarily by using the integral equations. In the papers previously cited in which second-order equations were integrated, the equations were equivalent to equations for the displacement differences, and the rigid-rotation solution furnished a particular solution with which the order of the system could be lowered.

Lee (1976) has used Fourier series to solve traction problems for circular sheets, and has also given a numerical method of solution for sheets of arbitrary shape. This work concerns sheets reinforced with two families of fibers not necessarily orthogonal to one another.

E. Crack and Tearing Problems

Because of the practical importance of failure analysis, some of the first nontrivial applications of the idealized theory dealt with fracture problems. In the case of plane stress of a sheet composed of two orthogonal families of fibers, it is more appropriate to call the defect a *tear* rather than a crack, but we shall use the term *crack* in all cases.

In papers, by England and Rogers (1973), Thomas and England (1974), and Thomas (1974), complete stress and displacement fields are worked out for some traction boundary value problems involving cracked or notched bodies. These solutions show that the fiber and normal line passing through the tip of the crack are singular, carrying finite forces. England and Rogers (1973) advocated fracture criteria based on the fiber force F_0 at the crack tip, the force N_0 in the normal line there, and the value of the maximum shearing stress.

The tip forces F_0 and N_0 are analogous to the stress intensity factors used in elastic fracture mechanics (Sih *et al.*, 1965; Rice, 1968). The characteristic $1/r^{1/2}$ behavior of the elastic stress is not found when the idealized theory is used, because the tip singularity is buried inside the stress concentration layers represented by the singular fiber and normal line (Section IV,G). However, the elastic stress intensity factors can be expressed in terms of F_0 and N_0 if the material is such that the idealized theory is a valid approximation (Section III,E), and in particular the infinite shearing stress at the tip is characterized by these parameters.

The tip forces F_0 and N_0 can be used for the correlation of data in the same way that stress intensity factors are, without introducing the elastic stress intensity factors explicitly. The advantage of using the tip forces as the basic parameters is that they are relatively easy to determine, and in fact general formulas for the crack-tip fiber force in terms of the boundary loading are available for some special classes of boundary shapes.

1. *Tip Forces*

Let us consider a body whose outer boundary is of the general form (2.11), with a crack of fairly arbitrary shape (Fig. 2). We suppose that finite surface tractions are prescribed on the external boundary and on the crack. We place the origin of coordinates at the crack tip, and suppose for simplicity that the crack lies in the quadrant where x and y are nonpositive, with no fiber or normal line crossing the crack more than once. We use the notation u_0, u_+, and u_- for the values of u in the regions $y \geq 0$, $y \leq 0$ to the right of the crack, and $y \leq 0$ to the left of the crack, respectively. For v we use the notation v_0, v_+, and v_- in the regions $x \geq 0$, $x \leq 0$ above the crack, and $x \leq 0$ below the crack, respectively.

In general u' will be discontinuous across the line $y = 0$ through the tip, and there will be a finite force $F(x)$ in this line. This force must vanish at the boundaries $x = L_+$ and $x = -L_-$ if the prescribed tractions at these points

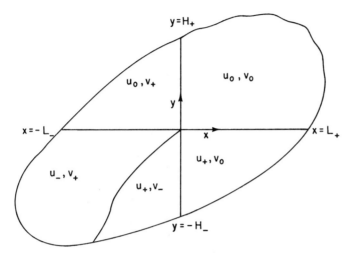

FIG. 2. Cracked sheet.

are finite. Hence, from (2.3) or (2.8), F is piecewise linear, with its maximum value F_0 at the crack tip:

$$F(x) = \pm G[u'_0(0) - u'_\pm(0)](L_\pm - |x|). \qquad (2.15)$$

Here $\pm$ is the sign of x. This gives two equal expressions for F_0 $[=F(0)]$ since F is continuous through $x = 0$. Similarly, the force $N(y)$ in the normal line $x = 0$ is

$$N(y) = \pm G[v'_0(0) - v'_\pm(0)](H_\pm - |y|), \qquad (2.16)$$

where $\pm$ is the sign of y, and $y = \pm H_\pm$ are the places where the normal line intersects the outer boundary.

Usually nonzero tractions are specified on the outer boundary and zero tractions on the crack faces, but this problem can be reduced by superposition to a problem with no crack, plus a residual problem in which there are tractions on the crack surfaces but not on the outer boundary. Sanchez-Moya and Pipkin (1977) considered the residual problem for some cases with a straight crack along the line $x = 0$, $y \leq 0$, loaded by arbitrary tractions. They found that for the problems considered, the tip force F_0 could be expressed in terms of the total opening force P and moment M (about the crack tip) of the prescribed normal tractions on one face of the crack. They found that if the boundary opposite to the one cut by the crack is a fiber $y = H_+$, then $\sigma_{xx} = 0$ in the region $y > 0$, except in the singular boundary fiber. Then F_0 can be deduced from elementary statics, by taking moments about the point $x = 0$, $y = H_+$:

$$F_0 = P + M/H_+. \tag{2.17}$$

A similar but more complicated result was found when the lower boundary is a straight line $y = -H_-$ but the upper boundary $y = y_+(x)$ is not necessarily straight. In this case

$$F_0 = P + M/D, \tag{2.18}$$

where D is a complicated mean depth of the uncracked part of the body, defined by an integral involving the body shape $y_+(x)$. Similar integrals have appeared in the crack and notch solutions previously cited.

Sanchez-Moya and Pipkin (1977) also considered the case of a body symmetrical about a fiber, with a symmetrical central crack or two symmetrically disposed edge cracks, symmetrically loaded. The symmetry ensures that $v = 0$, and it is then easy to show that $F_0 = P$ for an edge crack and $F_0 = P/2$ at each end of a central crack. This agrees with the results previously found by England and Rogers (1973) for bodies with two axes of symmetry.

2. Energy Release Rate

When a crack propagates, the total energy of the body and its loading system decreases, and the decrease per unit length of crack extension is the *energy release rate* $\mathscr{G}$. Sanchez-Moya and Pipkin (1977) calculated the energy release rate for straight cracks along the y-axis, by using an integral around the tip like that used in elasticity theory (Rice, 1968). Pipkin and Rogers (1978) observed that the rate of dissipation could be calculated more easily by considering the material passed over by the fiber and normal line through the tip, when the tip advances. The work done on this material is larger than the change of strain energy in it, and the discrepancy gives the energy release rate. This method makes it possible to compute the energy release rate for an infinitesimal extension of the crack in an arbitrary direction α (measured positive counterclockwise from the positive x-direction) that need not be related to the previous direction of the crack, α_0 say. The energy release rate was found to be

$$\mathscr{G} = (F_0^2/GL^*)(\sin \alpha)_+ + (N_0^2/GH^*)(\cos \alpha)_+, \tag{2.19}$$

where L^* and H^* are the harmonic mean distances defined by

$$2L^{*-1} = L_+^{-1} + L_-^{-1} \quad \text{and} \quad 2H^{*-1} = H_+^{-1} + H_-^{-1}. \tag{2.20}$$

The $+$ subscript on $\sin \alpha$ and $\cos \alpha$ means the positive part, i.e., f_+ is equal to f when f is positive and equal to zero when f is negative. Thus, with the crack oriented as in Fig. 2, the tip force F_0 contributes to the energy release rate if and only if the crack extension cuts the singular fiber carrying this force, and similarly N_0 contributes only if the normal line is cut. The direction of extension α_m that maximizes the energy release rate is given by

$$\tan \alpha_m = F_0^2 H^* / N_0^2 L^*, \tag{2.21}$$

and this direction is always in the quadrant $[0, \pi/2]$ in which the previous direction α_0 lies.

Since the energy release rate can also be expressed in terms of elastic stress intensity factors, the latter can be deduced from (2.19). We show how to do this in Section III,E.

3. *Crack Paths*

Many fracture criteria can be expressed in terms of a *fracture potential* $Q(\alpha; F_0, N_0, \alpha_0)$, which depends on the local state of stress at the crack tip through the parameters F_0, N_0, and α_0, and depends on the direction of incipient advance α. We suppose that Q increases as F_0 and N_0 do, until a critical value is reached for some direction α, and that the crack then propagates in the direction for which the critical value is first attained. This must be the direction α_m that maximizes the fracture potential. Let x_t and y_t be the coordinates of the tip. Then when the crack propagates, its trajectory is described by

$$dy_t/dx_t = \tan \alpha_m. \tag{2.22}$$

Pipkin and Rogers (1978) have pointed out that it is feasible to test a proposed fracture criterion by using it to calculate crack trajectories. They considered only quasistatic propagation, and calculated trajectories in a rectangular sheet under simple tension T on its ends $x = 0$ and $x = L$, with no traction on the boundaries $y = 0$ and $y = H$. For this problem it can be shown by elementary statics that

$$F_0 = T y_t + T y_t^2 / 2(H - y_t) \quad \text{and} \quad N_0 = T y_t^2 / 2(L - x_t), \tag{2.23}$$

for any crack of the general shape shown in Fig. 2.

One of the fracture criteria they considered is the maximum force criterion, in which the fracture progresses in the y-direction when F_0 reaches a critical value F_c, or in the x-direction when N_0 reaches a critical value N_c. This criterion is equivalent to the fracture potential

$$Q = \max[(F_0/F_c)(\sin \alpha)_+, (N_0/N_c)(\cos \alpha)_+], \tag{2.24}$$

with $Q = 1$ as the critical value. With this criterion, a crack started on the lower boundary either continues straight through the sheet on the same line

x = constant or, if started close enough to a corner, the crack eventually turns from the y- to the x-direction and the corner is torn off.

They also considered the maximum energy release rate criterion, for which $Q = \mathcal{G}$. By using (2.21) and (2.23) in (2.22), an equation for the crack path is obtained. Under this criterion, the paths are qualitatively like those for the maximum force criterion, but are smooth curves rather than straight or L-shaped.

F. Axisymmetric Problems

Several authors have considered deformations of tubes reinforced with helically wound fibers (see also Section V,H). Mulhern (1969) considered tubes with only one family of fibers, but most work concerns cases in which there are two families, symmetrically disposed. Smith and Spencer (1970) formulated the constitutive equations applicable to such cases (with plastic shearing response) and gave some solutions in which initially cylindrical surfaces remain cylindrical. Such solutions generally involve shearing in the axial direction, since the constraints prohibit pure radial expansion, and the boundary fibers then become singular because of the shearing stress discontinuity across the boundary. However, Spencer *et al.* (1974) have pointed out that infinitesimal pure homogeneous deformations are possible when the angle ϕ between the fibers and the axial direction satisfies $\tan^2\phi = 2$, and for this case the boundary fibers are not singular. This angle of winding is also exceptional for the pure bending solutions given by Spencer *et al.* (1975).

Troth (1976) has considered inflation, extension, and torsion of helically wound cylinders, using elasticity theory with extensible fibers, and has examined the limiting behavior of these solutions when the fiber extensibility grows small. Nicol (1978) has examined bending and flexure for cases in which the fibers are confined to discrete layers and are extensible, and has compared the limiting case of small extensibility with the results found by Spencer *et al.* (1975).

III. Plane Problems in Transversely Isotropic Elasticity

Possibly the most straightforward continuum model of a fiber-reinforced material is classical elasticity theory, with the material transversely isotropic about the fiber direction. The idealized theory that was discussed in Section II and the inextensible and stress concentration layer theories to be discussed in Section IV can be viewed as approximations to elasticity theory. In the present section we present the minimum amount of plane elasticity theory necessary to understand the nature of these approximations. We also record the basic results concerning crack-tip stress singularities, in a form that will facilitate comparison with the idealized and inextensible theories. The book by Green and Zerna (1954) is a standard reference on plane anisotropic elasticity, and the paper by Sih *et al.* (1965) gives the basic results concerning cracks in anisotropic bodies. Our notation differs from that of these authors.

After setting out the stress–strain relations for transversely isotropic materials, in Section III,B we consider the equation satisfied by Airy's stress function in plane problems. This equation contains two dimensionless parameters, the fiber extensibility ε and the normal compressibility c. We show that when ε and c are small, the singular fibers and normal lines encountered in the idealized theory are replaced by thin layers whose thicknesses are $O(\varepsilon)$ and $O(c)$, respectively. The result of the idealized theory that disturbances propagate along fibers and normal lines without attenuation is also modified. In elasticity theory, disturbances decay over distances of the order of $1/\varepsilon$ and $1/c$ along fibers and normal lines, respectively. We show this in Section III,D for some half-plane problems. At the same time, we illustrate the methods of approximation that will be discussed further in Section IV.

A. Stress–Strain Relations

We consider elastic materials that are transversely isotropic about the x_1 direction of a system of Cartesian coordinates x_1, x_2, x_3. The stress–strain relations for such a material have the form

$$\varepsilon_{11} = \sigma_{11}/E - v\sigma_{22}/E - v\sigma_{33}/E,$$

$$\varepsilon_{22} = -v\sigma_{11}/E + \sigma_{22}/E' - v'\sigma_{33}/E',$$

$$\varepsilon_{33} = -v\sigma_{11}/E - v'\sigma_{22}/E' + \sigma_{33}/E', \tag{3.1}$$

$$\varepsilon_{1\alpha} = \sigma_{1\alpha}/2G \ (\alpha = 2, 3),$$

$$\varepsilon_{23} = \sigma_{23}/2G',$$

where

$$G' = E'/2(1 + v'). \tag{3.2}$$

The strain energy density W has the form

$$W = (E/2)\varepsilon_{11}^2 + (K/2)(\varepsilon_{\alpha\alpha})^2 + 2G\varepsilon_{1\alpha}\varepsilon_{1\alpha} + G'e_{\alpha\beta}e_{\alpha\beta}, \tag{3.3}$$

where

$$e_{\alpha\beta} = \varepsilon_{\alpha\beta} - \tfrac{1}{2}\varepsilon_{\gamma\gamma}\delta_{\alpha\beta}. \tag{3.4}$$

Greek subscripts have the range 2, 3, and a repeated index implies summation over that range. In (3.3), K is defined by

$$1/K = 2(1 - v')/E' - 4v^2/E. \tag{3.5}$$

The strain energy is positive definite if and only if

$$E > 0, \qquad K > 0, \qquad G > 0, \qquad G' > 0. \tag{3.6}$$

In either plane stress or plane strain in the x, $y(x_1, x_2)$ plane, the relevant stress–strain relations can be written in the form

$$Gu_{,x} = A\sigma_{xx} - B\sigma_{yy'},$$

$$Gv_{,y} = -B\sigma_{xx} + C\sigma_{yy'}, \tag{3.7}$$

$$G(u_{,y} + v_{,x}) = \sigma_{xy}.$$

Relations of the same form are valid for orthotropic materials such as sheets with fibers in the x- and y-directions. For transversely isotropic materials, the coefficients in (3.7) are defined in terms of those in (3.1) in the case of plane stress by

$$A = G/E, \qquad B = vG/E, \qquad C = G/E', \tag{3.8}$$

and in plane strain by

$$A = (1 - v^2 E'/E)G/E, \qquad B = (1 + v')vG/E, \qquad C = (1 - v'^2)G/E'. \tag{3.9}$$

B. Plane Problems—Fundamental Parameters

For either plane stress or plane strain, the two-dimensional equilibrium equations imply that the stress can be represented in terms of Airy's stress function χ in the form

$$\sigma_{xx} = G\chi_{,yy}, \qquad \sigma_{xy} = -G\chi_{,xy}, \qquad \sigma_{yy} = G\chi_{,xx}. \tag{3.10}$$

Because the stress has been scaled with respect to the shear modulus G, χ has the dimensions of area.

When (3.10) is used in (3.7) and u and v are eliminated, an equation for χ is obtained, which can be put into the form

$$\left(\frac{\partial^2}{\partial x^2} + \varepsilon^2 \frac{\partial^2}{\partial y^2} \right) \left(c^2 \frac{\partial^2}{\partial x^2} + \frac{\partial^2}{\partial y^2} \right) \chi = 0. \tag{3.11}$$

Here $1/\varepsilon^2$ and c^2 are the larger and smaller roots λ of the equation

$$A\lambda^2 - (1 - 2B)\lambda + C = 0. \tag{3.12}$$

We wish to consider materials for which the tensile modulus E in the fiber direction is large in comparison to the moduli E', G, and G', and these latter moduli are of the same order of magnitude. Then for both plane stress and plane strain, A and B are small parameters, so the roots of (3.12) are approximately $1/A$ and C. It follows that ε is a small parameter approximately equal to $A^{1/2}$, or

$$\varepsilon \cong (G/E)^{1/2}. \tag{3.13}$$

We are also interested in cases in which the material is nearly incompressible. From the strain energy (3.3) we see that K, defined by (3.5), is a bulk modulus for deformations that do not involve fiber-stretching. With G/E small and G/E' of normal order, then G/K is small if v' is close to unity. It then follows from the third equation of (3.9) that in plane strain, c is a small parameter approximation equal to $C^{1/2}$, or

$$c \cong (G/K)^{1/2}. \tag{3.14}$$

But for plane stress, with C given by (3.8), the value of K or v is immaterial. In this case c is approximately

$$c \cong (G/E')^{1/2}, \tag{3.15}$$

and this is not small even if the material is exactly incompressible. Of course, for a material with strong fibers in both the x- and y-directions, E' is also large in comparison to G, so c is a small parameter even for plane stress.

The parameters ε and c are the same as those called ε_t and ε_c by Everstine and Pipkin (1971) and called $1/\alpha_1$ and α_2 by Green and Zerna (1954). They differ slightly from the parameters called ε and c by Spencer (1974a), becoming equal to Spencer's parameters asymptotically as $\varepsilon \to 0$. Values given by Green and Zerna for oak and spruce in plane stress are equivalent to $\varepsilon = 0.43$ and $\varepsilon = 0.24$, respectively. The former value is typical of hardwoods and the latter of softwoods, which are much more anisotropic. Although ε^2 is small in both cases, the value of ε itself is important, so even in the case of spruce we might expect errors of the order of 25% when using approximations based on the smallness of ε. Much smaller values of ε are obtained for high-strength composites. Spencer (1974a) uses data for carbon fibers in epoxy to compute $\varepsilon = 0.16$. The data of Folkes and Arridge (1975) for polystyrene fibers in a butadiene matrix suggest that ε is about 0.06 for this material.

For the materials just mentioned, c is of the order of unity except in the case of the styrene–butadiene composite. Small values of c should be expected for reinforced rubber, for which the ratio G/K is very small, and for fabrics, for which G/E' is as small as G/E.

Because of the importance of the parameters ε and c, it is useful to write the stress–strain relations (3.7) in a form that exhibits these parameters:

$$Gu_{,x} = a\varepsilon^2\sigma_{xx} - b\varepsilon^2\sigma_{yy'},$$
$$Gv_{,y} = -b\varepsilon^2\sigma_{xx} + ac^2\sigma_{yy}. \tag{3.16}$$

Here

$$a = (1 - 2b\varepsilon^2)/(1 + \varepsilon^2 c^2). \tag{3.17}$$

When ε is small, $a = 1$ approximately, and b is approximately v for plane stress and $v(1 + v')$ for plane strain.

C. GENERAL SOLUTION IN TERMS OF STRESS FUNCTIONS

The general solution of (3.11) has the form

$$\chi = \phi(x, cy) + \psi(\varepsilon x, y), \tag{3.18}$$

where ϕ and ψ are harmonic functions of their arguments. In terms of the two complex variables

$$z_1 = x + icy \quad \text{and} \quad z_2 = \varepsilon x + iy, \tag{3.19}$$

we may write

$$\phi(x, cy) = \operatorname{Re}\Phi(z_1) \quad \text{and} \quad \psi(\varepsilon x, y) = \operatorname{Re}\Psi(z_2), \tag{3.20}$$

where Φ and Ψ are analytic and Re denotes the real part. The expressions for the stress components (3.10) in terms of the complex potentials are

$$\sigma_{xx}/G = \operatorname{Re}(-c^2\Phi'' - \Psi''),$$
$$\sigma_{xy}/G = \operatorname{Re}(-ic\Phi'' - i\varepsilon\Psi''), \tag{3.21}$$
$$\sigma_{yy}/G = \operatorname{Re}(\Phi'' + \varepsilon^2\Psi'').$$

By using (3.21) in (3.16) and the third equation of (3.7) and integrating, and absorbing the integration constants into the potentials, we obtain the following expressions for the displacements:

$$u = -\frac{\partial}{\partial x}\operatorname{Re}[(1 - \varepsilon^2 k)\Psi + \varepsilon^2 k\Phi],$$
$$v = -\frac{\partial}{\partial y}\operatorname{Re}[(1 - \varepsilon^2 k)\Phi + \varepsilon^2 k\Psi]. \tag{3.22}$$

Here the parameter k is defined by

$$k = b + ac^2. \tag{3.23}$$

D. Half-Plane Problems

The stress function generally depends on the constants ε and c not only through the scaled coordinates εx and cy, but also parametrically. In the present section we consider some cases in which the parametric dependence can be determined explicitly, so that the behavior of the solution in the limit $\varepsilon \to 0$ can be determined.

We first consider a body in the half-space $y \geq 0$, with either u, v, σ_{xy}, or σ_{yy} prescribed and different from zero over the boundary $y = 0$, and with a complementary component prescribed as zero. We suppose that the data are consistent with the condition $\chi = 0$ at infinity, in order to avoid discussion of matters that are not relevant to our present purpose. For each of the eight types of problems mentioned, the stress function has the form

$$\chi = C_1 P(x, cy) + C_2 P(x, y/\varepsilon), \tag{3.24}$$

where $P(x_1, x_2)$ is a harmonic function determined by the inhomogeneous boundary condition, which is independent of the parameters ε and c. The constants C_1 and C_2 depend on the parameters.

1. *Normal Load Prescribed*

As a more specific class of examples, consider the boundary condition

$$\sigma_{yy}(x, 0)/G = n(x) \qquad (-\infty < x < \infty), \tag{3.25}$$

with either $\sigma_{xy} = 0$ or $u = 0$ over the whole boundary. For $\varepsilon = 0$, (3.21) shows that a stress function $\chi = P(x, cy)$ will satisfy (3.25) if

$$P_{/11}(x_1, 0) = n(x_1). \tag{3.26}$$

We take $P(x_1, x_2)$ to be defined by the conditions that it is harmonic in the half-plane, vanishes at infinity, and satisfies (3.26). Since these conditions do not involve the parameters ε and c, $P(x_1, x_2)$ is independent of these parameters.

The approximation $\chi \cong P(x, cy)$ will satisfy the second boundary condition $\sigma_{xy} = 0$ or $u = 0$ only accidentally if at all. To satisfy the second condition, it is necessary to add a contribution from the potential denoted ψ in (3.18), written as a function of x and y/ε since the scale of variation in the x-direction is set by the boundary condition (3.25). Although an iterative method of solution might be envisaged, it is soon apparent that the exact solution has the form (3.24), with ψ merely a differently scaled version of P. When this is realized, the coefficients C_1 and C_2 can be determined algebraically from the two boundary conditions. For example, if the second boundary condition is $\sigma_{xy} = 0$, then

$$C_1 = 1/(1 - \varepsilon c) \qquad \text{and} \qquad C_2 = -\varepsilon c C_1. \tag{3.27}$$

The second approximation that would have been obtained by the iterative method is

$$\chi \cong P(x, cy) - \varepsilon c P(x, y/\varepsilon). \tag{3.28}$$

2. Discussion

The approximation $\chi \cong P(x, cy)$ is exact when $\varepsilon = 0$, and this is the solution given by the inextensible theory (Section IV). In this theory, as in the idealized theory (Section II), the difference between $\sigma_{xy}(x, 0+)$ and the prescribed boundary value $\sigma_{xy} = 0$ produces a finite force in the singular fiber $y = 0$.

The solution given by the stress concentration layer theory, also discussed in Section IV, modifies the inextensible solution by adding a small but rapidly varying term that removes the shearing stress discontinuity at the boundary. In the present problem (3.28) is the stress concentration layer solution. From (3.22) we see that the new term yields a stretching displacement u that is $O(\varepsilon)$ rather than strictly zero, and the tensile stress σ_{xx} contributed by the new term is $-(c/\varepsilon)P_{,22}$, which is large of order $1/\varepsilon$ rather than infinite. Since $P(x_1, x_2)$ approaches zero as x_2 approaches infinity, we can regard it as approximately zero at $x_2 = L$, say, where L is a length scale set by the boundary condition (3.24). Then $P(x, y/\varepsilon)$ approaches zero within a distance of the order of $y = \varepsilon L$. Thus the $O(\varepsilon)$ stretching displacement and the $O(1/\varepsilon)$ tensile stress are mainly confined to a thin layer next to the boundary, whose thickness is $O(\varepsilon)$ rather than zero.

In the solution of the present problem according to the idealized theory, σ_{yy} is independent of y and thus the normal stress applied to the boundary penetrates along normal lines to infinity, with no attenuation. From the elastic solution we see that σ_{yy} is given in the lowest order of approximation by $P_{,11}(x, cy)$. Thus σ_{yy} actually decays to zero, but the characteristic length for this decay is L/c. This decay length becomes infinite as c approaches zero, and this is the limiting case considered in the idealized theory.

3. Layer Thicknesses and Penetration Distances

Solutions for the half-plane $x \geq 0$, with boundary conditions of uniform type over the whole boundary $x = 0$ (and one condition homogeneous) have the form

$$\chi = C_1 P(x/c, y) + C_2 P(\varepsilon x, y), \tag{3.29}$$

since the scale of variation in the y-direction is set by the inhomogeneous boundary condition. From solutions of this kind we find that stress concentration layers along normal lines have thicknesses that are $O(c)$ in comparison to the length scale L set by the boundary data, and prescribed normal data (σ_{xx} or u) penetrate along fibers to distances of the order of L/ε.

The complete duality between fibers and normal lines in the idealized theory is modified in elasticity theory. Every solution for a half-plane with fibers parallel to the boundary can be converted to a solution for a half-plane with fibers perpendicular to the boundary, merely by interchanging ε and c in the solution; *except* that in (3.22), $\varepsilon^2 k$ is not subject to this interchange.

The preceding results have the following implications concerning solutions obtained from the inextensible or idealized theories. Singular fibers represent thin layers whose thickness is $O(\varepsilon)$, in which the tensile stress is of the order of the fiber force given by the the inextensible theory, divided by the layer thickness. Disturbances propagate along fibers to a distance of order $1/\varepsilon$. Similarly, the thickness of a singular normal line is $O(c)$, the tensile stress in it is $O(1/c)$, and disturbances propagate along normal lines to distances that are $O(1/c)$.

We did not show explicitly that the tensile stress is of the order of $F(x)/L\varepsilon$, although it follows from the results given here. An explicit example in Section IV,F illustrates this. Particular examples illustrating stress penetration and boundary stress concentration layers have been given by Green and Taylor (1939) and Everstine and Pipkin (1971). The $O(1/\varepsilon)$ penetration distance has been demonstrated as a general result for slender bodies loaded on an end, in studies of Saint-Venant's principle by Horgan (1972a,b, 1974) and Choi and Horgan (1977, 1978). These results have the experimental implication that end effects are important at a considerable distance from the end of a sample, as observed by many experimentalists including Folkes and Arridge (1975).

E. Crack-Tip Stress Singularity

In Section IV,G we show how to connect the results concerning fracture, obtained from the idealized theory in Section II,E, with the results of elastic fracture analysis. The required results from elasticity theory are recorded in the present section. These results are in principle contained in the paper by Sih *et al.* (1965); we specialize their results.

The crack geometry is as in Section II,E, but since here we consider only the region infinitesimally close to the tip, we may regard the crack as straight. Let **t** be a unit vector tangential to the crack and directed toward its tip, and let **n** be normal to the crack:

$$\mathbf{t} = \mathbf{i}\cos\alpha + \mathbf{j}\sin\alpha, \qquad n = -\mathbf{i}\sin\alpha + \mathbf{j}\cos\alpha. \qquad (3.30)$$

The crack direction α is in the interval 0 to $\pi/2$. In a coordinate system with **t** and **n** as base vectors, the components of stress that are of primary interest are

$$\sigma_{nn}/G = (\mathbf{t} \cdot \nabla)^2 \chi \qquad \text{and} \qquad \sigma_{tn}/G = -(\mathbf{t} \cdot \nabla)(\mathbf{n} \cdot \nabla)\chi. \tag{3.31}$$

These components must vanish on the crack surfaces.

The physical radius r and angle θ are defined by $x + iy = r\exp(i\theta)$. However, the stress function will be expressed in terms of the complex variables z_1 and z_2 defined in (3.19). On the line ahead of the crack, tangential to it, $z_1 = rt_1$ and $z_2 = rt_2$, where

$$t_1 = \mathbf{t} \cdot \nabla z_1 = \cos\alpha + ic\sin\alpha,$$
$$t_2 = \mathbf{t} \cdot \nabla z_2 = \varepsilon\cos\alpha + i\sin\alpha. \tag{3.32}$$

On the crack surfaces,

$$z_1 = rt_1 e^{\pm i\pi} \qquad \text{and} \qquad z_2 = rt_2 e^{\pm i\pi}. \tag{3.33}$$

We consider only the terms of leading order, $O(1/r^{1/2})$, in the stress, and thus we consider stress functions that are linear combinations of $z_1^{3/2}$ and $z_2^{3/2}$. Two independent linear combinations satisfy the conditions of zero traction on the crack surfaces. We choose to use the two defined by

$$\chi_v = \tfrac{4}{3}\operatorname{Re}\big[t_1(z_1/t_1)^{3/2} - ct_2(z_2/t_2)^{3/2}\big] \tag{3.34}$$

and

$$\chi_u = \tfrac{4}{3}\operatorname{Re}\big[i\varepsilon t_1(z_1/t_1)^{3/2} - it_2(z_2/t_2)^{3/2}\big]. \tag{3.35}$$

Neither of these has the dimensions of a stress function. The actual stress function will be given near the tip by

$$\chi = K_u\chi_u + K_v\chi_v, \tag{3.36}$$

where K_u and K_v have dimensions of the square root of length.

We are particularly interested in the energy release rate in crack advance. This can be calculated from the stress on the line ahead of the crack and the crack displacement discontinuity. The components of the displacement discontinuity are

$$\Delta u_n = 2\mathbf{n} \cdot \mathbf{u} \qquad \text{and} \qquad \Delta u_t = 2\mathbf{t} \cdot \mathbf{u} \tag{3.37}$$

where **u** is the displacement on the surface $\theta = \alpha + \pi$. The energy release rate can be evaluated by carrying out an integral over the tip (Rice, 1968), with the result that

$$\mathscr{G} = (\pi/4)(\sigma_{nn}^* \Delta u_n^* + \sigma_{nt}^* \Delta u_t^*), \tag{3.38}$$

where the starred quantities are the algebraic coefficients of $1/r^{1/2}$ and $r^{1/2}$ in the expressions for the stresses and displacements, respectively.

The values of various quantities of interest, as computed from χ_u and χ_v, are presented in Table I. In the table, LA means values on the line ahead of the crack, $\theta = \alpha$, and Cr means the crack surface $\theta = \alpha + \pi$. The following abbreviations are used:

$$I_1 = 1 - \varepsilon c, \qquad I_2 = 1 - 2\varepsilon^2 k, \qquad J = \varepsilon c - \varepsilon^2 k(1 + \varepsilon c). \qquad (3.39)$$

TABLE I

STRESS AND DISPLACEMENT FACTORS
FOR TIP FIELD

Stress function		χ_u	χ_v
σ_{nn}^*/GI_1	(LA)	$\sin \alpha$	$\cos \alpha$
σ_{tn}^*/GI_1	(LA)	$-\cos \alpha$	$\sin \alpha$
u^*/J	(LA)	0	2
v^*/J	(LA)	2	0
u^*/I_2	(Cr)	-2ε	0
v^*/I_2	(Cr)	0	$2c$
$\Delta u_n^*/I_2$		$4\varepsilon \sin \alpha$	$4c \cos \alpha$
$\Delta u_t^*/I_2$		$-4\varepsilon \cos \alpha$	$4c \sin \alpha$
$\mathscr{G}/\pi GI_1 I_2$		ε	c

For the potential χ_u, the displacement on the crack surface is in the x-direction, and for χ_v it is in the y-direction. In either case, there is a combination of sliding and opening modes, depending on the direction of the crack.

For the general stress function (3.36), the energy release rate is easily computed from (3.38) and the table entries. It is found that the stress from χ_u does no work on the displacement from χ_v, and vice versa, so the energy release rate contains no cross term $K_u K_v$:

$$\mathscr{G} = \pi G(\varepsilon K_u^2 + c K_v^2)I_1 I_2. \qquad (3.40)$$

The technique used in computing the energy release rate here requires the crack to continue in the same direction, so in comparing (3.40) with the result (2.19) given by the idealized theory, α in (2.19) is restricted to be the present direction of the crack. Then the positive-part notation in (2.19) is irrelevant. We note that I_1 and I_2 approach unity when ε approaches zero. Then for (3.40) to agree with (2.19) to lowest order in ε and c, it is sufficient that

$$K_u^2 = F_0^2 \sin \alpha / \pi G^2 \varepsilon L^* \qquad (3.41)$$

and

$$K_v^2 = N_0^2 \cos \alpha / \pi G^2 cH^*. \qquad (3.42)$$

Although it does not follow from comparison of (2.19) and (3.40) that the two separate conditions (3.41) and (3.42) are necessary, this does follow from a more detailed comparison of the elastic displacement field with the displacement field to be determined from stress concentration layer theory in Section IV,G.

When ε is small but c is not, K_u is correctly given to lowest order in ε by (3.41), but the value of K_v given by (3.42) should not be expected to be accurate. The part of the tip field represented by χ_v should be found by using the inextensible theory, rather than the idealized theory, as Sanchez-Moya and Pipkin (1978) have done for the particular crack orientation $\alpha = \pi/2$.

When ε and c are both small, K_u and K_v are both known in terms of the tip forces F_0 and N_0, which are easily determined from the idealized theory, and thus the elastic tip field can be characterized completely in terms of parameters determined from a much simpler theory.

IV. Asymptotic Approximations to Elasticity Theory

For elastic materials for which the parameter ε is small, boundary value problems can be solved by ordinary or singular perturbation techniques. When all unknowns are expanded in powers of ε, the equations governing the $O(1)$ terms are equivalent to the theory of inextensible materials given by England *et al.* (1973) and Morland (1973). In the inextensible theory the displacement component u is a function of y alone, as in the idealized theory, but v is a harmonic function of x and cy. Thus, the solution of problems in the inextensible theory entails the solution of Laplace's equation, but not the much more complicated simultaneous determination of two harmonic functions that is necessary in elasticity theory.

Singular fibers, but not singular normal lines appear in the solutions of problems in the inextensible theory. As we have seen in Section III,D,2, a singular fiber actually represents a thin region of high tensile stress. Singular perturbation techniques are required in order to determine the stress distribution in the stress concentration layer (Section IV,D–G). In this theory it is again necessary to determine only one harmonic function, and in fact it is possible to write down general solutions covering most cases that are likely to be encountered.

A. The Inextensible Theory

By setting $\varepsilon = 0$ in the elastic stress–strain relations (3.16) we obtain

$$u_{,x} = 0 \quad \text{and} \quad \sigma_{yy} = (G/c^2)v_{,y}. \tag{4.1}$$

The shearing stress–strain relation (3.7c) is not altered. The condition $u_{,x} = 0$ is the fiber inextensibility condition, which yields the result that $u = u(y)$. The tensile stress σ_{xx} is the reaction to this constraint, and as such it no longer appears in the stress–strain relations. Instead, it is found by integrating the x-component of the equilibrium equation, as in Section II,B:

$$\sigma_{xx}/G = -xu''(y) - v_{,y}(x, y) + f(y). \tag{4.2}$$

By using the third equation of (3.7) and the second part of (4.1) in the y-component of the equilibrium equation, we obtain

$$v_{,xx} + c^{-2}v_{,yy} = 0. \tag{4.3}$$

In Section III we found that v is exactly a sum of two harmonic functions, one harmonic in x and cy and the other harmonic in εx and y. In the present approximation, the second function is absent. The form of the stress function implied by the present results is

$$\chi = \phi(x, cy) + g(y) - xu(y), \tag{4.4}$$

where f in (4.2) is related to g by $f = g''$. The displacement v is given in terms of ϕ or the complex potential Φ [see (3.20)] by

$$v = -\phi_{,y} = -c\phi_{,2}(x, cy) = -\operatorname{Re} ic\Phi'(z), \tag{4.5}$$

where $z = x + icy$ was denoted by z_1 in Section III.

B. Boundary Conditions in the Inextensible Theory

Except on boundaries that lie along a characteristic $y = \text{constant}$, the components of surface traction T_x and T_y are related to the interior stress in the usual way:

$$T_x/G = (-xu'' - v_{,y} + f)n_x + (u' + v_{,x})n_y, \tag{4.6a}$$

$$T_y/G = (u' + v_{,x})n_x + c^{-2}v_{,y}n_y. \tag{4.6b}$$

Here n_x and n_y are the components of the unit outward normal vector.

As in the idealized theory, the pure displacement boundary value problem is not well-posed because $u(y)$ should not be prescribed at more than one point on each fiber. However, v can be prescribed over the whole boundary, and u at one end of each fiber. Then u is determined trivially, and v is the

solution of a Dirichlet problem. With T_x prescribed at the other end of each
fiber, (4.6a) determines the function f in (4.2).

Morland (1973) discusses a number of boundary conditions for plane
problems, and gives a uniqueness proof for these conditions in a later paper
(Morland, 1975). Here we consider only one further case, the pure traction
boundary value problem. The procedure for obtaining boundary conditions
on v in this problem is applicable in other cases as well.

The stress–resultant relation (2.13a) remains valid in the inextensible
theory, although (2.13b) does not, since v now depends on y as well as x.
By eliminating u' from (2.13a) and (4.6b), we obtain

$$\left(v_{,x} - \frac{\Delta v}{\Delta x}\right)n_x + c^{-2}v_{,y}n_y = T_y/G - (f_x/\Delta x)n_x. \tag{4.7}$$

Here $f_x = F_x/G$, and F_x is the resultant traction explained in Section II,D.
The condition (4.7) takes a simpler appearance when v is expressed in
terms of a complex potential as in (4.5). To obtain this simpler form, we
first note that if the boundary is described parametrically in terms of arc
length s by $x = x(s)$ and $y = y(s)$, then $n_x = y'(s)$ and $n_y = -x'(s)$. Then with
$z = x + icy$, $z'(s)$ is $-n_y + icn_x$. By using these relations in (4.7), with (4.5),
we obtain

$$(d/ds)\text{Re}\,\Phi' + (\Delta v/\Delta x)n_x = (f_x/\Delta x)n_x - T_y/G. \tag{4.8}$$

Thus, the boundary condition involves the tangential derivative of the
harmonic function of x and cy that is conjugate to v. This is the same as the
normal derivative of v when $c = 1$, but not in general. Of couse, the derivative
involved in (4.7) is purely a normal derivative on any boundary $x = $ constant
or $y = $ constant.

The boundary condition (4.7) or (4.8) is not of a standard type except on
boundaries along fibers, where $n_x = 0$, because otherwise the boundary
condition involves the values of v at two different boundary points, through
the difference Δv. It can be presumed that standard existence and uniqueness
results for traction problems will be valid since the original problem can be
posed as an energy-minimization problem; in this sense the problem is
simpler before it is reduced to a problem involving v alone. The uniqueness
theorems for constrained materials with a positive definite strain energy,
given by Pipkin (1976), are in this spirit. However, Morland (1975) has erased
all doubt by proving existence directly.

Hayes and Horgan (1974) considered the three-dimensional displacement
boundary value problem for inextensible materials, and gave conditions
that are both necessary and sufficient for uniqueness. They later did the
same for a more general class of mixed boundary value problems, restricted
to be such that singular fibers cannot appear (Hayes and Horgan, 1975).

Singular fibers appear most commonly when part of the boundary of the body lies along a fiber. The tangential boundary condition (4.6a) need not be satisfied on a bounding fiber, and this simplifies matters. Since only the resultant tangential traction on a boundary fiber is used in determining v [through the stress-resultant (2.13a)], the shearing stress will generally be discontinuous across the boundary and there will be a finite force $F(x)$ in the boundary fiber, as in the idealized theory. The relation (2.8) remains valid. In particular, when the boundary $y = 0$ is a traction-free boundary with the body above it, the force in the boundary fiber is

$$F(x)/G = f - u'(0)x - v(x, 0)$$

$$= f - u'(0)x + \operatorname{Re} ic\Phi'(x),$$

where f is a constant.

C. SOME SOLVED PROBLEMS

Because only one harmonic function needs to be found in order to solve a problem in the plane theory of inextensible materials, all of the standard techniques for solving Laplace's equation are useful, even such prosaic methods as separation of variables.

England *et al.* (1973) discuss the solution in terms of complex potentials in some detail. They give the point force and point moment solutions for a body occupying the whole plane, and discuss the behavior of solutions at infinity in the whole plane and in semi-infinite regions. They give the solutions for a point force inside a half-plane and for edge tractions on a half-plane, and for tractions on the edges of an infinite strip bounded by fibers.

In addition to half-plane problems, Morland (1973) considers rectangular plates under various boundary conditions and computes the solutions numerically for various values of c. These results are of interest in showing how solutions from the inextensible theory approach solutions from the idealized theory as c approaches zero. We note that these solutions could equally well have been obtained by separation of variables. Morland also used complex variable methods to solve the problem of an elliptical hole in an infinite plate.

Spencer (1974a), in connection with stress concentration layer theory, gives the inextensible solutions for a finite crack parallel to the fibers in simple shear and a finite crack perpendicular to the fibers with a prescribed opening displacement. Spencer (1977) also gives a variety of interesting examples in a review article on the subject.

In all of these solutions, the boundary conditions are such that the two-point condition (4.7) need not be used. No published solution involves the two-point condition in a nontrivial way.

D. Stress Concentration Layers

In obtaining the stress–strain relations (4.1) of the inextensible theory from those of elasticity theory, (3.16), we implicitly treated all displacement and stress components and their derivatives as $O(1)$ with respect to ε. This procedure leads to solutions involving singular fibers. From elasticity theory (Section III,D,2) we realize that singular fibers represent layers whose thickness is $O(\varepsilon)$, in which the tensile stress is $O(1/\varepsilon)$. This high tensile stress causes a stretching $u_{,x}$ that is $O(\varepsilon)$ according to (3.16). These order estimates are the basis for the approximations used in determining the stress distribution in the stress concentration layer.

If a singular fiber lies along the line $y = y_0$, it is convenient to move the origin of y to that line, so that $y_0 = 0$. Then the rapid variation of σ_{xy} in the y-direction, represented by a discontinuity when the inextensible theory is used, can be described by taking σ_{xy} to be a function of the stretched coordinate

$$Y = y/\varepsilon. \tag{4.10}$$

Similarly, u and σ_{xx} are to be regarded as functions of Y. However, the well-behaved values of v and σ_{yy} given by the inextensible theory are still treated as functions of y; these functions are in effect constant through the thickness of a stress concentration layer.

If the layer lies on the interior of the body, it is convenient to add a rigid displacement to the inextensible solution $u(y)$ so that $u(0)$ is zero. Then the stretching displacement within the layer can be written as

$$u = \varepsilon U(X, Y). \tag{4.11}$$

Here $X = x$. If the layer lies along an external boundary of the body it is more convenient to treat the stretching displacement as an addition to the inextensible solution. We denote the latter by $u_0(y)$ now to distinguish it from u:

$$u = u_0(y) + \varepsilon U(X, Y). \tag{4.12}$$

We mean (4.11) or (4.12) to represent u only to first order in ε, rather than exactly. By using either of these in the stress–strain relation (3.16), we find that to leading order in ε,

$$\sigma_{xx}/G = (1/\varepsilon)U_{,X}. \tag{4.13}$$

This approximation neglects $O(1)$ terms; it represents only the part of the tensile stress that becomes infinite when ε approaches zero.

By using (4.13) in (3.16b) we find that to lowest order, σ_{yy} and v are still related as in the second equation of (4.1), the relation valid for the inextensible theory.

From the third equation of (3.7), the shearing stress is

$$\sigma_{xy}/G = u'_0 + U_{,Y} + v_{,x}. \tag{4.14}$$

Here we have used (4.12) and (4.10). The $O(\varepsilon)$ stretching displacement produces a shearing stress that is $O(1)$ because of the rapid variation of U in the y-direction.

When (4.13) and (4.14) are used in the x-component of the equilibrium equation, the $O(1/\varepsilon)$ terms yield

$$U_{,XX} + U_{,YY} = 0. \tag{4.15}$$

Thus U is harmonic in X and Y. In complex notation,

$$U(X, Y) = \operatorname{Re} W(Z), \tag{4.16}$$

where

$$Z = X + iY = x + iy/\varepsilon = z_2/\varepsilon, \tag{4.17}$$

z_2 being the variable (3.19) used in the exact elastic solution.

E. Boundary Conditions for Stress Concentration Layers

Only a limited number of types of boundary conditions on U need to be considered, if the conditions in the exact problem are compatible with the inextensible theory. We restrict attention to such conditions, since otherwise the order estimates we have used are not valid.

The domain of X and Y has a highly restricted form. First consider the ends of a layer, lying on boundaries of the form $x = x(y)$. In terms of the stretched variable Y, the boundary is $X = x(\varepsilon Y)$. Then for every finite value of Y, in the limit $\varepsilon \to 0$ the boundary is $X = x(0) = X_0$ (say). Boundaries at the ends of a layer always become straight vertical lines when viewed in terms of X and Y.

The domain of Y is always either infinite or semi-infinite. For an edge layer with the body in the region $y \geq 0$, at any fixed positive value of y, however small, y/ε approaches infinity as $\varepsilon \to 0$. Thus in the stretched coordinate, the edge layer occupies the region $0 \leq Y < \infty$. Similarly, for interior layers the domain of Y is $-\infty < Y < \infty$.

The types of conditions that U may satisfy are also highly restricted. First consider an end $X = X_0$ where u is specified. The inextensible solution will already satisfy that condition, so if the stretching displacement is added to the inextensible solution as in (4.12), the condition on U is

$$U(X_0, Y) = 0. \tag{4.18}$$

When u is expressed in the form (4.11) that is convenient for interior layers, we have introduced the convention that $u_0(0) = 0$. Then the prescribed boundary displacement has the form $u = u_0'(0)\varepsilon Y + O(\varepsilon^2)$, so U must satisfy the condition

$$U(X_0, Y) = u_0'(0)Y. \tag{4.19}$$

At an end where finite surface tractions are prescribed, the $O(1/\varepsilon)$ term in σ_{xx} must vanish in order to avoid becoming infinite:

$$U_{,X}(X_0, Y) = 0. \tag{4.20}$$

The surface tractions will then be in error by $O(1)$. The condition (4.20) is analogous to the condition in the inextensible theory that the finite force in a singular fiber must vanish at a boundary where finite surface tractions are prescribed. This is necessary but not sufficient to satisfy the traction boundary conditions.

For an edge layer with the body in the region $y \geq 0$, the shearing stress (4.14) is required to approach that given by the inextensible theory, for every fixed positive value of y, as $\varepsilon \to 0$. This condition is satisfied if

$$U_{,Y}(X, \infty) = 0. \tag{4.21}$$

Similarly, for interior layers the shearing stress matching conditions yield

$$U_{,Y}(X, \pm\infty) = u_0'(0\pm). \tag{4.22}$$

For an edge layer, the condition $\sigma_{xy} = \sigma(X)$ on $Y = 0$ finally yields a boundary condition with a function of arbitrary form in it. With (4.14), we find that $U_{,Y}$ must account for the difference between the applied shearing stress $\sigma(X)$ and that given by the inextensible theory:

$$U_{,Y}(X, 0) = \sigma(X)/G - u_0'(0) - v_{,x}(X, 0). \tag{4.23}$$

From (2.8), the stress difference here can be expressed in terms of the fiber force $F(X)$ given by the inextensible theory. Then (4.23) takes the form

$$U_{,Y}(X, 0) = F'(X)/G. \tag{4.24}$$

The right-hand member is replaced by $-F'/G$ if the body lies below the bounding fiber, rather than above it.

F. EDGE LAYERS

Because the variety of possible boundary conditions is not large, the problems of principal interest can be solved once and for all. In the present section we consider boundary layers, and in the following section, an example involving an interior layer.

Consider an edge layer with the body in the region $Y \geq 0$, with surface tractions prescribed on one end, $X = 0$ say. Then (4.20) gives

$$U_{,X}(0, Y) = 0. \tag{4.25}$$

Suppose that either the displacement is prescribed at $X = L$ or the traction is prescribed at $X = 2L$. Then

$$U(L, Y) = 0 \quad \text{or} \quad U_{,X}(2L, Y) = 0. \tag{4.26}$$

On $Y = 0$, we suppose that the shearing traction is prescribed, so that (4.24) is valid. The final boundary condition is (4.21).

Under the condition of the second part of Eq. (4.26), separation of variables yields

$$U(X, Y) = \sum_{1}^{\infty} U_n(\cos \lambda_n X)(\exp(-\lambda_n Y)), \tag{4.27}$$

where

$$\lambda_n = n\pi/2L. \tag{4.28}$$

The boundary fiber force may be expanded as

$$F(X) = \sum_{1}^{\infty} F_n \sin \lambda_n X, \tag{4.29}$$

and the series is uniformly convergent since $F(0) = F(2L) = 0$. Then by using (4.27) and (4.29) in (4.24), we find that $U_n = -F_n/G$. It then follows from (4.13) that the tensile stress is

$$\sigma_{xx} = \sum_{1}^{\infty} F_n(\sin \lambda_n X)(\lambda_n/\varepsilon) \exp(-\lambda_n Y). \tag{4.30}$$

Under the condition (4.26a), the solution has the same form with U_{2n} and F_{2n} equal to zero.

In (4.30), the y-dependent factors are approximately Dirac deltas:

$$(\lambda_n/\varepsilon) \exp(-\lambda_n y/\varepsilon) \cong \delta(y). \tag{4.31}$$

Thus, the tensile stress is approximately $F(X)\delta(y)$.

As an example we consider the boundary layer along the edge $y = 0$ of the cantilever considered in Section II,C, with the fixed and loaded ends reversed to fit the present formalism. Then the force in the lower boundary fiber has the form $F = \sigma X$. The coefficients F_n in (4.29) are easily found (with $F_{2n} = 0$), and when these coefficients are used in (4.30), the resulting series can be summed:

$$\sigma_{xx} = -(2\sigma/\pi\varepsilon) \operatorname{Re} \ln \tan \pi(Z - L)/4L. \tag{4.32}$$

Here Z is defined in (4.17). Further details of the cantilever problem are given by Everstine and Pipkin (1973).

G. INTERIOR LAYER THROUGH THE CRACK TIP

Spencer (1974a) has given a detailed analysis of the stress concentration layer passing through the tip of a finite crack perpendicular to the fibers in an infinite body, and Sanchez-Moya and Pipkin (1978) have examined the tip singularity for a crack in a finite body, again perpendicular to the fibers. The tip singularity has the same structure in both cases, and indeed in any case in which the crack cuts across the fibers, not necessarily perpendicular to them.

We use the general crack configuration and the notation explained in Sections II,E and III,E. The tip angle α is restricted to be different from zero; the crack is not parallel to the fibers.

We take $u = 0$ at the crack tip and $u'_0(0) = 0$ just ahead of the tip; the latter condition can be imposed because the solution involves an arbitrary rigid rotation. Then the condition (4.22), in the present notation, requires that

$$U_{,Y}(X, -\infty) = u'_-(0) \qquad (-L_- \leq X \leq 0),$$

$$U_{,Y}(X, -\infty) = u'_+(0) \qquad (0 \leq X \leq L_+), \tag{4.33}$$

$$U_{,Y}(X, \infty) = 0 \qquad (-L_- \leq X \leq L_+).$$

With traction boundary conditions on the crack surfaces and at the ends of the layer, U must satisfy (4.20) on these surfaces:

$$U_{,X}(0\pm, Y) = 0 \qquad (Y < 0)$$

$$U_{,X}(\pm L_\pm, Y) = 0 \qquad (-\infty < Y < \infty). \tag{4.34}$$

As explained in Section IV,E, the crack surfaces and the end boundaries appear to be vertical when viewed in terms of the stretched coordinate, so the analysis of a crack perpendicular to the fibers gives results correct to leading order for any crack orientation with $\alpha \neq 0$.

The displacement gradients $u'_\pm(0)$ in (4.33) can be expressed in terms of the force F_0 at the crack tip given by the inextensible or idealized solution, by using (2.15).

The problem is easily solved by conformal mapping (Sanchez-Moya and Pipkin, 1978), yielding a solution in implicit form with Z expressed as a function of W [defined in (4.16)]. Inversion to find W in terms of Z near the tip gives, to leading order,

$$W(Z) = (2iF_0/G)(\pi L^*)^{-1/2}(Z/i)^{1/2} + \cdots . \tag{4.35}$$

In order to compare this result with those discussed in Section III,E, we begin by noting that the displacement u here corresponds to a term from the exact elastic solution (3.22a) of the form

$$u = -\operatorname{Re} \varepsilon \Psi'(z_2). \tag{4.36}$$

With this and (4.11), (4.16), and (4.17), we see that W is related to Ψ by

$$W(Z) = -\Psi'(\varepsilon Z) \tag{4.37}$$

in this problem. Now, for the elastic crack-tip solution (3.36), we find that

$$-\Psi'(\varepsilon Z) = 2(iK_u + cK_v)(\varepsilon Z/t_2)^{1/2}. \tag{4.38}$$

From (4.37), (4.35) must be equal to (4.38), or rather to the term of leading order in ε in (4.38). This implies that K_u has the value (3.41) deduced less rigorously from the energy release rate.

In the present stress-concentration-layer approximation, the crack surfaces are cleared of traction only to $O(1/\varepsilon)$, and not to $O(1)$. The higher order terms needed to give zero traction on the crack surfaces are now easily found by a bootstrap procedure: with the value of K_u deduced from the present approximation, $K_u \chi_u$ is the part of the tip field associated with the stretching displacement, χ_u being defined in (3.35).

In the inextensible theory, the expression (3.42) for K_v is meaningless. Sanchez-Moya and Pipkin (1978) show, in effect, that the term $K_v \chi_v$ in the stress function for the tip singularity is given to lowest order in ε by terms arising from the displacement $v(x, cy)$ given by the inextensible theory.

V. Finite Deformations

The theory of materials reinforced with inextensible cords was originally formulated by Adkins and Rivlin (1955) and Adkins (1956a) for application to reinforced rubber. In this theory the fibers are restricted to lie in one or more discrete surfaces, rather than being continuously distributed throughout the material as in the theories considered here. Adkins (1958) considers materials with continuously distributed fibers, but with three families rather than only one.

Rivlin (1955, 1959, 1964) also developed a theory of networks formed from two families of inextensible cords, and Adkins (1956b) extended this theory to account for elastic shear resistance; these theories apply to fabrics and rubberized sheets. The finite plane stress theory for such materials is quite simple in some respects, but this theory remains largely undeveloped. Kydoniefs (1970) has considered axisymmetric deformations of such materials, and Lee (1976) has considered the plane-stress traction boundary value problem for small deformations of sheets reinforced with two families of fibers that are not necessarily orthogonal.

The theory of plane stress of sheets reinforced with only one family of fibers is relatively difficult. Adkins (1956b) and Mulhern *et al.* (1967) have discussed this theory very briefly.

Here we consider only plane strain of incompressible materials reinforced with one family of inextensible fibers, which are distributed throughout the material. Stress–strain relations for a material of this kind were given by Green and Adkins (1960), but most of the mathematical theory discussed in the present section originated in a paper by Mulhern *et al.* (1967). Although the latter paper concerns plastic shearing response while the former deals with elasticity, much of the mathematical structure of the theory is independent of the nature of the shearing stress response (Pipkin and Rogers, 1971a), and for simplicity we discuss only elastic response here.

A long review of the finite deformation theory has been given by Pipkin (1973), and shorter reviews have been given by Rogers (1975, 1977) and Pipkin (1977). The book by Spencer (1972a) also contains most of the theory discussed here.

A. Fiber Flux and Fiber Density

A finite deformation is described by specifying the final position $\mathbf{x}$ of a particle as a function $\mathbf{x}(\mathbf{X})$ of its initial position $\mathbf{X}$. The value of any function at a given particle can be regarded as a function of either $\mathbf{x}$ or $\mathbf{X}$, whichever may be convenient.

We consider materials composed of inextensible fibers. The configuration of the fibers when the body is undeformed is specified by a field of unit vectors $\mathbf{a}_0(\mathbf{X})$ that are tangential to the fibers. The field of unit tangent vectors after the deformation is denoted $\mathbf{a}(\mathbf{x})$. A material element $d\mathbf{X} = \mathbf{a}_0\,ds$, which lies along a fiber before the deformation, maps onto an element $d\mathbf{x} = \mathbf{a}\,ds$ after the deformation, still along a fiber and still of the same length.

We also suppose that the material is incompressible in bulk. Consider an infinitesimal prism in the undeformed material, with base area dS_0 oriented in the direction of the unit vector $\mathbf{v}_0$. Let the sides of the prism be along the fiber direction, with length ds. Then the volume of the prism is $dV_0 = \mathbf{v}_0 \cdot \mathbf{a}_0\,dS_0\,ds$. The deformation changes this prism into one with base $\mathbf{v}\,dS$ and volume $dV = \mathbf{v} \cdot \mathbf{a}\,dS\,ds$. Incompressibility implies that $dV = dV_0$, so

$$\mathbf{a}_0 \cdot \mathbf{v}_0\,dS_0 = \mathbf{a} \cdot \mathbf{v}\,dS. \tag{5.1}$$

Then the flux of the field of fiber directions through any material surface is conserved by the deformation (Mulhern *et al.*, 1967). By applying this result to an arbitrary closed material surface and using the divergence theorem, we find that

$$\mathbf{V}_0 \cdot \mathbf{a}_0 = \mathbf{V} \cdot \mathbf{a}, \tag{5.2}$$

where $\mathbf{V}_0$ is the gradient with respect to $\mathbf{X}$ and $\mathbf{V}$, is the gradient with respect to $\mathbf{x}$ (Pipkin and Rogers, 1971a).

Let $N_0(\mathbf{X})$ be the number density of fibers in the undeformed body, defined to be such that $N_0(\mathbf{X})\mathbf{a}_0 \cdot \mathbf{v}_0 \, dS_0$ is the (signed) number of fibers passing through the surface element $\mathbf{v}_0 \, dS_0$. The corresponding quantity in the deformed body is $N(\mathbf{x})\mathbf{a} \cdot \mathbf{v} \, dS$, where $N(\mathbf{x})$ is the number density after the deformation. With (5.1), conservation of fibers then implies that $N(\mathbf{x}) = N_0(\mathbf{X})$; the fiber density is conserved under deformation (Mulhern *et al.*, 1967).

It then follows that the quantity $\mathbf{V} \cdot (N\mathbf{a})$ is conserved under deformations. This represents the net number of fiber origins per unit volume, and it is zero when fibers do not begin or end inside the body of the material. If this is the case, and if in addition the material is formed by packing fibers as closely together as possible, so that $N(\mathbf{x})$ is uniform, then

$$\mathbf{V}_0 \cdot \mathbf{a}_0 = \mathbf{V} \cdot \mathbf{a} = 0. \tag{5.3}$$

B. Plane Deformations

In plane deformations the mapping $\mathbf{x}(\mathbf{X})$ has the form

$$\mathbf{x}(\mathbf{X}) = \mathbf{i}x(X, Y) + \mathbf{j}y(X, Y) + \mathbf{k}Z, \tag{5.4}$$

in terms of Cartesian coordinates. We consider cases in which the fibers initially lie in planes $Z = \text{constant}$ and have the same arrangement in each plane, so that the initial field of fiber directions has the form

$$\mathbf{a}_0(\mathbf{X}) = \mathbf{i}_0 \cos \theta_0(X, Y) + \mathbf{j}_0 \sin \theta_0(X, Y). \tag{5.5}$$

After a plane deformation, the fibers again lie in planes $z = \text{constant}$ and the fiber direction field has the form

$$\mathbf{a}(\mathbf{x}) = \mathbf{i} \cos \theta(x, y) + \mathbf{j} \sin \theta(x, y). \tag{5.6}$$

Thus in either case, the fiber direction is defined by a scalar function θ_0 or θ, the fiber angle.

Let $\mathbf{n}_0$ and $\mathbf{n}$ be fields of unit vectors orthogonal to $\mathbf{a}_0$ and $\mathbf{a}$, respectively, in the plane of deformation:

$$\mathbf{n}_0 = -\mathbf{i}_0 \sin \theta_0 + \mathbf{j}_0 \cos \theta_0, \qquad \mathbf{n} = -\mathbf{i} \sin \theta + \mathbf{j} \cos \theta. \tag{5.7}$$

The curves with $\mathbf{n}_0$ or $\mathbf{n}$ as tangent vectors, the orthogonal trajectories of the fibers, are called *normal lines*. Unlike fibers, normal lines are not material lines in general.

If we treat $\mathbf{a}(\theta)$ and $\mathbf{n}(\theta)$ as the functions of θ defined by (5.6) and (5.7), then

$$\mathbf{a}'(\theta) = \mathbf{n}(\theta) \qquad \text{and} \qquad \mathbf{n}'(\theta) = -\mathbf{a}(\theta). \tag{5.8}$$

The Serret–Frenet relations for the fibers and normal lines are easily derived by using (5.8):

$$(\mathbf{a} \cdot \mathbf{V})\mathbf{a} = \mathbf{n}/r_a, \qquad (\mathbf{a} \cdot \mathbf{V})\mathbf{n} = -\mathbf{a}/r_a,$$

$$(\mathbf{n} \cdot \mathbf{V})\mathbf{a} = \mathbf{n}/r_n, \qquad (\mathbf{n} \cdot \mathbf{V})\mathbf{n} = -\mathbf{a}/r_n. \tag{5.9}$$

Here the radii of curvature r_a and r_n are defined by

$$1/r_a = \mathbf{a} \cdot \mathbf{V}\theta \qquad \text{and} \qquad 1/r_n = \mathbf{n} \cdot \mathbf{V}\theta. \tag{5.10}$$

Since $\mathbf{V} \cdot \mathbf{a} = \mathbf{V}\theta \cdot \mathbf{a}'$, it follows from (5.8) and (5.10) that $\mathbf{V} \cdot \mathbf{a} = 1/r_n$. Thus, the important relation (5.2) implies that $r_n = r_{n0}$; the radius of curvature of a normal line is conserved in a plane deformation (Pipkin and Rogers, 1971a).

C. DEFORMATION GRADIENT—CONSTRAINTS— COMPATIBILITY CONDITIONS

The deformation can be described locally as a linear mapping of elements $d\mathbf{X}$ onto elements $d\mathbf{x}$:

$$d\mathbf{x} = (d\mathbf{X} \cdot \mathbf{V}_0)\mathbf{x}(\mathbf{X}). \tag{5.11}$$

The mapping is defined locally when the maps of three elements $d\mathbf{X} = \mathbf{a}_0 \, ds$, $\mathbf{n}_0 \, ds$, and $\mathbf{k}_0 \, ds$ have been specified.

Because the deformation is plane, an element $\mathbf{k}_0 \, ds$ initially in the axial direction maps onto an element $\mathbf{k} \, ds$ still in the axial direction. Also, $\mathbf{a}_0 \, ds$ maps onto $\mathbf{a} \, ds$ by the definitions of $\mathbf{a}_0$ and $\mathbf{a}$ and the inextensibility of fibers. The element $\mathbf{n}_0 \, ds$ maps onto an element of the form $(\mathbf{n}c_1 + \mathbf{a}c_2) \, ds$. Incompressibility requires that the volume spanned by $\mathbf{k} \, ds$, $\mathbf{a} \, ds$ and the latter vector be $(ds)^3$, so $c_1 = 1$. We write $c_2 = \gamma$ and call γ the *amount of shear*. Then

$$(\mathbf{a}_0 \cdot \mathbf{V}_0)\mathbf{x} = \mathbf{a}, \qquad (\mathbf{n}_0 \cdot \mathbf{V}_0)\mathbf{x} = \mathbf{n} + \gamma\mathbf{a}, \qquad (\mathbf{k}_0 \cdot \mathbf{V}_0)\mathbf{x} = \mathbf{k}. \tag{5.12}$$

Equivalently,

$$d\mathbf{x} = \mathbf{a}(\mathbf{a}_0 \cdot d\mathbf{X}) + (\mathbf{n} + \gamma\mathbf{a})\mathbf{n}_0 \cdot d\mathbf{X} + \mathbf{k}(\mathbf{k}_0 \cdot d\mathbf{X}). \tag{5.13}$$

For future reference we note that with the chain rule, (5.12) implies that

$$\mathbf{a}_0 \cdot \nabla_0 = \mathbf{a} \cdot \nabla, \qquad \mathbf{n}_0 \cdot \nabla_0 = \mathbf{n} \cdot \nabla + \gamma \mathbf{a} \cdot \nabla, \qquad \mathbf{n} \cdot \nabla = \mathbf{n}_0 \cdot \nabla_0 - \gamma \mathbf{a}_0 \cdot \nabla_0.$$

$$(5.14)$$

Given γ, θ, and θ_0, we may integrate (5.13) to determine the deformation $\mathbf{x}(\mathbf{X})$. To ensure that (5.13) is a perfect differential, the amount of shear and fiber angles must satisfy certain compatibility conditions, which are obtained by eliminating $\mathbf{x}$ from (5.12) by cross-differentiation. In doing this, it must be borne in mind that directional derivatives do not commute. Instead, by using (5.9) we find that

$$(\mathbf{a} \cdot \nabla)(\mathbf{n} \cdot \nabla) - (\mathbf{n} \cdot \nabla)(\mathbf{a} \cdot \nabla) = -(\mathbf{a}/r_a + \mathbf{n}/r_n) \cdot \nabla, \qquad (5.15)$$

and of course the same relation is valid for the initial fields. Then eliminating $\mathbf{x}$ from the first two equations of (5.12) and separating the resulting equation into components in the $\mathbf{a}$ and $\mathbf{n}$ directions, we obtain the compatibility conditions (Pipkin, 1973)

$$(\mathbf{a}_0 \cdot \nabla_0)(\gamma - \theta + \theta_0) + \gamma \mathbf{n}_0 \cdot \nabla_0 \theta_0 = 0 \qquad (5.16)$$

and

$$\gamma \mathbf{a}_0 \cdot \nabla_0 \theta - \mathbf{n}_0 \cdot \nabla_0 \theta + \mathbf{n}_0 \cdot \nabla_0 \theta_0 = 0. \qquad (5.17)$$

With the third equation of (5.14), (5.17) can be put into the simpler form $\mathbf{n} \cdot \nabla \theta = \mathbf{n}_0 \cdot \nabla_0 \theta_0$, or, with (5.10),

$$r_n = r_{n0}. \qquad (5.18)$$

Thus, (5.17) is the requirement that the radius of curvature of the normal line through a given particle be preserved in the deformation. The new compatibility condition (5.16) determines the variation of the amount of shear along a fiber. We return to these conditions in Section V,F, where we explain how they are used in solving boundary value problems.

D. Stress and Equilibrium

We assume that the material has reflectional symmetry in planes $Z = $ constant, so that in plane deformations there is no shearing stress on these planes. Then the stress has the form

$$\boldsymbol{\sigma} = T\mathbf{aa} - P(\mathbf{nn} + \mathbf{kk}) + S(\mathbf{an} + \mathbf{na}) + S_{33}\mathbf{kk}. \qquad (5.19)$$

We use dyadic notation; $\mathbf{aa}$ means the tensor with Cartesian components $a_i a_j$. T is the tensile stress in the fiber direction, and P is a pressure that is isotropic around the fiber direction. These stress components are reactions to the constraints of fiber inextensibility and bulk incompressibility, and neither is directly related to the deformation by a constitutive equation. The shearing stress S and the normal stress difference S_{33} must be specified by constitutive equations. We suppose that the response of the material is elastic, so far as these functions are concerned, but it will be evident that the theory is largely independent of this assumption. For elastic response, S and S_{33} are functions of the local deformation gradient, and since the local deformation is defined apart from orientation by the amount of shear, S and S_{33} are functions of the amount of shear. The strain energy density W is also a function of the amount of shear, related to S by $S(\gamma) = W'(\gamma)$. Spencer (1972a) gives the full three-dimensional form of the constitutive equation, which is not needed here (see also Pipkin, 1973).

We suppose that the material is homogeneous in the z-direction. Then the stress equation of equilibrium, $\mathbf{V} \cdot \boldsymbol{\sigma} = \mathbf{0}$, yields

$$\mathbf{a} \cdot \mathbf{V}T + (P + T)/r_n = 2S/r_a - \mathbf{n} \cdot \mathbf{V}S, \qquad (5.20\text{a})$$

$$\mathbf{n} \cdot \mathbf{V}P - (P + T)/r_a = 2S/r_n + \mathbf{a} \cdot \mathbf{V}S, \qquad (5.20\text{b})$$

and $\partial P/\partial z = 0$ (Pipkin and Rogers, 1971a).

At a boundary point where the unit outward normal is $\mathbf{v}$ we find that

$$\boldsymbol{\sigma} \cdot \mathbf{v} = \mathbf{a}(T\mathbf{a} \cdot \mathbf{v} + S\mathbf{n} \cdot \mathbf{v}) + \mathbf{n}(-P\mathbf{n} \cdot \mathbf{v} + S\mathbf{a} \cdot \mathbf{v}). \qquad (5.21)$$

If the boundary is not a fiber or a normal line, this must be equal to the surface traction $\mathbf{T}$ (say) on the boundary. Then

$$T\mathbf{a} \cdot \mathbf{v} = \mathbf{T} \cdot \mathbf{a} - S\mathbf{n} \cdot \mathbf{v} \qquad \text{and} \qquad -P\mathbf{n} \cdot \mathbf{v} = \mathbf{T} \cdot \mathbf{n} - S\mathbf{a} \cdot \mathbf{v}. \quad (5.22)$$

As in the infinitesimal theory, it is sometimes possible to determine the deformation purely kinematically without use of the equilibrium equations. In such cases all the quantities in (5.20) and (5.22) except P and T are known when the deformation has been determined. Then the relations (5.20) are partial differential equations in characteristic form, with the fibers and normal lines as characteristics. The first determines the variation of T along a fiber, and the second governs the variation of P along a normal line. With boundary values of the form (5.22) to give T at one end of each fiber and P at one end of each normal line, the stress field is determinate.

Because the system is hyperbolic, solutions can involve singular fibers and normal lines as in the infinitesimal theory. On boundaries along fibers or normal lines, (5.22) need not be satisfied. Instead, the boundary line becomes singular, carrying a finite force that equilibrates the discrepancy between the applied tangential traction and the shearing stress inside the body. This force can be found by regarding S in the final member of (5.20a) or (5.20b) as a step function, whose derivative across the boundary is a Dirac delta. Then if the boundary is curved, the surface tension causes a discontinuity in the normal stress across it, so the normal component of (5.22) also need not be satisfied.

Singular fibers are common in the finite deformation theory, because a fiber that lay along a boundary before the body was deformed still lies along a boundary after the deformation, and it should be expected that this fiber will be singular. However, normal lines are not material lines in general, and even if the undeformed body were bounded by a normal line, this need not be the case for the deformed body. This is one of the reasons that singular normal lines are less usual in the finite deformation theory.

E. Lines of Discontinuity—Energetic Admissibility

It is kinematically admissible for fibers to change direction discontinuously. If fibers change from the direction $\mathbf{a}_-$ to the direction $\mathbf{a}_+$ as they pass through a surface element $\mathbf{v}\,dS$, continuity of the flux $\mathbf{a} \cdot \mathbf{v}\,dS$ implies that $\mathbf{v} \cdot \mathbf{a}_- = \mathbf{v} \cdot \mathbf{a}_+$ (Spencer, 1972a). In two dimensions, this implies that the line of discontinuity bisects the angle between the fibers on the two sides of it (Pipkin and Rogers, 1971a).

Let $\Delta\theta$ and $\Delta\gamma$ be the changes in θ and γ across a discontinuity line in two dimensions. When two fibers separated by the distance ds pass through the line, one travels further than the other, so to speak, by the distance $2\tan(\Delta\theta/2)\,ds$. This implies that (Pipkin, 1978)

$$\Delta\gamma = 2\tan(\Delta\theta/2). \tag{5.23}$$

Since (5.23) is valid for any line crossed by fibers, and since $\Delta\theta = 0$ for any normal line, there can be no discontinuity in γ across a normal line. This is a second reason why singular normal lines are unusual in the finite-deformation theory.

Let $\mathbf{v}$ and $\boldsymbol{\tau}$ be unit vectors normal and tangential to a line of discontinuity, and define

$$\sigma_{vv} = \mathbf{v} \cdot \boldsymbol{\sigma} \cdot \mathbf{v} \qquad \text{and} \qquad \sigma_{v\tau} = \mathbf{v} \cdot \boldsymbol{\sigma} \cdot \boldsymbol{\tau}. \tag{5.24}$$

The equilibrium conditions at a discontinuity are that σ_{vv} and $\sigma_{v\tau}$ are continuous across it. With (5.19), these continuity conditions yield expressions for T and P on one side of the line in terms of their values on the other side (Pipkin and Rogers, 1971a). These conditions are used in place of (5.20) when the characteristics change direction discontinuously.

In most problems, the kinematic conditions of the problem allow more than one kinematically admissible solution, some of them having lines of discontinuity. Pipkin and Rogers (1971b) discuss a particular case in which the continuous solution can be shown to be the correct one because the discontinuous solution does not satisfy the traction boundary conditions of the problem. However, there are problems in which there are infinitely many solutions, all kinematically and statically admissible, satisfying all constraint and equilibrium conditions and all of the boundary conditions. Everstine and Rogers (1971) encountered a case of this kind in a machining problem, and chose to use the one solution that had no discontinuity line. The physical reason for choosing the smooth solution has only recently been discovered.

Pipkin (1978) analyzed two solutions of a bending problem, one continuous and the other having a discontinuity line, and showed that both solutions satisfied all kinematic and equilibrium requirements. However, the discontinuous solution did not minimize the total energy of the body and its loading system, nor even render it stationary. In a virtual deformation, some material passes through the discontinuity line, and the displacement gradient (amount of shear) in this material changes by a finite amount even though the displacement is infinitesimal. Then the change of energy per unit volume for this material exceeds the work done on it by the amount $\Delta W - \sigma_{\tau v}\Delta\gamma$. This may be negative, but the point is that it need not be zero, so the energy is not stationary in small virtual deformations.

Uniqueness is restored by requiring solutions to minimize the energy. It appears that discontinuity lines are energetically admissible only when they are imposed by displacement boundary conditions; in such cases the discontinuity line cannot move in any admissible virtual deformation. As a rule of thumb, solutions with discontinuity lines need not be considered when a continuous alternative is available.

F. PARALLEL FIBERS

When the material is formed by packing fibers as closely together as possible and no fiber terminates in the interior of the body, the divergence of the field $\mathbf{a}_0$ is zero (Section V,A). If $\mathbf{a}_0$ is also a plane field, this divergence is equal to the curvature $1/r_{n0}$ (Section V,B), so the normal lines are *straight*. The fibers are then parallel in the sense that concentric circles are parallel; the perpendicular distance between two fibers is constant along their length.

The compatibility condition (5.18) implies that if normal lines are straight initially then they remain straight in every state of deformation. That is, the fibers remain parallel in every state of deformation, and the distance between two fibers never changes. Because of the simplicity of these conditions, most published solutions of plane problems involve materials with initially parallel fibers.

Problems in which one of the boundaries lies along a fiber are especially simple. If the deformed shape of the fiber is given, the straight normal lines perpendicular to it can be constructed, and the orthogonal trajectories of these straight lines are the loci of the remaining fibers. When the position of one particle on each fiber is given, the positions of the others along the known locus are determinate from the inextensibility condition. Thus the deformation can be determined completely, if the normal lines through the boundary fiber cover the whole region in which the body might lie.

When the deformation can be found geometrically in this way, it is unnecessary to integrate the basic kinematic relation (5.13) explicitly. The integrability conditions (5.16) and (5.18) are central to the theory, but an explicit analytical expression for the deformation $\mathbf{x}(\mathbf{X})$ usually is not needed.

The determination of the stress field is particularly simple when the fibers are parallel. If the fibers in the deformed body are straight as well as parallel, then both $1/r_n$ and $1/r_a$ are zero, and (5.20) yields

$$T_{,x} = -S_{,y} \quad \text{and} \quad P_{,y} = S_{,x}, \tag{5.25}$$

where x and y are coordinates along the fiber and normal directions [compare with (2.3) and (2.4)]. These may be integrated directly, as in the infinitesimal case.

When the deformed fibers are parallel but not straight, the fiber angle θ is constant along normal lines but variable from one normal line to another, so θ can be used as a characteristic coordinate. As a second coordinate, constant along fibers, we use the distance ξ of the fiber from the boundary fiber $\xi = 0$. The fiber radius of curvature r_a varies linearly along normal lines because the fibers are parallel:

$$r_a(\xi, \theta) = r_0(\theta) - \xi. \tag{5.26}$$

Then the characteristic coordinates ξ and θ are related to the position $\mathbf{x}$ by

$$d\mathbf{x} = \mathbf{a}(\theta)r_a(\xi, \theta)\, d\theta + \mathbf{n}(\theta)\, d\xi + \mathbf{k}\, dz, \tag{5.27}$$

which yields

$$\mathbf{x} = \int_0^\theta \mathbf{a}(\theta')r_0(\theta')\, d\theta' + \xi\mathbf{n}(\theta) + \mathbf{k}z + \mathbf{x}_0, \tag{5.28}$$

where $\mathbf{x}_0$ is a constant of integration.

In these coordinates the equilibrium relations (5.20) are

$$\partial T/\partial \theta = 2S - r_a\, \partial S/\partial \xi, \qquad \partial(r_a P)/\partial \xi = T + \partial S/\partial \theta. \tag{5.29}$$

In this case T must be determined first, since the equation for P involves T as well.

Even when the deformation is known, the amount of shear usually is not obvious by inspection. With parallel fibers, the final member of (5.16) is zero, and integration yields

$$\gamma = \theta - \theta_0 + k(\xi), \qquad (5.30)$$

where $k(\xi)$ is constant along each fiber, to be determined by evaluating (5.30) at a boundary point where γ, θ, and θ_0 are all known. We give an example of this in Section V,G.

Examples illustrating the determination of T and P have been given by Mulhern *et al.* (1967), Pipkin and Rogers (1971a,b), Everstine and Rogers (1971), Rogers and Pipkin (1971b), Kao and Pipkin (1972), Spencer (1972a), and Pipkin (1973, 1978).

G. STRESS RESULTANTS

As in the infinitesimal theory (Section II,D), the method of shearing stress resultants can sometimes be used to determine the deformation in problems with specified tractions. A trivial example is the shearing deformation shown in Fig. 3. The kinematic constraints imply that the fibers remain straight and parallel, and the amount of shear γ is the displacement gradient $u'(y)$, which is now finite. Then the resultant shearing stress on a line $y = $ constant is $LS(u')$, and this must be equal to the shearing force F_x. This implies that u' is constant, and determines its value when the form of the function $S(\gamma)$ is specified (Pipkin, 1973).

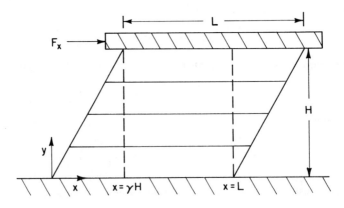

FIG. 3.   Sheared slab.

The preceding example is exceptional because fibers in the deformed body generally are not straight, and the total shearing force on a curved fiber has no simple relation to the applied force. However, normal lines are always straight if they were straight when the body was undeformed, so it is often possible to make use of the resultant shearing stress on normal lines.

As an example, consider finite deformation of the cantilever described in Section II,C. The fibers are initially straight and parallel to the X-direction, so that $\theta_0 = 0$. To determine the amount of shear from (5.30), we observe that an element $d\mathbf{X} = \mathbf{n}_0\,dY$ that is along the wall $X = 0$ before the deformamation is still along the wall after the deformation, so that $d\mathbf{x} = d\mathbf{X}$; *but* $d\mathbf{x} = (\mathbf{n} + \gamma\mathbf{a})\,dY$ since the element was along a normal line. Hence $\gamma = 0$ on the end, and $\theta = 0$ there. It then follows from (5.30) that $\gamma = \theta$ everywhere in the deformed body if the deformation is smooth.

With this preliminary kinematic information, we know that the shearing stress $S(\theta)$ is constant along normal lines in the deformed body, so the resultant shearing stress on a normal line that passes completely across the body is $HS(\theta)$. Now, suppose that the body is deformed by a dead load $\mathbf{F}$ that is distributed in some way over the end $X = L$, and that the sides $Y = 0$ and $Y = H$ are free from traction. Then

$$HS(\theta) = \mathbf{F} \cdot \mathbf{n}(\theta). \tag{5.31}$$

This determines one value $\theta = \alpha$ (say), or at most a few discrete values of α, and it implies that throughout most of the deformed beam the fibers are parallel to the direction α and the beam is in a state of simple shear with $\gamma = \alpha$.

Since the end $x = 0$ is a normal line with $\theta = 0$, there must be a region near this end in which the assumptions leading to (5.31) are not satisfied. For $\alpha > 0$, all conditions are satisfied by letting normal lines near the end meet at the common point $x = 0$, $y = H$; in this way these normal lines neither have two free ends, so that (5.31) would have to be satisfied, nor do they intersect inside the body, which is kinematically impossible. The fibers in this fan region lie on circles centered at the point where the normal lines meet (Fig. 4).

Most of the papers mentioned at the end of Section V,F involve deformations composed of simply sheared regions and fan regions. Rogers and Pipkin (1971b) have used the method of stress resultants in a more complicated case in which the fibers initially lie in concentric circles, reinforcing a pressurized tube. The tube is deformed into an oval shape by squeezing it between two rigid parallel plates. The amount of shear is not constant along normal lines in this example, but it varies in a simple way. This problem is further complicated by the fact that because part of the surface traction is

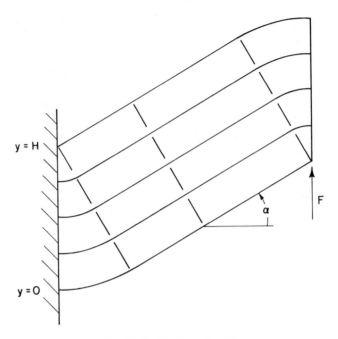

FIG. 4.   Finite deflection of cantilever.

the internal pressure in the tube, the resultant traction on the part of the body to one side of a normal line is known only implicitly in terms of the unknown boundary shape. Nevertheless, the problem can be reduced to a single equation for the radius of curvature $r_0(\theta)$ in (5.26), which defines the shape of the deformed tube. Rogers (1975) has used similar methods in a crack-inflation problem.

In passing, we note that for the deformations shown in Figs. 3 and 4, the infinitesimal theory would yield singular normal lines along the boundaries $X = 0$ and $X = L$. In the shearing problem, these singular normal lines are replaced by the triangular regions in which $x < \gamma H$ or $x > L$, and it is easy to verify that the total tensile stress σ_{yy}, integrated across the triangle, is the force that appears in the infinitesimal theory as a finite force carried by the boundary normal line. In the cantilever problem, the singular normal line is replaced by the fan region.

As in the infinitesimal theory, problems are simpler when the body is bounded by a fiber. Craig and Hart (1977) have solved a pure traction boundary value problem for a region not bounded by characteristics, by using perturbation methods for relatively small deformations, and they have also proved uniqueness for the perturbation equations for the general traction boundary value problem. Aside from this, no pure traction boundary value problems have been solved for bodies of general shape.

H. AXISYMMETRIC AND OTHER PROBLEMS

Certain finite deformations of isotropic (unreinforced) incompressible materials have such a high degree of symmetry that they automatically yield equilibrium stress fields. Some of these deformations are still possible when the material is reinforced by fibers, if the fibers do not spoil the symmetry of the problem (Pipkin, 1970, 1973). Beskos (1973) has considered some problems of this kind for incompressible materials, and has shown that some of these solutions remain valid when the material is compressible (Beskos, 1972).

Problems that are two-dimensional by virtue of axial symmetry are nearly as simple as plane problems, but complete analytical solutions are less easy to obtain. Mulhern (1969) and Smith and Spencer (1970) considered tubes helically wound with one and two families of fibers, respectively, and solved some problems in which initially cylindrical surfaces remain cylindrical. Beskos and Jenkins (1975) have encountered one of Mulhern's solutions, involving extension with twist, in a model of muscle fibers.

Arcisz (1973) considered tubes with fibers parallel to the axial direction, under general axisymmetric deformations with twist, and obtained a special solution in which cylindrical surfaces do not remain cylindrical, the only known complete analytical solution of this kind. Pipkin (1975) considered axisymmetric deformations without twist, for bodies with fibers along meridianal curves satisfying $\mathbf{V} \cdot \mathbf{a} = 0$, and showed that displacement problems could be reduced to quadratures. In the axisymmetric case, the condition $\mathbf{V} \cdot \mathbf{a} = 0$ implies that normal lines are catenaries rather than straight lines. The special solution found by Arcisz (1973) is a case in which the normal lines are congruent.

All published plane-strain solutions involve parallel fibers, except the special generalized-plane-strain solutions discussed by Pipkin (1974). In these problems, distortion of cross sections is caused by extension in the z-direction. The extension can be supported by an axial tension alone, if and only if the fibers initially form a Hencky–Prandtl net. In this case they again form such a net after the deformation, and the deformation has a relatively simple description in terms of characteristic coordinates.

ACKNOWLEDGMENT

This paper was prepared under Grant MCS76-08808 from the National Science Foundation. We gratefully acknowledge this support.

48   *Allen C. Pipkin*

REFERENCES

ADKINS, J. E. (1956a). Cylindrically symmetrical deformations of incompressible elastic materials reinforced with inextensible cords. *J. Ration. Mech. Anal.* **5**, 189–202.

ADKINS, J. E. (1956b). Finite plane deformation of thin elastic sheets reinforced with inextensible cords. *Philos. Trans. Ry. Soc. London, A Ser.* **249**, 125–150.

ADKINS, J. E. (1958). A three-dimensional problem for highly elastic materials subject to constraints. *Q. J. Mech. Appl. Math.* **11**, 88–97.

ADKINS, J. E., and RIVLIN, R. S. (1955). Large elastic deformations of isotropic materials X. Reinforcement by inextensible cords. *Philos. Trans. Ry. Soc. London, A Ser.* **248**, 201–223.

AIFANTIS, E. C., and BESKOS, D. E. (1976). Dynamic universal solutions for fiber-reinforced incompressible isotropic elastic materials. *J. Elast.* **6**, 353–367.

ALTS, T. (1976). Thermodynamics of thermoelastic bodies with kinematic constraints. Fiber-reinforced materials. *Arch. Ration. Mech. Anal.* **61**, 253–289.

ARCISZ, M. (1973). Finite, axially symmetric deformation of plastic fibre-reinforced materials. *Arch. Mech. Stosow.* **25**, 883–894.

BESKOS, D. E. (1972). Universal solutions for fiber-reinforced compressible isotropic elastic materials. *J. Elast.* **2**, 153–168.

BESKOS, D. E. (1973). Universal solutions for fiber-reinforced incompressible isotropic elastic materials. *Int. J. Solids Struct.* **9**, 553–567.

BESKOS, D. E., and JENKINS, J. T. (1975). A mechanical model for mammalian tension. *J. Appl. Mech.* **42**, 755–758.

CHEN, P. J., and GURTIN, M. E. (1974). On wave propagation in inextensible elastic bodies. *Int. J. Solids Struct.* **10**, 275–281.

CHEN, P. J., and NUNZIATO, J. W. (1975). On wave propagation in perfectly heat conducting inextensible elastic bodies. *J. Elast.* **5**, 155–160.

CHOI, I., and HORGAN, C. O. (1977). Saint-Venant's principle and end effects in anisotropic elasticity. *J. Appl. Mech.* **44**, 424–430.

CHOI, I., and HORGAN, C. O. (1978). Saint-Venant end effects for plane deformation of sandwich strips. *Int. J. Solids Struct.* **14**, 187–195.

CRAIG, M. S., and HART, V. G. (1977). "The Stress Boundary Value Problem for Finite Plane Deformations of a Fibre-reinforced Material." Dept. Math., University of Queensland, Brisbane, Australia.

ENGLAND, A. H. (1972). The stress boundary value problem for plane strain deformations of an ideal fibre-reinforced material. *J. Inst. Math. Appl.* **9**, 310–322.

ENGLAND, A. H. (1975). An inclusion in a strong anisotropic material. *J. Elast.* **5**, 259–271.

ENGLAND, A. H., and ROGERS, T. G. (1973). Plane crack problems for ideal fibre-reinforced materials. *Q. J. Mech. Appl. Math.* **26**, 303–320.

ENGLAND, A. H., FERRIER, J. E., and THOMAS, J. N. (1973). Plane strain and generalized plane stress problems for fibre-reinforced materials. *J. Mech. Phys. Solids* **21**, 279–301.

EVERSTINE, G. C., and PIPKIN, A. C. (1971). Stress channelling in transversely isotropic elastic composites. *Z. Angew. Math. Phys.* **22**, 825–834.

EVERSTINE, G. C., and PIPKIN, A. C. (1973). Boundary layers in fiber-reinforced materials. *J. Appl. Mech.* **40**, 518–522.

EVERSTINE, G. C., and ROGERS, T. G. (1971). A theory of machining of fiber-reinforced materials. *J. Compos. Mater.* **5**, 94–106.

FOLKES, J. M., and ARRIDGE, R. G. C. (1975). The measurement of shear modulus in highly anisotropic materials: The validity of St. Venant's principle. *J. Phys. D* **8**, 1053–1064.

GREEN, A. E., and ADKINS, J. E. (1960). "Large Elastic Deformations." Oxford Univ. Press (Clarendon), London and New York.

GREEN, A. E., and TAYLOR, G. I. (1939). Stress systems in aelotropic plates. I. *Proc. Ry. Soc. London, Ser. A* **173**,

GREEN, A. E., and ZERNA, W. (1954). "Theoretical Elasticity." Oxford Univ. Press (Clarendon), London and New York.

GURTIN, M. E., and PODIO GUIDUGLI, P. (1973). The thermodynamics of constrained materials. *Arch. Ration. Mech. Anal.* **51**, 192–208.

HAYES, M., and HORGAN, C. O. (1974). On the displacement boundary-value problem for inextensible elastic materials. *Q. J. Mech. Appl. Math.* **27**, 287–297.

HAYES, M., and HORGAN, C. O. (1975). On mixed boundary-value problems for inextensible elastic materials. *Z. Angew. Math. Phys.* **26**, 261–272.

HORGAN, C. O. (1972a). On Saint-Venant's principle in plane anisotropic elasticity. *J. J. Elast.* **2**, 169–180.

HORGAN, C. O. (1972b). Some remarks on Saint-Venant's principle for transversely isotropic composites. *J. Elast.* **2**, 335–339.

HORGAN, C. O. (1974). The axisymmetric end problem for transversely isotropic circular cylinders. *Int. J. Solids Struct.* **10**, 837–852.

JONES, N. (1976). Dynamic behavior of ideal fibre-reinforced rigid-plastic beams. *J. Appl. Mech.* **43**, 319–324.

KAO, B., and PIPKIN, A. C. (1972). Finite buckling of fiber-reinforced columns. *Acta Mech.* **13**, 265–280.

KYDONIEFS, A. D. (1970). Finite axisymmetric deformations of an initially cylindrical membrane reinforced with inextensible cords. *Q. J. Mech. Appl. Math.* **23**, 481–488.

LEE, D. A. (1976). The plane-stress boundary-value problem for a material reinforced by two families of strong fibres. *Q. J. Mech. Appl. Math.* **29**, 277–293.

MORLAND, L. W. (1973). A plane theory of inextensible transversely isotropic elastic composites. *Int. J. Solids Struct.* **9**, 1501–1518.

MORLAND, L. W. (1975). Existence of solutions of plane traction problems for inextensible transversely isotropic elastic solids. *J. Aust. Math. Soc. B* **19**, 40–54.

MULHERN, J. F. (1969). Cylindrically symmetric deformations of a fibre-reinforced material. *Q. J. Mech. Appl. Math.* **22**, 97–114.

MULHERN, J. F., ROGERS, T. G., and SPENCER, A. J. M. (1967). A continuum model for fibre-reinforced plastic materials. *Proc. R. Soc. London, Ser. A* **301**, 473–492.

NICOL, D. A. C. (1978). Bending and flexure of a reinforced tube. *J. Inst. Math. Appl.* **21**, 117–126.

PARKER, D. F. (1975). "Dynamic Flexural Deformations in an Ideal Fibre-reinforced Beam." Dept. Theor. Mech., University of Nottingham, England.

PIPKIN, A. C. (1970). Non-linear phenomena in continua. *In* "Non Linear Continuum Theories in Mechanics and Physics and Their Applications" (R. S. Rivlin, ed.), pp. 51–150. Edizioni Cremonese, Rome.

PIPKIN, A. C. (1973). Finite deformations of ideal fiber-reinforced composites. *In* "Composite Materials, Vol. 2: Micromechanics" (G. P. Sendeckyj, ed.), pp. 251–308. Academic Press, New York.

PIPKIN, A. C. (1974). Generalized plane deformations of ideal fiber-reinforced materials. *Q. Appl. Math.* **32**, 253–263.

PIPKIN, A. C. (1975). Finite axisymmetric deformations of ideal fibre-reinforced composites. *Q. J. Mech. Appl. Math.* **28**, 271–284.

PIPKIN, A. C. (1976). Constraints in linearly elastic materials. *J. Elast.* **6**, 179–193.

PIPKIN, A. C. (1977). Finite deformations in materials reinforced with inextensible cords. *In* "Finite Elasticity" (R. S. Rivlin, ed.), pp. 91–102, ASME AMD 27. Am. Soc. Mech. Eng., New York.

PIPKIN, A. C. (1978). Energy changes in ideal fiber-reinforced composites. *Q. Appl. Math.* **35**, 455–463.

PIPKIN, A. C., and ROGERS, T. G. (1971a). Plane deformations of incompressible fiber-reinforced materials. *J. Appl. Mech.* **38**, 634–640.

PIPKIN, A. C., and ROGERS, T. G. (1971b). A mixed boundary-value problem for fiber-reinforced materials. *Q. Appl. Math.* **29**, 151–155.

PIPKIN, A. C., and ROGERS, T. G. (1978). Crack paths in sheets reinforced with two families of inextensible fibers. *Mech. Today* **5** (in press).

PIPKIN, A. C. and SANCHEZ-MOYA, V. (1974). Existence of solutions of plane traction problems for ideal composites. *SIAM J. Appl. Math.* **26**, 213–220.

RICE, J. R. (1968). Mathematical analysis in the mechanics of fracture. *In* "Fracture, Vol. II", pp. 191–311. Academic Press, New York.

RIVLIN, R. S. (1955). Plane strain of a net formed by inextensible cords. *J. Ration. Mech. Anal.* **4**, 951–974.

RIVLIN, R. S. (1959). The deformation of a membrane formed by inextensible cords. *Arch. Ration. Mech. Anal.* **2**, 447–476.

RIVLIN, R. S. (1964). Networks of inextensible cords. *In* "Nonlinear Problems of Engineering" (W. F. Ames, ed.), pp. 51–64. Academic Press, New York.

ROGERS, T. G. (1975). Finite deformations of strongly anisotropic materials. *In* "Theoretical Rheology" (J. F. Hutton, J. R. A. Pearson, and K. Walters, eds.), pp. 141–168. Applied Science Publ., London.

ROGERS, T. G. (1977). Deformations of strongly anisotropic materials. *Rheol. Acta* **16**, 123–133.

ROGERS, T. G., and PIPKIN, A. C. (1971a). Small deflections of fiber-reinforced beams or slabs. *J. Appl. Mech.* **38**, 1047–1048.

ROGERS, T. G., and PIPKIN, A. C. (1971b). Finite lateral compression of a fibre-reinforced tube. *Q. J. Mech. Appl. Math.* **24**, 311–330.

SANCHEZ-MOYA, V., and PIPKIN, A. C. (1977). Energy release rate for cracks in ideal composites. *Int. J. Solids Struct.* **13**, 571–578.

SANCHEZ-MOYA, V., and PIPKIN, A. C. (1978). Crack-tip analysis for elastic materials reinforced with strong fibers. *Q. J. Mech. Appl. Math.* **31**, 349–362.

SCOTT, N., and HAYES, M. (1976). Small vibrations of a fibre-reinforced composite. *Q. J. Mech. Appl. Math.* **29**, 467–486.

SHAW, L., and SPENCER, A. J. M. (1977). Impulsive loading of ideal fibre-reinforced rigid-plastic beams. *Int. J. Solids Struct.* **13**, 823–854.

SIH, G. C., PARIS, P. C., and IRWIN, G. R. (1965). On cracks in rectilinearly anisotropic bodies. *Int. J. Fract. Mech.* **1**, 189–203.

SMITH, G. E., and SPENCER, A. J. M. (1970). A continuum theory of a plastic-rigid solid reinforce by two families of inextensible fibres. *Q. J. Mech. Appl. Math.* **23**, 489–504.

SPENCER, A. J. M. (1972a). "Deformations of Fibre-reinforced Materials." Oxford Univ. Press (Clarendon), London and New York.

SPENCER, A. J. M. (1972b). Plane strain bending of laminated fibre-reinforced plates. *Q. J. Mech. Appl. Math.* **25**, 387–400.

SPENCER, A. J. M. (1974a). Boundary layers in highly anisotropic plane elasticity. *Int. J. Solids Struct.* **10**, 1103–1123.

SPENCER, A. J. M. (1974b). Dynamics of ideal fibre-reinforced rigid-plastic beams. *J. Mech. Phys. Solids* **22**, 147–159.

SPENCER, A. J. M. (1976). A note on an ideal fibre-reinforced rigid-plastic beam brought to rest by transverse impact. *Mech. Res. Commun.* **3**, 55–58.

SPENCER, A. J. M. (1977). "Simple Methods of Stress Analysis for Highly Anisotropic Materials." *In* "Fibre Reinforced Materials: Design and Engineering Applications," pp. 21–27. Institute of Civil Engineers, New York.

SPENCER, A. J. M., ROGERS, T. G., and MOSS, R. L. (1974). An optimal angle of winding for pressurized fibre-reinforced cylinders. *Mech. Res. Commun.* **1**, 27–32.

SPENCER, A. J. M., MOSS, R. L., and ROGERS, T. G. (1975). Pure bending of helically wound ideal fibre-reinforced cylinders. *J. Elast.* **5**, 287–296.

THOMAS, J. N. (1974). A stress boundary value problem for an ideal fibre-reinforced rectangular beam with a longitudinal notch. *Z. Angew. Math. Phys.* **25**, 553–563.

THOMAS, J. N., and ENGLAND, A. H. (1974). The stress boundary value problem for an ideal fibre-reinforced rectangular plate with a hole. *J. Inst. Math. Appl.* **14**, 347–374.

TRAPP, J. A. (1971). Reinforced materials with thermomechanical constraints. *Int. J. Eng. Sci.* **9**, 757–773.

TROTH, M. R. (1976). Materials reinforced by almost inextensible fibres. *J. Inst. Math. Appl.* **18**, 265–278.

WEITSMAN, Y. (1972). On wave propagation and energy scattering in materials reinforced by inextensible fibers. *Int. J. Solids Struct.* **8**, 627–650.

WESTERGAARD, H. M. (1938). A problem of elasticity suggested by a problem in soil mechanics: Soft material reinforced by numerous strong horizontal sheets. *In* "Contributions to the Mechanics of Solids Dedicated to Stephen Timoshenko," pp. 268–277. Macmillan, New York.

ADVANCES IN APPLIED MECHANICS, VOLUME 19

Theory of Water-Wave Refraction

R. E. MEYER

Department of Mathematics
University of Wisconsin
Madison, Wisconsin

Dedicated to J. B. Keller

I. Introduction

Gravity waves on the surface of oceans and lakes have a propagation speed dependent on the water depth. For long waves, that dependence may be strong and such waves are directly modulated by the changes in water depth they encounter during propagation over natural water bodies. For shorter waves, the dependence of propagation speed on the depth may be tenuous, but natural seabed slopes are generally very small and as the waves travel over long distances, for instance, above a continental shelf, the locally minute modulation by changes in the depth often accumulates to bring about important, even dominant effects. Slow modulation, moreover, can culminate in drastic changes of seaway over surprisingly short distances, and disaster mechanisms may result.

The understanding and prediction of such effects is simplified by two salient facts. The sedimentary nature of the seabed assures great gentleness of almost all natural depth changes encountered. For all waves but the tides, the process is a modulation in which the wave propagates virtually unchanged over distances of a few wavelengths; over many wavelengths, however, gradual changes in the wave pattern take place. The modulation is therefore characterized by a small parameter, denoted by ε throughout this article; it characterizes the ratio of the wavelength to the horizontal scale of the wave pattern as a whole. The second fact is that natural waves have small amplitude during most of their travel. The main concern of refraction theory is to describe the process to a first approximation in both these small parameters. Within this framework, the last two decades have brought very substantial advances, and this article aims to outline some of the main ones among them.

Such processes of modulation are of interest in many branches of science, indeed, most of our physical notions about waves involve them intimately; for without slowness of change characterized by a small modulation parameter ε, even a term like wavelength loses its meaning. Water-wave refraction is therefore a member of a large family of wave modulation subjects in science, and many members of the family interact. For the sake of unity of this article, however, no such cross-connections with other branches of science will be discussed. For the same reason, even water-wave refraction by mechanisms other than depth changes will be excluded.

As a distinct field of oceanography, refraction theory was introduced by Munk and Traylor (1947) in an article on the ray method of wave tracing, which has since been perfected and widely applied in coastal engineering. An important further step was made by Keller (1958) in placing the intuitive ray method into a rational framework of asymptotic approximation to the classical, linear theory of small-amplitude surface waves. This geometrical optics framework (Section IV) turned out to be not quite the right one, but it identified a most fruitful, rational direction and sparked nearly all the advances to be reported in this article.

As is usual when a new idea opens a field, it set in train simultaneous developments in opposite directions. On one hand, the geometrical optics approximation was used to improve the efficiency of the ray method drastically and to adapt it to far-reaching extensions of its practical application. In this process, the traditional emphasis on the rays in oceanographical work underwent a revision. They have physical meaning, but their role is more technical than substantive. Nor is a recourse to rays always necessary to obtain the information primarily desired by oceanographers and engineers; more direct avenues have been discovered in some cases and may find extension to others.

These changes were accelerated by the discovery of wave trapping. Less than a generation ago, it was well-established that oceans differ basically from bounded basins, such as harbors and lakes, in regard to wave resonance. Except in regard to the tides, an ocean is effectively unbounded, so that any temporary concentration of wave energy will be dispersed promptly by radiation to far distant parts of the ocean and resonant wave modes are impossible in the open ocean, much in contrast to harbors and lakes. But that established truth turned out to be a misconception derived from a preoccupation with two-dimensional water motion. The first resonant wave modes of unbounded water bodies were discovered by Ursell in 1951, and it was soon recognized that their mechanism, in many cases, is precisely that of refraction.

A very simple, physical description of it was given by Longuet-Higgins (1967). Since the phase velocity of water waves increases with the local water depth, the parts of a wave crest lying over deeper water travel faster than the parts of the same wave crest lying over shallower water. In the course of its propagation, such a wave front therefore turns gradually toward the shallows. This is at the root of the common observation on beaches that the crests end up almost parallel to the shore line, even when they had been approaching the coast at an oblique angle from the sea. But if a wave be then reflected from shore, the same mechanism will make its crest turn gradually away from the direction facing straight out to sea, except under very special circumstances. The possibility therefore arises that some wave crests may be turned back toward the coast before they can reach the deep sea. They are then trapped between the coast and an invisible barrier; edge waves are an example. Once generated, such trapped waves must sweep again and again over the zone between shore and barrier, and if the phases match, then largely self-sustained, trapped wave modes become a plausible possibility, and a potential for resonances is indicated.

This challenge could not be met by the original ray methods. The geometrical optics approximation, on the other hand, proved equal to it (Section VII) by leaving the rays behind in daring steps toward simple, practical resonance conditions.

On the other hand, Keller's (1958) conjecture that the ray method constructs the first term of an asymptotic expansion of the linear water wave solutions with respect to the modulation parameter ε set in train a succession of researches aming to understand the foundation of refraction theory by proving the conjecture. The first obstacle encountered was the fact that the original, geometrical optics approximation (Section IV) reproduces the failure of the standard ray methods at caustics and shores. This obstacle was overcome by Shen and Keller by the construction of a uniformized approximation free of the defect (Sections V, VIII). As a by-product, quantitative estimates of maximal wave amplitudes were obtained (Section V), which had frustratingly eluded the earlier ray methods for nearly 30 years.

The next obstacle was discovered in the work on wave trapping. The geometrical optics approximation to this phenomenon indicates the occurrence of radiation damping and of direct harmonic excitation of resonance, which are nonetheless inaccessible to geometrical optics, as originally formulated, and remain inaccessible to it even after uniformization. To overcome this obstacle required a reexamination of the basic formulation of refraction, which led to a new conjecture: the first asymptotic approximation to the solutions of the classical linear surface-wave theory appears to be governed by a partial differential equation of Helmholtz' type (Section III). That equation implies not only the first, uniformized geometrical optics approximation, but also yields predictions of energy leakage and resonant excitation and even of wave reflection by underwater slopes (Section IX). These are subtle effects and their prediction represents a notable theoretical success, but full confidence in them must depend on mathematical proof from the classical linear theory or on experimental confirmation.

The latter appears at this time to be an even more difficult task than the former. In part, this derives from scale effects which impede observation of the predicted phenomena on a scale much smaller than the natural ocean scale. One aim of this article is to collect the theoretical developments into a coherent basis for experiment and observation.

The mathematical proof in the framework of the classical linear theory of surface waves is also still outstanding, except under very restricted circumstances (Appendix III). The voyage toward it, however, has already led to enough discoveries of potential, practical benefit to prompt this article. It starts with a brief outline of the framework of the classical linear theory of surface waves (Section II) and the introduction and discussion of the refraction equation (Section III). The original ray theory is then derived from it in the elegant and efficient geometrical optics version (Section IV), and its extension to a uniform ray theory is described in Section V.

The main difficulty still besetting the practical application of refraction theory concerns wave reflection at the shore. This is not a refraction effect, even on the linear theory of water waves. An appropriate specification of wave reflection at the shore is needed, however, as a boundary condition for the refraction equation whenever shores are relevant. Section VI comments on what appears known to date.

Section VII and VIII outline the results of the last decade on wave trapping by seabed topography. The first describes what can be found out already from the simple ray theory, the second describes further predictions derived from the refraction equation by an analysis relying strongly on the symmetry of strictly round islands. The extension to real islands appears still unclear, but the qualitative features of the predictions have general relevance and appear to be of immediate practical interest. The reader may be disappointed by the absence of direct, quantitative answers to the practical questions of ocean engineering, but the theory already offers much unsuspected physical insight to guide educated guesses.

Continental slopes and submarine ridges and valleys should plausibly reflect some part of waves traveling over them, especially in the case of long waves, but credible estimates of this refraction effect have eluded mathematical oceanography for generations; they are now available (Section IX).

Most of refraction theory concerns established, steady wave patterns, but ray theory has also been extended to the treatment of the earlier stages, at least, of the time evolution of wave patterns from initial conditions (Section X).

Finally, the advances of refraction theory sparked by Keller have stimulated applications and extensions to a multitude of related phenomena in oceanography, such as internal waves, Coriolis effects, currents, planetary waves, and many others, which will not be touched upon here because any attempt to do so would immediately lead far afield.

II. Linear Surface Waves

Mass conservation for a moving fluid can be expressed by

$$\partial \rho / \partial T + \operatorname{div}(\rho \mathbf{v}) = 0, \tag{2.1}$$

where $\mathbf{v}$ denotes the vector of fluid velocity and ρ, the density, which will be assumed constant in this article. Internal waves due to density stratification are thereby excluded from consideration, and the equation reduces to

$$\operatorname{div} \mathbf{v} = 0 \tag{2.2}$$

[For more details and background for this section, the reader may find Stoker (1957) or Meyer (1971a) useful.] For convenience, grad and div will be used throughout to denote three-dimensional vector operation; by contrast, two-dimensional operation in the horizontal plane of the undisturbed water surface will be denoted throughout by the ∇ symbol.

The basic notions of fluid kinematics imply that the boundary of any body of fluid is convected with the fluid motion: the normal component of fluid velocity relative to the boundary must vanish. Let X, Y, Z denote Cartesian coordinates fixed with respect to the earth, with Z measured vertically upward from the undisturbed water surface, and U, V, W, the corresponding components of the fluid velocity. If the water surface at time T be represented by the equation

$$Z = \zeta(X, Y, T),$$

the kinematic condition implies

$$D\zeta/DT = W(X, Y, \zeta, T),\qquad (2.3)$$

where

$$D/DT = \partial/\partial T + \mathbf{v} \cdot \text{grad}$$

is the "convective" rate of change seen by an observer moving with the local fluid velocity. The seabed will be assumed impermeable and hence is another such boundary segment. If it is represented by

$$Z = -H(X, Y),$$

then the kinematic condition implies

$$U\,\partial H/\partial X + V\,\partial H/\partial Y + W = 0 \qquad \text{at}\quad Z = -H(X, Y).\qquad (2.4)$$

The gravitational acceleration g will be taken constant and other body forces, such as Coriolis', will be ignored; their effects on refraction are relatively minor at wave periods of less than a few hours. The effects of the viscous stresses will also be neglected; they are relatively minor, except close to shore. The conservation of momentum can then be expressed by

$$\rho\, D\mathbf{v}/DT = -\text{grad}(p + \rho g Z).\qquad (2.5)$$

The interaction of waves and wind through the normal stress at the water surface is also of minor significance for refraction and will be neglected, so that

$$p(X, Y, \zeta, T) = \text{const.} = 0\qquad (2.6)$$

represents the dynamic boundary condition at the water surface.

Except close to shore (Section VI), water waves usually have a small amplitude in the sense that the slope of the water surface is small. Refraction can increase the amplitude markedly, but the present aim of the theory is more to identify the circumstances in which this occurs than to describe really large waves. It is therefore customary to linearize the dynamical conditions (2.5) and (2.6) by neglecting terms that are formally of second order in the velocity and surface elevation. Since the effect of currents has been reviewed recently in a volume of this publication (Peregrine, 1976) they will not be considered here, and the linearization then replaces $D\mathbf{v}/DT = \partial\mathbf{v}/\partial T + (\mathbf{v} \cdot \text{grad})\mathbf{v}$ by $\partial\mathbf{v}/\partial T$ in (2.5), which now shows the motion to have a velocity potential,

$$v = \text{grad}\,\Phi$$

governed by Laplace's equation,

$$\text{div grad}\,\Phi = 0,\qquad (2.7)$$

by (2.2).

Moreover, (2.5) now shows $\partial \Phi / \partial T + p/\rho + gZ$ to be a function only of the time, which may be absorbed into $\partial \Phi / \partial T$ without change to the velocity field grad Φ, and the surface condition (2.6) then becomes

$$\partial \Phi / \partial T + g\zeta = 0 \qquad \text{at} \quad Z = \zeta(X, Y, T).$$

The small-amplitude assumption together with (2.7) assures a high degree of smoothness for Φ even at the surface, and with $D\zeta/DT$ in (2.3) also replaced by $\partial \zeta / \partial T$, for consistency, the last equation may be written as

$$\partial^2 \Phi / \partial T^2 + gW = 0 \qquad \text{at} \quad Z = \zeta.$$

In turn, this may be compared with its value at $Z = 0$; by the mean-value theorem of calculus, the difference is also formally of second order in amplitude and hence is neglected so that (2.6) is finally replaced by

$$\partial^2 \Phi / \partial T^2 + g\, \partial \Phi / \partial Z = 0 \qquad \text{at} \qquad Z = 0. \tag{2.8}$$

Similarly, the surface elevation ζ is computed from

$$\zeta = -g^{-1}\, \partial \Phi / \partial T \qquad \text{at} \quad Z = 0. \tag{2.9}$$

Equations (2.7), (2.8), and (2.4) are collected below.

$$\frac{\partial^2 \Phi}{\partial X^2} + \frac{\partial^2 \Phi}{\partial Y^2} + \frac{\partial^2 \Phi}{\partial Z^2} = 0 \qquad \text{for} \quad 0 > Z > -H(X, Y),$$

$$\frac{\partial^2 \Phi}{\partial T^2} + g\frac{\partial \Phi}{\partial Z} = 0 \qquad \text{for} \quad Z = 0, \tag{2.10}$$

$$\frac{\partial \Phi}{\partial X}\frac{\partial H}{\partial X} + \frac{\partial \Phi}{\partial Y}\frac{\partial H}{\partial Y} + \frac{\partial \Phi}{\partial Z} = 0 \qquad \text{for} \quad Z = -H(X, Y).$$

Equations (2.10) together with appropriate initial, shore, and radiation conditions, form a complete system for the determination of the velocity potential Φ. They are the best-known system of water-wave equations and are often referred to as the "exact linear" equations because their analytical difficulty is still too formidable for most purposes.

Further approximation is therefore the norm, and the best-known is the "long-wave" (or "shallow-water") approximation which neglects vertical acceleration and other details of the vertical structure of the motion so that $\mathbf{v} = \mathbf{v}(X, Y, T)$. Many accounts of diverse motivations for such a formal approximation have been given (e.g., Peregrine, 1972; Meyer and Taylor, 1963). Typically, the waves described by (2.3) to (2.7) are *skin waves* of exponential decay as $Z \to -\infty$ (Longuet-Higgins, 1953) and are therefore significant only in a surface layer. Their penetration depth is seen from (2.7) to be the reciprocal of the wave number, that is, to be the scale on which the surface elevation ζ varies horizontally. When the water depth is small compared with this penetration depth, very little of the decay can actually take place, and $\mathbf{v} \approx \mathbf{v}(X, Y, T)$.

The equations governing such a long-wave approximation can be obtained as follows. By (2.2),

$$\frac{\partial}{\partial X} \int_{-H}^{\zeta} U \, dZ + \frac{\partial}{\partial Y} \int_{-H}^{\zeta} V \, dZ = \left(U \frac{\partial}{\partial X} + V \frac{\partial}{\partial Y} \right)(H + \zeta) - \int_{-H}^{\zeta} \frac{\partial W}{\partial Z} \, dZ$$

$$= -\partial \zeta / \partial T,$$

by (2.3) and (2.4). If the vertical variation of U and V is now neglected,

$$-\partial \zeta / \partial T = \frac{\partial}{\partial x} [U(\zeta + H)] + \frac{\partial}{\partial y} [V(\zeta + H)]$$

results and the velocity potential Φ may also be approximated by a function $\Phi_h(X, Y, T)$. Finally, terms formally of second order in the amplitude are again neglected, so that (2.9) may be used to approximate the last equation by

$$g^{-1} \partial^2 \Phi_h / \partial T^2 = \mathrm{div}(H \, \mathrm{grad} \, \Phi_h)$$

or

$$(\nabla^2 - (gH)^{-1} \partial^2 / \partial T^2)\Phi_h + H^{-1}(\nabla H) \cdot \nabla \Phi_h = 0, \qquad (2.11)$$

a wave equation for the horizontal variation of Φ, and similarly of ζ, by (2.9).

While the long-wave equation (2.11) has played an important historical role in the discovery of refraction phenomena, it is strictly appreciable only to the tides. Even for tsunamis, dispersion effects are applicable over longer propagation distances (Van Dorn, 1961; Carrier, 1966). For many waves of practical interest, the water depth exceeds the penetration depth over the continental shelf; it is much smaller near shore, of course, but even edge waves (Ursell, 1952) are not closely covered by (2.11). Refraction theory is therefore better founded on (2.10). However, the analytical difficulties of the "exact" linear theory for essentially three-dimensional motions are formidable, and further approximation is required for practical progress.

III. Refraction Equations

Most theoretical work on refraction has been based directly or indirectly on the geological fact (Shepard, 1963) that seabed slopes are small. Indeed, slopes appreciably in excess of 1/100 are rare, and surface waves on which the seabed has an influence are therefore generally characterized by a small parameter ε representing the seabed slope, with a typical value of $1/1000$. Moreover, the sedimentary origin of all but geologically very recent seabeds implies that the slope varies gently, that is, if X, Y are referred to a scale representing the surface-wave pattern, then the undisturbed water depth H depends smoothly on εX, εY, rather than on X, Y.

It is worth emphasis that the units of length, etc., must be intrinsic ones related to the wave pattern, if a nondegenerate analytical description is to result. Thus the vertical scale must represent the penetration depth of the water motion, which is proportional to the wavelength. The horizontal scale, by contrast, should be related to the scale of the wave pattern, i.e., to the scale on which the wavelength changes with distance or to the horizontal extent of the wave pattern, in the case of trapped waves (Meyer, 1971b). This is a little awkward, however, not only because the correct, intrinsic scales are not known *a priori*, but also because the same analysis may be designed to cover more than one set of scales. Most analyses therefore use frankly nonphysical, topographic scales to introduce nondimensional co-ordinates and water depth by

$$X = Lx, \qquad Y = Ly, \qquad T = (L/g)^{1/2}t,$$
$$Z = \varepsilon Lz, \qquad H(X, Y) = \varepsilon Lh(x, y). \tag{3.1}$$

In regard to the true scales of the motion, it may be noted right away that any intrinsic, horizontal scale could be at most of the order of L, but might turn out to be smaller. *A fortiori*, the wavelength must be anticipated to be usually very small in these units, and the solutions, therefore, to be so rapidly oscillating that they fail to tend to any functions at all as $\varepsilon \to 0$. The penetration depth must be anticipated to be similarly small compared with L, but not necessarily as small as the vertical scale εL. The need for an *a priori* choice of some intrinsic scaling is therefore only postponed by (3.1), not avoided. Indeed, an effective modification of the time scale will become necessary presently.

Since (2.10) is a linear system on a domain, and with coefficients, inde-pendent of time, its solutions may be Fourier-transformed with respect to t, and a considerable clarification results from concentrating attention on the individual Fourier components

$$\Phi(X, Y, Z, T) = (L^3 g)^{1/2} e^{-i\omega t} \phi(x, y, z) \tag{3.2}$$

of (dimensional) frequency $\omega(g/L)^{1/2}$ and surface elevation

$$\zeta = iL\omega\phi(x, y, 0).$$

For a direct interpretation as physical wave modes, the real part may be taken everywhere. Equations (2.10) now become

$$\frac{\partial^2 \phi}{\partial x^2} + \frac{\partial^2 \phi}{\partial y^2} + \varepsilon^{-2} \frac{\partial^2 \phi}{\partial z^2} = 0 \qquad \text{for} \quad 0 > z > -h(x, y),$$

$$\partial \phi / \partial z = \varepsilon \omega^2 \phi \qquad \text{at} \quad z = 0, \tag{3.3}$$

$$\frac{\partial \phi}{\partial x} \frac{\partial h}{\partial x} + \frac{\partial \phi}{\partial y} \frac{\partial h}{\partial y} + \varepsilon^{-2} \frac{\partial \phi}{\partial z} = 0 \qquad \text{at} \quad z = -h(x, y).$$

The asymptotic theory of such a system in the limit $\varepsilon \to 0$ is called a mathematical short-wave theory because it stands in an analogous relation to the "exact" linear theory of surface waves as short-wave optics stands to the theory of Maxwell's equations and the "quasi-classical approximation" stands to quantum mechanics. In an oceanographical sense, however, it is neither a short-wave nor a long-wave theory, but includes both, being merely an asymptotic theory for gentle seabed topography.

A definite asymptotic theory of (3.3), however, is seen to require a decision on the frequency scale. The most general choice is shown by (3.3) to be that permitting

$$\varepsilon \omega^2 = \eta$$

to be a parameter of order unity, and this will be assumed in the following. The real frequency is then

$$\omega (g/L)^{1/2} = [g\eta/(\varepsilon L)]^{1/2} \gg (g/L)^{1/2}$$

and the real time scale is only $(\varepsilon L/g)^{1/2}$. Smaller frequencies may still be covered by the limit $\eta \to 0$ of the results.

It may be observed that even the formal argument for linearizing the equations of motion (Section II) breaks down at the shore in this general refraction limit. The local depth there becomes smaller than the surface elevation, and the mean-value argument then applies to the seabed condition as well as the surface condition, giving in (3.3)

$$(\partial \phi / \partial x) \partial h / \partial x + (\partial \phi / \partial y) \partial h / \partial y = -\varepsilon^{-2} \eta \phi$$

at $z = 0$ so that $\phi^{-1} |\text{grad } \phi| = O(\varepsilon^{-2})$, contrary to the assumption of small amplitude. The shore will be discussed further in Section VI. The refraction limit $\varepsilon \to 0$ with η/ε^2 bounded, in which this particular difficulty disappears, is covered as a special "long-wave" limit by the general refraction approximation.

Analogies with optics are suggested already by the long-wave equation (2.11). Like all wave equations, it is soluble by the help of characteristics or rays, and such procedures were adapted (Munk and Traylor, 1947) to approximate forms of (2.10) by heuristic or ad hoc arguments to indicate constructions for rays and statements of energy conservation along them. This approach soon found acceptance in oceanography (Section IV). A major step forward, nonetheless, occurred with the introduction (Keller, 1958) of the mathematical theory of geometrical optics into surface-wave refraction.

In its original form, this consists in solving (3.3) by an asymptotic expansion

$$\phi \sim \cosh[k(z + h)] \, e^{iS(x,y)/\varepsilon} \sum_{r=0}^{\infty} A_r(x, y, z)\varepsilon^r \qquad (3.4)$$

analogous to the asymptotic expansion of short-wave optics. The striking feature of this expression is the rapidly varying exponential factor of local wavelength $2\pi\varepsilon/|\nabla S|$, which describes the pattern of crests and troughs of the water surface. Substitution in (3.3) shows the factor $\cosh[k(z + h)]$ to describe the vertical variation of the first approximation, which is of a form familiar from the simple solutions of (2.10) valid in water of uniform depth (Stoker, 1957). Accordingly, A_0 in (3.4) is independent of z and $A_0 \cosh(kh) = A_s(x, y)$ describes the main distribution of amplitude over the water surface. The higher order terms in the expansion describe corrections to both the surface pattern and vertical structure. Thus (3.4) is a physically well-motivated and plausible form of description for surface-wave patterns.

It can hardly be emphasized rapidly enough, however, that the "asymptotic expansion" (3.4) is *false*, like so many formal expansions that appear "to work" because they set up an apparently consistent procedure for the determination of successive terms. Nonetheless, (3.4) has been signally helpful, and one of its most important services has been to set refraction into a rational framework of *asymptotic approximation* to the solutions of the "exact" linear equations (2.10). It put an end to individual, heuristic approaches by focusing attention on the proper aim of correct asymptotic approximation: it changed the questions asked.

The new viewpoint motivates the definition of the object of refraction *theory* as that of determining the *first asymptotic approximation* to the solution of the classical linear surface-wave equations (2.10) as $\varepsilon \to 0$ for smooth, gentle seabed topography.

This frees us automatically of the incorrect expansion in (3.4): in view of the smallness of ε in nature, there is little practical interest in higher order corrections to the true, first approximation. If an approximation be unsatisfactory, on the other hand, attempts to improve it by higher order corrections based on it only throw good money after bad. The rational definition of refraction theory also removes from it complications arising from local exceptions to the gentleness of the seabed topography, as at harbor works or rocky promontories. These give rise to *diffraction* effects which demand in principle a reexamination of the limit $\varepsilon \to 0$ for regular $h(x, y)$ implied in (3.3) (see Appendix to this section).

A more immediate question arises within refraction theory proper: if (3.4) be false, what approach can lead to the true, first asymptotic approximation? Two main reasons are known for the failure of (3.4). The first (Keller, 1958) is recognized by the fact that direct determination of the function $A_0(x, y)$ from (3.3) (Section IV) shows it to develop singularities. These turn out to be local failures of the asymptotic quality of the approximation scheme (3.4). Originally, they were bridged conjecturally by drawing on the experience from exact solutions of (2.10) for specially simple, "canonical" problems. Later, a much more sophisticated approximation scheme (Section V) was proposed by Shen and Keller (1975) which contains (3.4) but is free from such flaws. This *uniform refraction approximation* appears likely to yield indeed a first asymptotic approximation to a large class of solutions of (2.10).

A second reason for the failure of (3.4) arises for waves which grow or decay in time so that the frequency ω is complex. Then (3.3) is found to imply that $S(x, y)$ in (3.4) must also be complex. If $S_i(x, y)$ denotes its imaginary part, the factor $\exp(-S_i(x, y)/\varepsilon)$ will change so much in magnitude over short distances as to dwarf any changes described by the asymptotic expansion in (3.4). Indeed, a unit difference in S_i corresponds to a ratio $\exp(-1/\varepsilon)$ in the corresponding values of $|\phi|$, which is smaller than every term in the expansion (3.4)! In such circumstances, any asymptotic scheme involving powers of ε becomes useless and the form of the solution must be altogether different from (3.4).

The mathematical approach to the asymptotics of (3.3) for complex frequency remains to be explored, but a consensus appears to be developing on what may be a promising start. It derives from the basic refraction notion that the depth $h(x, y)$ is a smooth function of the topographic coordinates while the surface elevation is a function of coordinates $x/\varepsilon = x'$ and $y/\varepsilon = y'$ measured on the wavelength scale. The vertical structure of the motion, i.e., the dependence of the velocity on z at fixed x, y, is a property of the wave on the scale of the penetration depth, which is comparable to the wavelength. Its nondegenerate description in the limit $\varepsilon \to 0$ must therefore be in terms of x', y'. In this frame, $h = h(\varepsilon x', \varepsilon y')$ tends to a constant as $\varepsilon \to 0$ and it is therefore plausible that the first approximation to the *vertical structure depend on* this constant—which is *the local depth* $h(x, y)$—but not on derivatives of h, which tend to zero with ε in the relevant x', y'-frame. This basic conjecture cannot, of course, be justified by assuming some formal expansion, but a closer motivation, based on averaging (3.3) over the water depth, has been given by Lozano and Meyer (1976), and for long waves a partial proof has been given by Harband (1977) (see the Appendix to this section).

The conjecture implies immediately that

$$\phi(x, y, z) \sim \phi(x, y, 0) F_0(z),$$

where F_0 is the function predicted by (2.10) for the case of uniform water depth [equal to the local value of $h(x, y)$ so that F_0 depends on x, y parametrically through h]. But, for uniform water depth, the first and last of Eqs. (3.3) imply $F_0''/F_0 = $ const. and $F_0'(-h) = 0$, whence

$$F_0(z) = \cosh[k(z + h)]/\cosh(kh)$$

with

$$k \tanh(kh) = \varepsilon \omega^2. \qquad (3.5)$$

Now multiply the first of (3.3) by F_0 and integrate over z from $-h$ to 0 to obtain for $\phi(x, y, 0) = \Psi(x, y)$

$$\int_{-h}^{0} F_0 \nabla^2(\Psi F_0)\, dz + \varepsilon^{-2} k^2 G\Psi = 0, \qquad G(x, y) = \int_{-h}^{0} F_0^2\, dz$$

because $F_0 = F_0(z; x, y)$ with $\partial^2 F_0/\partial z^2 = k^2 F_0$. The symbols ∇ and ∇^2 denote the gradient and Laplace operators with respect to x, y. This may also be written

$$\nabla(G\nabla\Psi) + \varepsilon^{-2} k^2 G\Psi = F_0^2\Big|_{z=-h} (\nabla h)\cdot\nabla\Psi - \Psi\int_{-h}^{0} F_0\nabla^2 F_0\, dz,$$

because $\nabla G = 2\int F_0 \nabla F_0\, dz + F_0^2 \nabla h$, and the last of (3.3) is in terms of the present notation

$$\Psi F_0(\nabla F_0)\cdot\nabla h + F_0^2(\nabla\Psi)\cdot\nabla h = 0 \qquad \text{at} \quad z = -h,$$

because $\partial F_0/\partial z = 0$ at $z = -h$, leaving

$$\nabla(G\nabla\Psi) + \varepsilon^{-2}\Psi\left\{k^2 G + \varepsilon^2\int_{-h}^{0} F_0\nabla^2 F_0\, dz + \varepsilon^2 F_0\nabla F_0\Big|_{z=-h}\cdot\nabla h\right\} = 0.$$

The smoothness of $h(x, y)$ implicit in refraction theory is inherited by $F_0(z; x, y)$ via (3.5). As $\varepsilon \to 0$ therefore, the last brace tends to $k^2 G$, uniformly in x, y, and to the first asymptotic approximation that we are seeking in refraction theory, the surface pattern $\Psi(x, y) = \phi(x, y, 0)$ is likely to be governed by

$$\nabla(G\nabla\Psi) + \varepsilon^{-2} k^2 G\Psi = 0,$$
$$G = \int_{-h}^{0} F_0^2\, dz = [\sinh(2kh) + 2kh]/[4k\cosh^2(kh)]. \qquad (3.6)$$

(Berkhoff, 1973; Jonsson *et al.*, 1976; Lozano and Meyer, 1976).

The auxiliary function $G(x, y)$ is a measure of the local water depth relative to the vertical wave structure or, very loosely, the local depth as the wave sees it. It may also be written as $\varepsilon|\mathbf{cg}|$ in terms of the classical phase and group velocities for linear waves on water of uniform depth. Equation (3.6) is seen to be a form of Helmholtz, or reduced wave, equation of large wave number k/ε for the surface pattern. The wave-number function $k(x, y)$ is determined by the full, classical dispersion relation (3.5) covering the whole oceanographic range from the long-wave approximation $\eta h = \varepsilon \omega^2 h \to 0$ to the short-wave limit $\eta \to \infty$. For water of finite, nonzero depth, (3.6) may also be written

$$\nabla^2 \psi + [\varepsilon^{-2} k^2 - G^{-1/2} \nabla^2 (G^{1/2})] \psi = 0 \qquad (3.7)$$

for $\psi = G^{1/2} \psi$, $\phi(x, y, z) = F_0 G^{-1/2} \psi$.

While a proper justification of (3.6) as the equation governing the first asymptotic approximation to (2.10) as $\varepsilon \to 0$ for gentle seabed topography remains to be given, it can be seen to agree with the appropriate exact solutions and to contain all the established approximations. For water of uniform depth, for instance, $\phi = F_0 \Psi$ satisfies (2.10) exactly, if Ψ satisfies (3.6), which then reduces to the standard form

$$\nabla^2 \Psi + (k/\varepsilon)^2 \Psi = 0$$

of the reduced wave equation. In the limit of deep water, moreover, $kh \to \infty$, $F_0 \sim \exp(\eta z)$, $k/\varepsilon \sim \omega^2$.

For shallow water, $\eta h \to 0$, (3.5) gives $k^2 \sim \eta/h$ so that $G \sim h$ and (3.6) tends to (2.11) for frequency ω and hence governs long-wave refraction as a special case. Conversely, the similarity between (3.6) and (2.11) is noteworthy: both are Helmholtz equations and therefore present the same type and degree of difficulty. The added flexibility and precision of (3.6) derives formally from the replacement of the local water depth H by the wave-depth function G computed from H by the dispersion relation (3.5).

The exact solution of (2.10) is known for a plane sloping beach (Friedrichs, 1948; Peters, 1952; Roseau, 1952) and (3.6) reproduces (Lozano and Meyer, 1976) Friedrichs' (1948) asymptotic approximation to it as $\varepsilon \to 0$ for finite, positive depth h and two-dimensional motion. Furthermore, if a solution of form (3.4) exists, then (3.6) reproduces (Lozano and Meyer, 1976) Keller's (1958) $S(x, y), A_0(x, y)$. It follows that the asymptotic approximation (Shen et al., 1968) as $\varepsilon \to 0$ to Ursell's (1952) exact edge-wave solutions on a plane sloping beach is also reproduced. Application of (3.6) to round islands (Lozano and Meyer, 1976) showed it also to reproduce the typical results of uniform ray theory (Section V), but without limitation to real frequency.

Other ray approximations are reproduced by (3.5) and (3.6), if further restrictive or ad hoc assumptions be added. All these comparisons indicate that (3.5) and (3.6) are the best general refraction equations known at the time of writing.

Appendix III. **Comments on Diffraction**

To complement the outline of a rational and fruitful theory of refraction just attempted, it will be useful to add some remarks illustrating its limitations. Realistic circumstances may involve local exceptions to the gentleness of the seabed slope and the effects arising therefrom are the main object of diffraction theory. To divide the difficulties, consider first those exceptions which arise at the lateral boundaries of the water body, and to divide further, begin with those among them not involving shores of natural, small beach slope.

Parts of the water boundary may be representable as clear-cut geometrical surfaces because they concern drastic changes in water depth over only a small fraction of a wavelength. Notions of refraction are then clearly inapplicable locally, but if the seabed variations are gentle otherwise and if the radius of curvature of the boundary surface is comparable to (or even larger than) the scale of the gentle seabed variations, then the effect of such parts of the water boundary may be very plausibly represented by boundary conditions for the refraction equations (3.6) or (3.7). Essentially, such a boundary condition specifies how a wave incident on the boundary gives rise to a reflected wave. A notable analysis of this reflection process is due to Kajiura (1961). Of course, it does not justify the rather common expedient of using "seawall" boundaries of this type to shirk the problems of natural shores.

Complications may arise, but some of them can still be accommodated similarly within the framework of refraction. For instance a rocky promontory may be adequately representable as a corner of a boundary surface and can then be interpreted as a point-source of reflected waves which can themselves be treated by refraction theory, given the character of the point source. The formalism for this can be taken over from electromagnetic theory (Lewis and Keller, 1964), and will not be discussed here. Again, a smooth boundary surface may give rise to a wave shadow and to creeping waves within it, which have their source on the shaded part of the boundary and can be treated by refraction theory (Lewis and Keller, 1964). Of course, the shadow region of one wave may be illuminated by another wave, and higher order refraction corrections to the latter—which are not discussed in the following—may then be larger than the amplitude of the former. The nature of a shadow boundary will be discussed in Section V.

Other complications may require a more genuine diffraction analysis. For instance, if the radius of curvature of the boundary surface be comparable to the wavelength, then its representation by a mere boundary condition is inadequate because it gives rise to genuine diffraction processes in the interior of the water body. Similarly, if such a boundary casts a shadow of width comparable to the wavelength, refraction theory cannot treat the diffraction phenomena encountered in and near the shadow.

A related difficulty of much more pervasive impact on oceanographic wave refraction arises from natural shores with beaches of small slope. The gentleness of the seabed slope is here of no avail, since the percentage change in water depth over a wavelength is not small near shore, and ease of physical access has made the difficulties of understanding wave shoaling well appreciated. The hope that shoaling could be represented by a boundary condition for the refraction equation (3.6), or even the classical equations (2.10), has not been well realized and the problem will be discussed in Section VI. In relation to diffraction, it should be noted here that, even if we assume the refraction equation (3.6) to hold up to the shore, its solutions have near the shore a character quite different from refraction. The reason for it is that the large parameter ε^{-2} in (3.6) is drowned locally by the singularity of $G^{-1}\nabla G$; the latter therefore dominates the shore character of the solutions. It will be seen in Sections V and VIII how solutions of the refraction equation (3.6) uniformly valid up to the very shore can be obtained nonetheless, but this mathematical achievement should not be permitted to obscure the critique of it in Section VI.

Related problems arise, of course, from banks shallow enough to give rise to shoaling.

Apart from these, diffraction effects beyond the scope of refraction theory arise also far from the boundaries of a water body in the rather rare instances of local exceptions to the gentleness of the seabed variation. For the case of a sharp underwater escarpment, reference may again be made to the work of Kajiura (1961).

Another analysis (Harband, 1977) has succeeded in proving the basic refraction hypothesis at least under limited circumstances and providing, moreover, a method for the quantitative estimation of the errors of the refraction approximation. This was achieved by investigating the perturbation to a plane, progressive, long wave on water of uniform depth that is caused by a mound or depression of the seabed of bounded lateral extent. To be specific, Harband (1977) starts from the hypotheses (i) that the unperturbed wave amplitude is small enough, by comparison to both wavelength and water depth, for validity of the classical linear equations (2.10) and (ii) that the height of the seabed mound or depression is small enough, by comparison again to the water depth and wavelength, for validity of a linear perturbation of the seabed boundary condition. These two *a priori* limits indicate equations amenable to solution by Fourier transform. Determination of the perturbation to the basic plane wave also requires radiation conditions. An alternative approach, used by Harband, is to start with the unperturbed plane wave over water of uniform depth, let the mound grow slowly and smoothly to its final, arbitrary shape and wait with the inversion of the transform until the transients have disappeared to leave the permanent wave perturbation unobscured. Of course, the difference between the two approaches is more apparent than real. For a more transparent interpretation of the solution, the water depth is now permitted to tend to zero by comparison with the wavelength of the unperturbed wave so that the linear long-wave limit results and the water volume directly above the mound tends to zero by comparison to (wavelength)^3. This is a reasonable step in that long waves may be expected to be particularly sensitive to seabed perturbations. The resulting limit, finally, is approximated by the stationary-phase principle for gentle seabed variation: the shape function of the mound is kept fixed, and so is the ratio of the earth volume in the mound to the water volume under an area (wavelength)^2, but the horizontal scale of the mound is taken large compared with the wavelength. To the first approximation, the result confirms the basic refraction hypothesis that the local wave number depends only on the local depth. Higher order corrections, moreover, could be extracted from Harband's (1977) representation of the solution.

IV. Simple Ray Theory

A. Ray Equations

The most efficient form of the ray methods familiar in coastal ocean-ography has been developed by Keller and his associates in their application of the geometrical optics approximation to short-wave physics. Substitution of (3.4) into (3.6) [or into (3.3), which is shown in Appendix IV,a to this section to lead to the same result (Keller, 1958)], noting that

$$\nabla(G\nabla\Psi) = \Psi\nabla(G\Psi^{-1}\nabla\Psi) + G\Psi|\Psi^{-1}\nabla\Psi|^2,$$

yields, when $A_0 \neq 0$,

$$0 = \varepsilon^{-2}(k^2 - |\nabla S|^2) + \frac{i}{\varepsilon G}\nabla(G\nabla S)$$

$$+ \frac{2i}{\varepsilon}(\nabla S)\{A_0^{-1}\nabla A_0 + \tanh(kh)\nabla(kh)\} + \Sigma_{15},$$

with Σ_{15} bounded as $\varepsilon \to 0$. As $\varepsilon \to 0$, therefore,

$$|\nabla S|^2 = k^2, \tag{4.1}$$

$$\nabla(A_s^2 G\nabla S) = 0 \tag{4.2}$$

become necessary, where $A_s(x, y) = A_0 \cosh(kh)$ represents the first approximation to the surface amplitude. Σ_{15} is ignored (Appendix IV,a).

Given the seabed topography $h(x, y)$, (3.5) thus determines $k(x, y)$ and (4.1) is a differential equation for the "phase function" $S(x, y)$, the "eiconal equation" of optics. The "transport equation" (4.2) is a "conservation law" for the first approximation to energy flux $\int|\phi|^2 \, dz$ in the gradient direction of S because the divergence theorem recasts it as

$$\int_C A_s^2 G \frac{\partial S}{\partial n} \, ds = 0 \tag{4.3}$$

around any (simple, rectifiable) closed curve C in the x-, y-plane.

The eiconal equation (4.1) is a first-order partial differential equation and may therefore be solved efficiently by the method of characteristics (Courant-Hilbert, 1962) as follows. The lines $S(x, y) = \text{const.}$ are called "phase lines"; they include the crest lines and trough lines of the surface pattern, by (3.4). Their orthogonal trajectories are called the "rays" of (4.1); if they be represented by $\mathbf{x} \equiv (x, y) = \mathbf{x}(\sigma)$ in terms of the arclength σ along them, then $d\mathbf{x}/d\sigma \parallel \nabla S$ and therefore $1 = |d\mathbf{x}/d\sigma|^2 \propto |\nabla S|^2 = k^2$, whence

$$d\mathbf{x}/d\sigma = k^{-1} \nabla S. \tag{4.4}$$

It follows from (4.1) that

$$dS/d\sigma = (\nabla S) \cdot d\mathbf{x}/d\sigma = k^{-1}|\nabla S|^2 = k, \tag{4.5}$$

which is an ordinary differential equation for S along the rays and thus shows them to be the characteristic curves of (4.1) (Courant-Hilbert, 1962). Anyway, if we know a ray $\mathbf{x}(\sigma)$, then

$$S(\mathbf{x}(\sigma)) = S(\mathbf{x}(\sigma_0)) + \int_{\sigma_0}^{\sigma} k(\mathbf{x}(\sigma')) \, d\sigma'$$

gives us the explicit variation of the phase function along it.

The conservation law (4.2) also relates to the rays. Choose any connected segment of a phase line $S(x, y) = S(\sigma_0)$, let $l(\sigma_0)$ denote its arclength and follow the rays through its endpoints to another phase line $S(x, y) = S(\sigma)$. They will there mark off a segment of arclength $l(\sigma)$, and if $|\sigma - \sigma_0|$ is not too large, these ray segments and phase line segments form a simple, rectifiable closed curve. But $\partial S/\partial n = 0$ on the rays and $\partial S/\partial n = dS/d\sigma = k$ on the phase lines, by (4.5), if the normal is taken in the sense of σ increasing. Hence, (4.3) becomes

$$\int_{l(\sigma)} A_s^2 G k \, ds = \text{const.}$$

independent of σ. Some accounts of ray methods are based on the *a priori* assumption of a conservation statement of this type. For a more local statement of energy flux invariance along rays, an expansion ratio of the ray family may be defined as

$$\Xi(\sigma) = \lim_{l(\sigma_0)\to 0} [l(\sigma)/l(\sigma_0)], \qquad (4.6)$$

and then

$$A_s^2 k G \Xi \equiv A^{*2} = \text{const. along a ray.} \qquad (4.7)$$

Mathematically, all is thus reduced to finding the rays $\mathbf{x}(\sigma)$ in the first place, for which we need to eliminate S from (4.4). That may be done by noting, in Cartesian coordinates x_i $(i = 1, 2)$,

$$\frac{d}{d\sigma}\frac{\partial S}{\partial x_i} = \sum_j \frac{dx_j}{d\sigma}\frac{\partial^2 S}{\partial x_j \partial x_i} = \frac{1}{k}\sum_j \frac{\partial S}{\partial x_j}\frac{\partial^2 S}{\partial x_j \partial x_i} = \frac{1}{2k}\frac{\partial}{\partial x_i}\sum_j \left(\frac{\partial S}{\partial x_j}\right)^2,$$

by (4.4), so that (4.1) and (4.4) give

$$\frac{d}{d\sigma}\left(k\frac{dx_i}{d\sigma}\right) = \frac{\partial k}{\partial x_i}, \qquad (i = 1, 2), \qquad (4.8)$$

which is a system of two ordinary differential equations for $x\sigma()$, $y(\sigma)$, in which $k = k(x, y)$ is given by (3.5).

The system (4.8) is actually very familiar from elementary mechanics. The vector $d\mathbf{x}/d\sigma$ is the unit tangent vector to the ray curve, and if $k\,d\mathbf{x}/d\sigma$ is thought of as corresponding to the momentum of a point particle, then (4.8) is Newton's equation governing its motion on an orbit corresponding to the ray under a force field corresponding to ∇k. This analogy can be helpful in solving the ray equation (4.8), for instance, when the seabed topography is axially symmetrical (Appendix IV,b).

Naturally, special ideal topographies have been sought out for which (4.8)—or at least its long-wave approximation resulting from use of $k^2 h = 1$ in the place of (3.5)—have explicit solutions in terms of elementary functions (Appendix IV,b). Natural topographies require numerical treatment to which references may be found, e.g., in Chao (1974). Figure 1 (from Skovgaard *et al.*, 1976) shows a set of rays computed with the full dispersion relation (3.5). Figures 2 and 3 (from Wilson *et al.*, 1965) show long-wave ray patterns computed for Monterey Bay; thus the rays are here constructed for (2.11), rather than for (3.5) and (3.6).

These figures illustrate that the finer details of seabed topography are not conspicuously reflected in the ray patterns because (4.8) is a second-order differential equation. Starting from the topography $h(x, y)$, ray construction therefore involves a double averaging process. The striking features of Figs. 2 and 3 arise from very pronounced underwater canyons.

Figure 1 illustrates the relevance of the expansion ratio (4.6), proportional to normal distance between adjacent rays, to the evaluation of amplitude

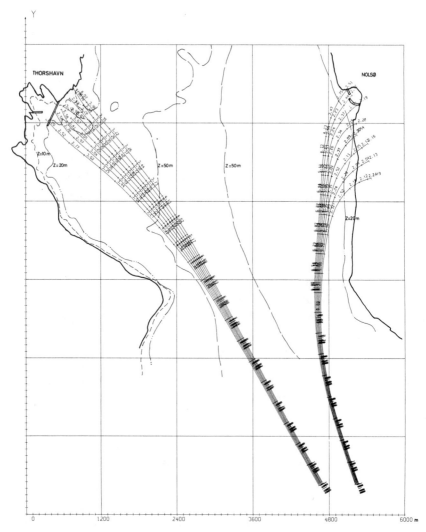

FIG. 1.   Ray diagram for Nolso Fjord on Faroe Islands, from Fig. 8, O. Skovgaard, I. G. Jonsson, and J. A. Bertelsen, Computation of wave heights due to refraction and friction. Reprinted from *J. Waterw., Harbors, Coastal Eng. Div., Am. Soc. Civ. Eng.* **102**, 103, by permission of the American Society of Civil Engineers. © 1976 by the American Society of Civil Engineers, New York.

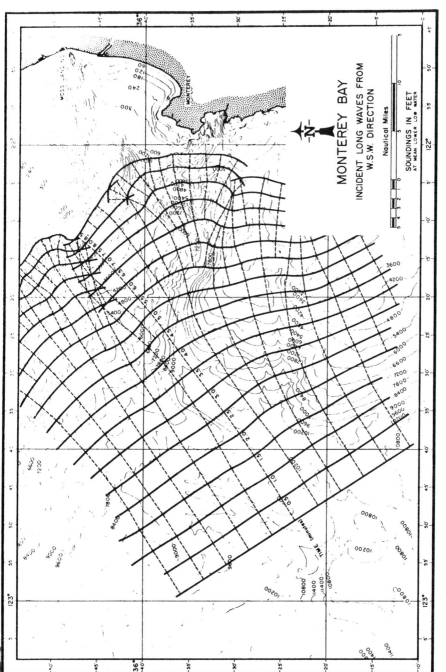

FIG. 2. Ray diagram for longwaves entering Monterey Bay, from Fig. 44, B. W. Wilson, J. A. Hendrickson, and R. E. Kilmer, "Feasibility Study for a Surge-action Model of Monterey Harbor, California." Reprinted from Contract Rep. No. 2-136, p. 104 (1965), by permission of the Waterways Experiment Station, Corps of Engineers, U. S. Army, Vicksburg, MI 39180.

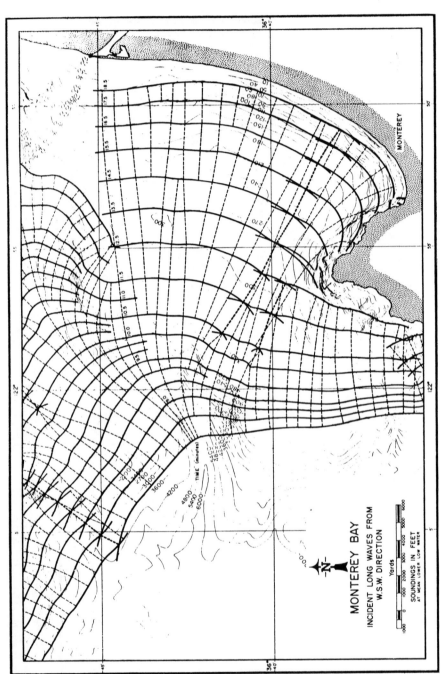

FIG. 3. Ray diagram for longwaves in Monterey Bay, from Fig. 47, B. W. Wilson, J. A. Hendrickson, and R. E. Kilmer, Feasibility Study for a Surge-action Model of Monterey Harbor, California. Reprinted from Contract Rep. No. 2-136, p. 108 (1965), by permission of the Waterways Experiment Station, Corps of Engineers, U.S. Army, Vicksburg, MI 39180.

change along rays from (4.7). On that basis alone, the amplitude would seem to decrease as a ray is followed toward shore in Fig. 1, but the factor kG varies in the opposite sense. In fact, even without consideration of nonlinear shoaling effects, (4.7) breaks down near the shore because k and G become singular.

Another difficulty in the direct use of (4.7) for the computation of local wave amplitudes arises from the fact that more than one ray may pass through a point of interest, even though the wave train incident from the open sea be plane and monochromatic. This possibility is illustrated in Figs. 2 and 3 and the realism of such complicated refraction patterns is shown strikingly by the aerial photograph of Fig. 4 (from Pierson *et al.*, 1955). If several rays thus pass through the same point, the respective waves interfere there.

FIG. 4. U.S. Coast and Geodetic Survey aerial photograph of waves off Ocracoke, North Carolina. An underwater ridge extends from the coast into the ocean. The ocean swell approaches in a direction near-normal to the right-hand edge of the figure, but the waves are then refracted toward the ridge crest from both sides of the ridge, demonstrating how the seabed turns the wave crests toward the shallows. A complicated wave interference pattern results over the ridge. Photo by U.S. Coast and Geodetic Survey, from Fig. 6.4, p. 190, of W. J. Pierson, G. Neumann, and R. W. James, "Practical Methods for Observing and Forecasting Ocean Waves," Publ. HOP-603, U.S. Naval Oceanographic Office, Washington, D.C., 1955.

An important study of ray tracing for a complicated, natural shelf topography is due to Chao (1974). It includes account of the sphericity of the earth, which has a quantitatively noticeable effect on waves traveling across large continental shelves. Careful computation indicated arrival of up to seven distinct rays at the target point even for a single wave period and swell direction in the ocean (Fig. 5). This does not, however, necessarily imply a marked amplitude enhancement at such a point. One corollary is that computation cannot generally be simplified by the procedure of tracing rays backward from the target point. Another is that even great effort cannot guarantee quite precise quantitative amplitude predictions by ray tracing

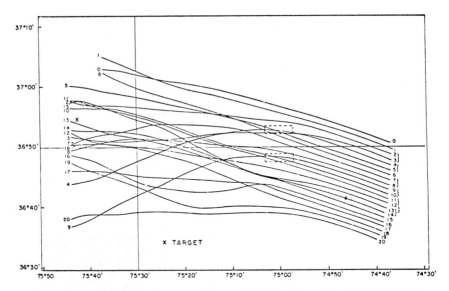

FIG. 5. Ray diagram for waves near Chesapeake Light Tower, from Fig. 3, Y.-Y. Chao, "Wave Refraction Phenomena Over the Continental Shelf Near the Chesapeake Bay Entrance," Rep. TM-47, p. 17. Dept. Meteorol. Oceanogr., New York University, New York, 1974. Reprinted from Document AD/A-002 056, Natl. Tech. Inf. Serv., U.S. Dept. of Commerce, Springfield, VA 22161.

for natural topographies. This approach may have passed a point of diminishing returns.

B. CAUSTICS

While ray patterns (Figs. 1–5) are helpful and illuminating, they are not of *direct* oceanographical interest. Typical information of primary interest would be maximal amplitudes and their location, and here (3.4) begins to fail us because (4.7) must be anticipated to predict infinite amplitudes in many cases. This happens where $kG \to 0$ or where the ray expansion ratio $\Xi \to 0$. The former occurs where the depth $h \to 0$, as at shores, by (3.5) and (3.6). The latter occurs where the rays form an envelope, called caustic by analogy with optics; some examples may be seen in Figs. 2 and 3.

A striking demonstration of the caustic effect has been given by Chao and Pierson (1972). A square wave tank consists of a shallower portion and a deeper one, connected by a bank in the tank floor running along a diagonal of the tank. One side of the tank, adjacent to the shallow portion, is formed by a wavemaker, and the other sides by vertical walls or wave absorbers; in the bird's-eye view of Fig. 6 (from Pierson, 1972b) the bank runs parallel

WAVE BEHAVIOR NEAR CAUSTICS

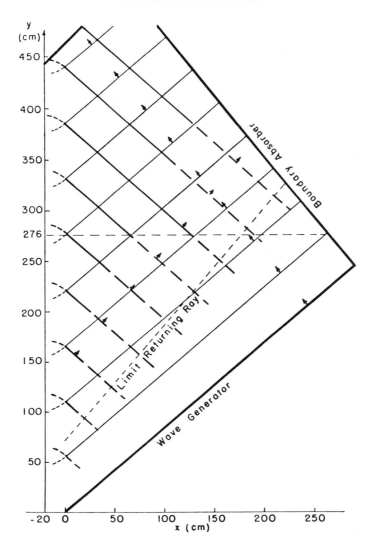

FIG. 6. Waves in a tank with an underwater bank (running along the *y*-axis of the figure). The wave crests seen on a snapshot of the motion are shown. One set of crests, parallel to the wave maker, belongs to the waves directly generated by it. The other, belongs to the waves traveling, after refraction over the bank, toward the absorber on the upper right-hand side. From Fig. 4, W. J. Pierson, Wave behavior near caustics in models and in nature. Reprinted from "Waves on Beaches" (R. E. Meyer, ed.), p. 171. © 1972 by Academic Press, New York.

to the left edge of the figure, the wavemaker occupies the lower edge. There is a range of wavemaker periods for which the waves are observed to cross the tank in a straightforward way, but over another range of periods, they are found unable to cross over the bank to the deeper portion! Instead, their amplitude peaks markedly over the bank and they are turned back, as from an invisible barrier, to recross the shallow portion in a different direction (Fig. 6). They are trapped over the shallow portion of the tank.

The same effect can occur over natural underwater banks and can be equally notable, not only for the local amplitude enhancement, but also for the cross-wave field it generates (Fig. 4). A vessel sailing into such a field may encounter a sudden aggravation of sea conditions much more serious than mere steepening of plain swell, and Pierson (1972a) has documented this as a probable cause of marine disasters by a close study of windfields and refraction patterns at the time and site of two trawler losses.

The analysis of refraction over a straight bank and more generally, over a straight continental slope, is simple and instructive. Suppose that the water depth depends only on distance x from a straight shore $x = 0$ and for simplicity let $h(x)$ be strictly monotone-increasing. The general refraction hypothesis of gentle seabed topography implies an upper bound on dh/dx, but no further specification will be necessary because there will be no need of actual rays. By (3.5), $k = k(h)$ decreases strictly monotonely from large values $k \sim h^{-1/2}$ near shore to a limit $k_\infty \geq \varepsilon \omega^2$ as $k \to \infty$. From (4.8), $k\, dy/d\sigma = $ const. and if c denotes its absolute value, then from (4.4) and (4.1),

$$\partial S/\partial y = \pm c, \qquad \partial S/\partial x = \pm (k^2 - c^2)^{1/2}, \qquad c \geq 0, \qquad (4.9)$$

whence (4.4) gives $dx/d\sigma$ and

$$\frac{dy/d\sigma}{dx/d\sigma} = \pm c(k^2 - c^2)^{-1/2} \qquad (4.10)$$

describes the ray slopes (Shen et al., 1968). The role of the parameter c is seen by picking an arbitrary reference point P (not at the shore, but also not too far from it). Then k depends at P only on the local depth and c determines the direction in which the ray passes through P. (At first acquaintance, it may seem strange that an infinity of ray directions are possible at a single point P, but this is clearly needed for the description of a point source, e.g. This shows also why the differential equation (4.8) of the rays must be at least of second order.) For a given physical wave, the value of c is determined by the manner in which the wave is generated.

It can be observed right away that (4.10) allows two basically different cases: if $c^2 < k_\infty^2$, then dy/dx is real for all $x \geq 0$, but if $c^2 > k_\infty^2$, then the ray slope is real only for $0 < x < a$ with a defined by $k(a) = c$, and no real rays can exist beyond $x = a$. This second case occurs whenever the initiation parameter c is in the range of $k(h)$ defined directly by (3.5). There is then a line $x = a > 0$ parallel to shore marking the outer border of the wave region. It traps the waves in $0 < x < a$; they are edge waves. At $x = a$, (4.10) gives $dx/dy = 0$ on the rays and they are refracted back into $x < a$ because (4.10) shows

$$\frac{d^2x}{dy^2} = \frac{dx}{dy}\frac{d}{dx}\frac{dx}{dy} = \frac{k}{c^2}\frac{dk}{dx} < 0,$$

if $dh/dx > 0$, since $k(h)$ is strictly monotone-decreasing, by (3.5). Therefore $x = a$ is a ray envelope or caustic. It is the only one because the ray slope depends only on h and is nonzero and finite for $0 < x < a$. Hence, $x = 0$ and $x = a$ mark the singularities of the amplitude A, by (4.7), (3.6), and (3.5). To determine a, incidentally, we need only compute the number $(1/c)\,\mathrm{arctanh}(1/c) = h(a)$, by (3.5).

When waves are thus refracted back from a caustic barrier, we encounter cross-waves, as in Figs. 4 and 6. Since the rays are defined as the orthogonals to the phase lines, we also encounter cross-rays, as allowed by the double sign in (4.10). Through each point of the wave region in Fig. 6, therefore, pass two distinct rays with the same value of c. It is important to realize that this source of confusion in ray patterns is normal. At a vertical tank wall, for instance, waves are reflected; the incident and reflected waves each satisfy (3.3) and each therefore has its own ray pattern. Through each point near the wall, accordingly, passes both an "incident" and a "reflected" ray. Other reflections may occur, e.g., at shore lines, and to simplify the text, caustic refraction will also be referred to as caustic reflection. Generally, then, the complete wave pattern will arise from the superposition of a finite number N of individual waves, each of which satisfies (3.3), and (3.4) must be replaced by

$$\phi \sim \sum_{j=1}^{N} \cosh[k(z + h)]\,A_j(x, y)\exp[iS_j(x, y)/\varepsilon]. \tag{4.11}$$

The wave number function $k(x, y)$ depends only on $h(x, y)$, by (3.5), not on j. To avoid confusion in the discussion of the associated ray families, moreover, it is helpful to associate with each wave j a separate sheet of the "wave region j," meaning that part of the (undisturbed) water surface which is actually reached by rays of the wave j.

The notion of sheets helps to clarify the nature of physical wave patterns. Over a continental slope, for instance, a wave striking the shore $x = 0$ is expected to be reflected, even if perhaps only partially, so that the x-component of the wave number vector changes sign, but the y-component does not. Now, the unit wave number vector (4.4) defines the unit ray-tangent vector, so shore reflection changes the sign of the ray slope dy/dx in (4.10). The wave region must therefore be covered by two ray "congruences," one representing the wave incident on the shore and the other, the reflected wave. When a separate "sheet" of the wave region is allotted to each ray congruence, it is also natural to regard the two sheets as connected at the shore, where the reflected wave and its rays are generated by the incident wave and its rays.

A quite similar situation arises at the caustic $x = a$, where the rays turn so that one ray congruence is continued from the caustic by the other ray congruence with the same c. The two ray sheets are therefore also connected at the caustic and form, in fact, a surface topologically equivalent to a cylinder round which the waves travel in helical fashion, unable ever to reach the open sea. A simple kinematical explanation of such partial wave trapping was sketched in the Introduction.

By contrast, if c is not in the range of $k(h)$, then there is no caustic, (4.10) predicts real ray slope for all $x > 0$. Such waves can therefore reach the open sea, or come from it, and are called progressive waves. There is still shore reflection and the wave region $x > 0$ is therefore still covered by two sheets of rays, representing respectively the incident and reflected waves, but the sheets are connected only at the shore.

The example illuminates the nature of ray patterns further, if the continental slope unbounded alongshore be replaced by a channel with plane, parallel sidewalls, say, at $y = \pm b$. If c is in the range of $k(h)$, the wave is now trapped altogether in the sense that it cannot escape the bounded wave region $0 < x < a$, $-b < y < b$. This will be taken up in Section VII.

For more general, natural seabed topography, the qualitative understanding of the ray pattern may need computer assistance with (4.8), but the main task—to convince oneself that he knows the rough positions of all the relevant caustics—is not basically computational. Those, and the shores, are the locations of the possible extreme wave amplitudes. An estimate of their magnitude, however, requires the uniform ray theory; (4.7) only predicts a singularity of the surface amplitude A_s because $\Xi = 0$ at a caustic and $kG = 0$ at a shoreline.

The singularity at shores and caustics also prevents a proper discussion of the *reflection conditions* by which the incident wave determines the reflected wave. Those are readily formulated for a vertical wall (e.g., channel side wall or seawall) where the normal velocity $\partial \phi/\partial n = 0$. Then in (4.11), since just two waves participate in the reflection process,

$$\sum_{j=+,-} \left(\frac{i}{\varepsilon} A_j \,\partial S_j/\partial n + \partial A_j/\partial n \right) \exp(iS_j/\varepsilon) = 0,$$

and without loss of generality, we may choose $S_+ = S_-$. As $\varepsilon \to 0$, that requires $A_+ \, \partial S_+/\partial n = -A_- \, \partial S_-/\partial n$ for the reflected $(+)$ wave in terms of the incident $(-)$ wave; the further requirement $\partial A_+/\partial n = -\partial A_-/\partial n$ will come to be satisfied automatically by what follows. The conservation law (4.3) adds the requirement $A_+^2 \, \partial S_+/\partial n = -A_-^2 \, \partial S_-/\partial n$, whence

$$S_+ = S_-, \quad A_+ = A_-, \quad \partial S_+/\partial n = -\partial S_-/\partial n \qquad \text{at a wall.} \qquad (4.12)$$

Of course, that is just Snell's law for standard optical reflection.

This argument collapses at a caustic or shore because A does not exist there. Uniform ray theory will be shown in the next section to fill also this gap by the reflection condition (5.5), (5.7) at a caustic and, if the condition of a bounded solution be adopted, also at a shore. However, reflection conditions are local in nature, and simple ray theory needs to know what they look like from a distance sufficient for it to apply. That will be shown in Section V to be

$$A_+^* = A_-^*, \quad S_+ = S_- - \tfrac{1}{2}\pi, \quad \partial S_+/\partial n = -\partial S_-/\partial n. \qquad (4.13)$$

If these conditions are added to simple ray theory, it can predict correctly phenomena (e.g., Section VII) that do not depend on local wave behavior or maximal amplitudes.

Appendix IV

a. Derivation of (4.2) from (3.3). Substitution of (3.4) in (3.3) and letting $\varepsilon \to 0$ implies a sequence of equations of which the first three are (3.5), (4.1), and, with $k(z + h) = \alpha$ for temporary abbreviation,

$$i \operatorname{sech} \alpha \, \frac{\partial}{\partial z}\left[\left(\cosh^2 \alpha\right)\frac{\partial A_1}{\partial z}\right] = 2\, \nabla S \cdot \nabla(A_0 \cosh \alpha) + A_0 \cosh^2 \alpha \, \nabla^2 S,$$

$$\partial A_1/\partial z = 0 \text{ at } z = 0 \text{ and } = -i A_0 \nabla h \cdot \nabla S \text{ at } z = -h.$$

To see that this has a vertical average independent of A_1, integrate once to get

$$i \cosh^2 \alpha \, \frac{\partial A_1}{\partial z} = (2\, \nabla S \cdot \nabla A_0 + A_0 \nabla^2 S + A_0 \nabla S \cdot \nabla) \int_0^z \cosh^2 \alpha \, dz.$$

The integral is $(4k)^{-1}[\sinh(2\alpha) - \sinh(2kh) + 2kz]$. For $z = -h$, in particular, $\alpha = 0$ and the left-hand side of the last equation is given by the boundary condition; on the right-hand side, the gradient of $(4k)^{-1}\sinh(2\alpha)$ is merely $\tfrac{1}{2}\nabla h$ when $\alpha = 0$, and the whole equation is now seen to reduce there to (4.2).

In the derivation of (4.1) and (4.2), the apparent additional condition $\Sigma_{15} = 0$ has been ignored for the following reason. Even on the assumption that the vertical structure of the motion be adequately approximated by the classical function $F_0(z; x, y)$, (3.3) was seen to imply

$$\nabla(G\nabla\Psi) + \varepsilon^{-2}\Psi\{k^2 G + O(\varepsilon^2)\} = 0,$$

rather than (3.6). On substitution of (3.4), this error term must be expected to make a contribution of the same order in ε as Σ_{15}. Hence, (4.1) and (4.2) are necessary for a first approximation, but Σ_{15} is relevant only to higher order refraction approximations which are outside the scope of this article.

b. Special topographies. The analogy between the ray equation (4.8) and Newton's equation can help in solving the ray equation when the seabed topography is axially symmetrical (Kriegsmann, 1979). If the center of symmetry is taken as the origin, the depth then depends only on the distance $r = |x|$ from the origin, and similarly $k = k(r)$, by (3.5), so that the corresponding particle moves in a central force field. It may be concluded right away that the "angular momentum" $L = x \wedge k\,dx/d\sigma$ should be constant along the ray; indeed, ∇k is now parallel to x, whence $dL/d\sigma = (dx/d\sigma) \wedge k\,dx/d\sigma + x \wedge \nabla k$, by (4.8), and both terms on the right are products of parallel vectors. If κ and n, moreover, denote respectively the curvature of, and unit normal to, the ray, then $n \cdot dx/d\sigma = 0$ and $\kappa n = d(dx/d\sigma)/d\sigma$, and therefore from (4.8), $n \cdot \nabla k = k\kappa$. Here $\nabla k = (dk/dr)x/r$, and since $x \wedge dx/d\sigma$ and $n \cdot x$ agree in magnitude,

$$k^{-2}\,dk/dr = \pm \kappa r/|L|,$$

with sign dependent on the orientation of n. Along a ray, therefore, the ray curvature κ is a ray constant $|L|$ times a function only of r that is determined readily from $h(r)$ via (3.5).

One of the simplest special cases arises for rays of constant curvature κ, that is, circles, and it occurs if, and only if, $k = (c_1 r^2 + c_2)^{-1}$ with constants c_2 and $2c_1 = \mp \kappa/|L|$ (Kriegsmann, 1979). The case $c_1 = c_2 = 1$ corresponds to Maxwell's fish eye lense of optics. A simpler seabed topography results in the long-wave limit $k^2 h \to \varepsilon\omega^2 = 1$ of the dispersion relation (3.5); with $c_1 = -c_2 = d^2 - 1$, the depth becomes

$$h = (r^2 - 1)^2/(d^2 - 1)^2,$$

and if that be adopted for $1 \leq r \leq d$ and complemented by $h = 1$ for $r \geq d$, an island of shore radius 1 and base radius d in an ocean of otherwise constant depth is described [Fig. 7, (from Kriegsmann, 1979)]. Figure 8 (from

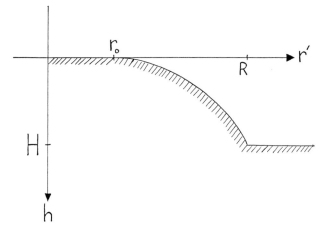

FIG. 7. Water depth vs. radius for Kriegsmann's island with circular rays, from Fig. 1, G. A. Kriegsmann, An illustrative model describing the refraction of long water waves by a circular island. Reprinted from *J. Phys. Oceanogr.* **9**, 608, by permission of the American Meteorological Society, © 1979 by the American Meteorological Society, Boston.

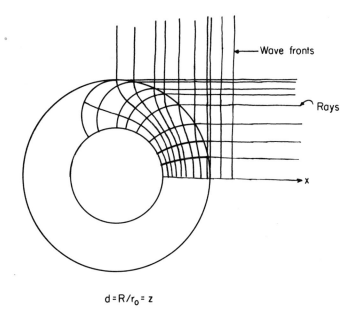

FIG. 8. Wave crests and rays for plane waves incident on Kriegsmann's island (cf. Fig. 7), from Fig. 2, G. A. Kriegsmann, An illustrative model describing the refraction of long water waves by a circular island. Reprinted from *J. Phys. Oceanogr.* **9**, 609, by permission of the American Meteorological Society. © 1979 by the American Meteorological Society, Boston.

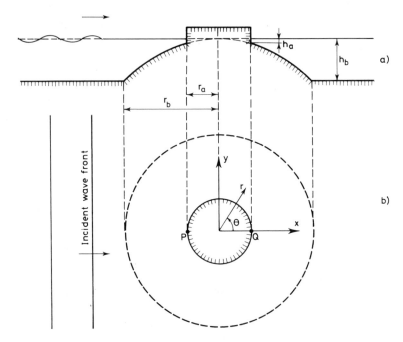

FIG. 9.   Water depth vs. radius for Homma's island with logarithmically spiraling rays, from Fig. 1, I. G. Jonsson and O. Skovgaard, A mild-slope wave equation and its application to tsunami calculations. Reprinted from *Mar. Geodesy* **2**, 42, by permission of Crane, Russak & Co., Inc. © 1978 by Crane, Russak & Co., Inc., New York.

Kriegsmann, 1979) shows the rays and wave fronts arising in the refraction of incident plane waves by such an island.

A popular island, since Homma (1950), has been that of paraboloidal shape, $h(r)/r^2$ = const. (Fig. 9) (from Jonsson and Skovgaard, 1978), with the awkward singularity at $r = 0$ traditionally evaded (at least, in part) by a vertical seawall at $r > 0$. While the island shape looks similar to that of Fig. 7, the rays are now logarithmic spirals and look quite different (Fig. 10) (from Jonsson and Skovgaard, 1978). An attraction of explicit ray formulas is the opportunity for computation of detailed wave fields without prohibitive expense; Fig. 11 (from Jonsson and Skovgaard, 1978) shows such a field for Homma's island. Further convex island shapes admitting explicit ray formulas have been explored by Christiansen (1976). It will be clear from the examples just sketched, that such explicit ray formulas tend to be bought at the expense of restriction to artificial topographies with at least some unnatural refraction properties. The explicitness can serve to demon-

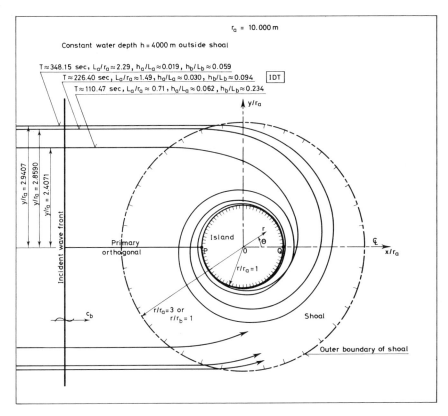

FIG. 10. Long-wave rays for Homma's island (cf. Fig. 9), from Fig. 9, I. G. Jonsson and O. Skovgaard, A mild-slope wave equation and its application to tsunami calculations. Reprinted from *Mar. Geodesy* **2**, 53, by permission of Crane, Russak & Co., Inc. © 1978 by Crane, Russak & Co., Inc., New York.

strate some interesting features, but are they among the realistic or the unnatural ones?

c. Cusped caustics. The notion of sheets also helps in the description of the local geometry of ray patterns at caustics. The normal case is that encountered over a continental slope (Section IV,B) where $dx/dy = 0$, $d^2x/dy^2 < 0$ on the ray curves $x = x(y)$ at the straight caustic $x = \text{const.} = a$. Here only the region $x < a$ on one side of the caustic is covered by rays, and that is covered twice, corresponding to two sheets joined at the caustic. More generally, a caustic is a curve similarly separating a region covered doubly by the rays from a region where there are no rays at all (Fig. 12).

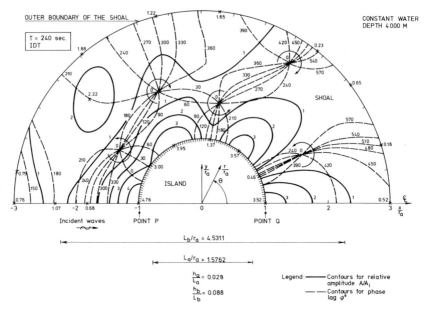

FIG. 11.   Plot of amplitude and phase lag (dashed) for the refraction pattern of plane waves incident on Homma's island (cf. Fig. 9). From Fig. 2, I. G. Jonsson and O. Skovgaard, A mild-slope wave equation and its application to tsunami calculations. Reprinted from *Mar. Geodesy* **2**, 46, by permission of Crane, Russak & Co., Inc. © 1978 by Crane Russak & Co., Inc., New York.

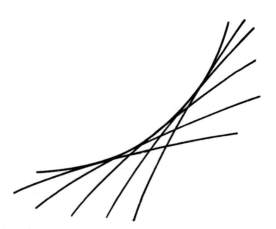

FIG. 12.   Caustic curve as envelope of a family of rays.

Ray envelopes encountered in computational ray tracing are liable to be cusped curves, as in Figs. 2 and 3, and this complicates the pattern. Locally, however, in a close neighborhood of any one caustic point P (other than the very cusp point), the difference arises merely from the simultaneous presence of two ray families one of which forms the envelope, while the other sweeps over the neighborhood of P. For full clarity, we recognize three sheets: two of them are connected at the envelope, and each of these covers only a half-neighborhood of P as at a normal caustic point (Fig. 12), while the third sheet has no local association with the envelope at all and covers the full neighborhood of P.

The cusped caustic is then seen to consist of two branches each of which has the normal character indicated in Fig. 12. The branches join at the cusp point, which is a geometrically exceptional point of the envelope and connects all three sheets.

d. Exact edge waves. Comparison of the ray theory of waves trapped along a continental coast with exact edge wave solutions of (2.10) known (Ursell, 1952) for the case of uniform beach slope illustrates another aspect of refraction theory. For small beach slope ε, the surface elevations $\zeta_n(x, y) = iL\omega_n\phi_n(x, y, 0)$ of the first three of Ursell's (1952) edge wave modes are given by

$$\phi_0 \sim e^{-x \cos \varepsilon} \cos y,$$

$$\phi_1 \sim \tfrac{1}{2}\phi_0 - \tfrac{1}{2}e^{-x \cos 3\varepsilon} \cos y,$$

$$\phi_2 \sim \phi_1 - \tfrac{1}{6}\phi_0 + \tfrac{1}{6}e^{-x \cos 5\varepsilon} \cos y.$$

Thus $\zeta_0(x, 0)$ decays monotonely with distance x from shore, while $\zeta_1(x, 0)$, $\zeta_2(x, 0)$ vanish at the shore but show a maximum further out, whence they also decay exponentially with increasing x. For these lowest modes, however, $\omega_n = \mathcal{O}(\varepsilon^{1/2})$ and ray theory based on (3.4) with $\varepsilon\omega^2 = \eta = \mathcal{O}(1)$ is not directly applicable. Nonetheless, simple ray theory is found (Shen *et al.*, 1968) to predict the correct edge wave frequencies even for these modes when $\varepsilon \ll 1$.

V. Uniform Ray Theory

To overcome the shortcomings of simple ray theory at shores and caustics, Shen and Keller (1975) propose that the surface potential $\Psi = \phi(x, y, 0)$ be approximated by

$$\Psi \sim \tfrac{1}{2}e^{i\theta/\varepsilon} \cosh(kh) R\left\{(A_+ + A_-)V(\rho) - \frac{i}{Q}\varepsilon^{1-\mu}(A_+ - A_-)V'(\rho)\right\}, \quad (5.1)$$

in which $\theta = \theta(x, y)$, $\rho = \varepsilon^{-\mu} v(x, y)$ with constant $\mu \geq 0$, $R(v)$ and $Q(v)$ are auxiliary functions to be chosen later, and $V(\rho)$ is a solution of

$$V'' + \varepsilon^{\mu} P V' + \varepsilon^{2\mu - 2} Q^2 V = 0, \qquad (5.2)$$

where $P = d \log(R^2/Q)/dv$.

Some of the motivation for the replacement of (3.4), viz.,

$$\Psi \sim A_0(x, y) \exp[i S(x, y)/\varepsilon],$$

will become apparent presently. Indeed, if $Q \equiv R \equiv \mu = 1$, then (5.2) reduces to $V'' + V = 0$ and without loss of generality,

$$V = e^{iv/\varepsilon} + e^{-iv/\varepsilon}$$

in (5.1), which becomes

$$\text{sech}(kh)\Psi \sim A_+ \exp[i(\theta + v)/\varepsilon] + A_- \exp[i(\theta - v)/\varepsilon]. \qquad (5.3)$$

This differs from (3.4) only by a pairing of waves, as is natural near a shoreline or caustic where (Section IV,B) an "incident" wave must always be expected to generate a "reflected" wave. Moreover, (5.3) shows that the single phase function $S(x, y)$ of (3.4) has been replaced in (5.1) by two "phase coordinates" $\theta(x, y)$ and $v(x, y)$ to anticipate a natural difference between surface pattern variation parallel and normal to a singular line C representing a shore or caustic. The variation in the direction tangential to C is represented by $\exp i\theta(x, y)/\varepsilon$ in (5.1), in close analogy to (3.4). By contrast, $v(x, y)$ is a phase coordinate transversal to C and greater flexibility for the representation of surface pattern variation in that direction is provided by the function $V(\rho)$ in which ρ is a stretched phase coordinate, but with a stretch exponent μ dependent on the character of C. Since the time-dependent potential $\Phi \propto \Psi \exp(-i\omega t)$, the choices $\omega > 0$ and $v = 0$ at C make the first (second) term in (5.3) represent a wave propagating in the direction of v increasing (decreasing). However, they are not (Appendix V,a) individual waves in the sense of (4.11) that each wave separately approximates a solution of (2.10).

The paper of Shen and Keller (1975) is elegant and of sweeping generality, it covers general unsteady propagation of not only surface waves, but also internal and acoustic waves, in rotating oceans and atmospheres, and the analysis is accordingly formidable. It may therefore help to sketch here the much simpler analysis pertinent to the physical scope of this article; this will also prove that the refraction equation (3.6) does govern the uniform ray theory. It is shown in Appendix V,a that if (5.1), (5.2) are to describe a first refraction approximation then—regardless of the choice of $Q(v)$ and $R(v)$—the functions

$$S_\pm(x, y) = \theta \pm \int_0^v Q(t)\, dt \qquad (5.4)$$

must each satisfy the eiconal equation (4.1) and the functions $A_+(x, y)\cosh(kh)$ must each satisfy the conservation law (4.2). Hence, the rays of the simple theory (Section IV) remain the rays of the uniform theory. As anticipated, there are two rays through each point and on choosing $Q \geq 0$ without loss of generality, those associated with the $+$ subscript again represent propagation in the direction of v increasing, which may in turn be chosen as that pointing toward the side of C on which there are rays (Fig. 12). The $+$ subscript then denotes propagation away from C, the $-$ subscript denotes a wave incident upon C. The notion of sheets is again convenient for distinguishing the waves, and then it follows as in the simple theory (Section IV) that the respective values of S_+ and A_+ along any ray (on the respective sheet) are given explicitly by (4.5) and (4.7), once they are known at one point of that ray. To fix the ideas in a physically natural way, we take them to be specified at one point of each incident ray at a positive distance from C independent of ε. For the specification of S_+, there is no loss of generality, and some convenience, in the choice

$$S_+ = S_- = 0, \quad v = 0 \qquad \text{on } C, \tag{5.5}$$

whence (4.5) gives $S_+(x, y)$ everywhere on the $+$ sheet. But (4.7) still implies that A_+ do not exist at C. On the other hand, they do not by themselves represent amplitudes in (5.1), and the object of uniform ray theory is to specify $Q(v)$ and $R(v)$ so that the expression (5.1) is bounded even at C. This specification is different for caustics and shorelines.

For a caustic C, the appropriate choice is known from the work of Kravtsov (1964) and Ludwig (1966) to be

$$\mu = 2/3, \qquad Q = R^2 = v^{1/2},$$

so that (5.2) becomes $V'' + \rho V = 0$, with $\rho = \varepsilon^{-2/3}v$ and

$$v = [\tfrac{3}{4}(S_+ - S_-)]^{2/3},$$

by (5.4) and (5.5). An appropriate solution is the Airy function $V(\rho) = \text{Ai}(-\rho)$ (Fig. p. 393 in Olver, 1974), and then the surface potential (5.1) becomes

$$\Psi \sim \tfrac{1}{2}e^{i\theta/\varepsilon}\cosh(kh)\{(A_+ + A_-)v^{1/4}\text{Ai}(-\rho)$$
$$+ i\varepsilon^{1/3}(A_+ - A_-)v^{-1/4}\text{Ai}'(-\rho)\}. \tag{5.6}$$

It is shown in Appendix V,b that $v^{1/4}A_-$ has a finite limit at the caustic C, if a regular, incident wave has been specified, and that the condition

$$\lim_{v \to 0} v^{1/4}(A_+ - A_-) = 0,$$

or equivalently,

$$A_+^* = A_-^* \qquad \text{at the caustic}, \tag{5.7}$$

is necessary and sufficient to make (5.6) bounded there and to determine a regular, reflected wave on the $+$ sheet. It follows that (5.6) and (5.7) also make the surface elevation $\zeta = iL\omega\Psi$ bounded at the caustic.

It should be mentioned that the analysis relies on the assumption that the caustic is smooth and strictly convex relative to the rays (Ludwig, 1966). A cusp point of a caustic (Appendix IV,c) remains a singular point at which a more involved *local* approximation to Ψ (Ludwig, 1966) is needed.

For large $\rho = \varepsilon^{-2/3}v > 0$, $\mathrm{Ai}(-\rho)$ and its derivative have asymptotic approximations (Olver, 1974) which give as first approximation to (5.6),

$$\Psi \sim \frac{1}{2}\pi^{-1/2}\varepsilon^{1/6}\cosh(kh)\left\{A_+\exp i\left(\frac{S_+}{\varepsilon}-\frac{\pi}{4}\right)+A_-\exp i\left(\frac{S_-}{\varepsilon}+\frac{\pi}{4}\right)\right\} \quad (5.8)$$

by (5.4). This is again just a pair of waves (3.4), so that uniform ray theory is seen to reproduce simple ray theory (Section IV) at sufficient distance from a caustic on the side where $v > 0$.

Comparison with (5.3), moreover, shows (5.8) to contain two new pieces of information. The total phase effect of the caustic refraction is a phase shift $-\pi/2$ from the incident to the refracted wave. Second, comparison of (5.6) and (5.8) permits us to make an explicit estimate of *amplitude enhancement* due to caustic refraction. At a point P away from the caustic, (5.8) gives the *incident* wave amplitude as

$$|\Psi_p| \sim \tfrac{1}{2}\pi^{-1/2}\varepsilon^{1/6}A_-\cosh(kh) = \tfrac{1}{2}\varepsilon^{1/6}(\pi kG\Xi)_p^{-1/2}A_-^*,$$

by (4.7). This may be related by (4.7) to the amplitude at the point Π of contact of the ray through P with the caustic C. Very close to the caustic, the second term in the brace of (5.6) is relatively small for sufficiently small ε, and if it be disregarded, the amplitude there is

$$|\Psi| \sim \cosh(kh)\lim_{v=0}(v^{1/4}A_-)|\mathrm{Ai}(-\rho)|,$$

by (5.7). By (4.7), therefore

$$|\Psi|_{\max}/|\Psi_p| \sim 2\varepsilon^{-1/6}\pi^{1/2}\max|\mathrm{Ai}|[(kG)_p/(kG)_\pi]^{1/2}\lim_{v=0}(v^{1/2}\Xi_p/\Xi)^{1/2}, \quad (5.9)$$

where kG is a function only of the local water depth h, by (3.5) and (3.6); the maximum of $\pi^{1/2}|\mathrm{Ai}(x)|$ is about unity. The last factor depends mainly on the ray convergence from P to the general neighborhood of the caustic (Appendix V,c). For sufficiently small ε, the most significant enhancement factor in (5.9) is seen to be $\varepsilon^{-1/6}$. For practical applications of (5.9), however, it should be borne in mind that not all waves are covered by the limit $\varepsilon \to 0$ of (3.3), with $\varepsilon\omega^2 = \mathcal{O}(1)$ (Appendix IV,d). On the other hand, the realism of the predictions of uniform ray theory has been illustrated by the experiments of Chao and Pierson (1972).

Uniform ray theory also revises the notion of the caustic as a sharp boundary of the surface wave pattern. It remains a sharp envelope of the rays as mathematical characteristics of (4.1), but $\text{Ai}(-\rho)$ in (5.6) does not vanish beyond the envelope $\rho = 0$ (Fig. 12). However, Ψ decays rapidly with distance beyond the caustic: from the asymptotic approximations of $\text{Ai}(x)$ (Olver, 1974), and (5.6) and (5.7),

$$\Psi \sim \tfrac{1}{2}\pi^{-1/2}\varepsilon^{1/6}\cosh(kh)\lim(v^{1/4}A_-)e^{i\theta/\varepsilon}v^{-1/4}\exp(-\tfrac{2}{3}|v|^{3/2}/\varepsilon) \qquad (v < 0),$$

so that the decay is exponentially rapid in ε. The one-sided limit of $v^{1/4}A_-$ defined by the ray pattern is here understood to be extended by continuity to the side $v < 0$ of C over the exponentially small distances involved. Such decay is called evanescence, in fact, the surface-wave behavior described by (5.6) is quite analogous to optical wave behavior at a shadow boundary. In the case of a cusped caustic, of course, the evanescent wave is negligible by comparison with the higher order refraction corrections to the wave on the third sheet. The wave behavior beyond the caustic also explains the choice of the solution $V(\rho) = \text{Ai}(-\rho)$ of Airy's equation $V'' + \rho V = 0$ for the representation (5.8). Any other solution would show exponential growth for $v < 0$, which is physically quite implausible and by making Ψ unbounded, would defeat the very object of the uniform theory.

For a shoreline, a more direct approach to the determination of $V(\rho)$ in (5.1) offers itself since $k^2 h \to \eta = 1$ and $h^{-1}G \to 1$ in (3.5) and (3.6) as the depth $h(x, y) \to 0$ and therefore (3.6) approaches the long-wave equation

$$\nabla(h\nabla\Psi) + (k/\varepsilon)^2 h\Psi = 0.$$

It will be assumed that the shoreline is smooth and typical in that the normal derivative of the depth is nonzero. Let θ and n denote orthogonal coordinates parallel and normal to the shoreline, respectively, scaled so that $\nabla\theta$ and ∇n are independent of ε, with $n = 0$ at the shore and increasing seaward. For an approximate solution of the general form of (5.1),

$$\Psi \sim e^{i\theta/\varepsilon}\cosh(kh)N(n),$$

the long-wave equation implies

$$hN'' + \alpha N' + \varepsilon^{-2}N = 0,$$

to the first approximation as $\varepsilon \to 0$, $h \to 0$, with $\alpha = \partial h/\partial n > 0$. For

$$\rho = 2h^{1/2}/(\alpha\varepsilon), \qquad V(\rho) = N(n),$$

this becomes

$$V'' + \rho^{-1}V' + V = 0,$$

which is (5.2) with

$$Q \equiv \mu = 1, \quad v = \varepsilon\rho = 2h^{1/2}/\alpha, \quad P = v^{-1}, \quad R = v^{1/2}.$$

It is also Bessel's equation of zero order so that

$$V = A_1 H_0^{(1)}(\rho) + A_2 H_0^{(2)}(\rho)$$

in terms of the Hankel functions. For $v > 0$, $\varepsilon \to 0$, their asymptotic approximations give

$$\Psi \sim (2\varepsilon/\pi v)^{1/2} \cosh(kh)\{A_1 \exp[i(\theta + v)/\varepsilon - \pi/4] + A_2 \exp[i(\theta - v)/\varepsilon + \pi/4]\},$$

which is again analogous to (5.3) and (5.8), with

$$v^{1/2} A_+ = (2\varepsilon/\pi)^{1/2} A_1, \quad v^{1/2} A_- = (2\varepsilon/\pi)^{1/2} A_2$$

and a *phase shift* $-\pi/2$ from the incident to the reflected wave. [This is of course analogous to standard, tidal theory (Lamb, 1932).] The form of (5.1) at a shore is therefore

$$\Psi \sim (\pi/2\varepsilon)^{1/2} \cosh(kh)e^{i\theta/\varepsilon}v^{1/2}[A_+ H_0^{(1)}(\rho) + A_- H_0^{(2)}(\rho)], \qquad (5.10)$$

and away from the close neighborhood of the shore, $\rho = 2h^{1/2}/(\alpha\varepsilon) \to \infty$ and (5.10) is in turn approximated by

$$\Psi \sim \cosh(kh)\{A_+ \exp[i(\theta + v)/\varepsilon - \pi/4] + A_- \exp[i(\theta - v)/\varepsilon + \pi/4]\}. \qquad (5.11)$$

It may be noted that $v^{1/2}A_\pm$ tends to a finite limit at the shore (Shen and Keller, 1975) because

$$A_\pm \cosh(kh) = A_\mp^*(kG\Xi)^{-1/2},$$

by (4.7), and $kG \sim h^{1/2} = \alpha v/2$ as $h \to 0$, by (3.5) and (3.6), while the ray expansion ratio Ξ tends to a nonzero limit (Appendix V,d). This is not unexpected, in view of the familiar constancy of $h^{1/4}$ times amplitude in tidal theory.

The choice of A_+/A_- at a shore is more in the nature of a boundary condition (Section VI), by contrast with the caustic, where it was seen to be implicit in the general objective of uniform ray theory to construct a bounded approximation to the potential. However, if the conventional shore condition of bounded surface elevation $\zeta = iL\omega\Psi$ be adopted, it follows from (5.10) that

$$\Psi \sim (\pi/2\varepsilon)^{1/2} \cosh(kh)e^{i\theta/\varepsilon}v^{1/2}(A_+ + A_-)J_0(\rho),$$

$$\lim_{v \to 0} v^{1/2}(A_+ - A_-) = 0 \qquad (5.12)$$

(Shen and Keller, 1975).

The last condition is just (5.7) again and permits us to estimate the amplification by comparing the *incident* wave amplitude $|\Psi_p|$ at a point P away from shore with the *total* amplitude $|\Psi_s|$ at the shorepoint on the incident ray through P. By (4.7), (5.11), and (5.12)

$$|\Psi_s/\Psi_p|^2 = 4\pi(\varepsilon\alpha)^{-1}kG\Xi, \tag{5.13}$$

where $kG\Xi$ is to be evaluated at P with the expansion ratio normalized to unity at the shore. The amplification is thus seen to be proportional to the reciprocal root of the beach slope $\varepsilon\alpha$, in keeping with a general result (Meyer and Taylor, 1972) of the linear theory of surface waves normally incident on a beach. Comparison with (5.9) shows that shore amplification is likely to be more important than caustic amplification and also easier to compute. Since kG is given by (3.5) and (3.6) directly in terms of the local depth h at P, the main influence of the ray pattern (Fig. 1) is to select the shore point corresponding to a point P at which the incident wave amplitude is specified. A second, but relatively minor, influence arises from the expansion ratio.

Since the common formalism of (5.1) and (5.2) covers the simple ray theory (Section IV) as well as the neighborhood of a caustic or a shoreline, it can be used for the whole wave region, even if several shores and caustics be present (Shen and Keller, 1975). In the case of two caustics, for instance, the choice $\mu = 1/2$, $Q^2 = v(v_1 - v)$, with constant $v_1 > 0$ and $R = Q^{1/2}$, gives

$$\rho = \varepsilon^{-1/2}v, \qquad V'' + \rho(\rho_1 - \rho)V(\rho) = 0$$

for (5.2), which is Weber's (parabolic cylinder) equation. It has a solution approximated near $\rho = 0$ and $\rho = \rho_1 = v_1/\varepsilon^{1/2}$ by the Airy functions of $-(v_1/\varepsilon^2)^{1/3}v$ and $(v_1/\varepsilon^2)^{1/3}(v - v_1)$, respectively. In the case of a shoreline at $v = 0$ and a caustic at $v = v_1 > 0$, such as often occurs around islands, the choice $\mu = 2/3$, $P = v^{-1}$, $Q = (v_1 - v)^{1/2}$ gives

$$\rho = \varepsilon^{-2/3}v, \qquad V'' + \rho^{-1}V' + (\rho_1 - \rho)V(\rho) = 0$$

for (5.2), and $R(v) = v^{1/2}(v_1 - v)^{1/4}$. Its solutions are approximated near $\rho = 0$ by Bessel functions of $v_1^{1/2}v/\varepsilon$, and one solution is approximated near ρ_1 by $Ai(\rho - \rho_1)$. For $0 < v < v_1$, they are approximated by (4.11).

Appendix V

a. Eiconal and transport equations near a singular line. It is convenient to regard $v(x, y)$ as a coordinate normal to the singular line C, so that $\nabla\theta \cdot \nabla v = 0$. The approximation (5.1) is based on the assumption that C is smooth and strictly convex relative to the rays (in particular, its radius of curvature is bounded from zero). Such a regular, orthogonal coordinate system (θ, v) then exists in a sufficiently narrow strip about C.

   Since the waves are necessarily paired near C, whether it be a caustic or shoreline, it is not obvious that the simplest representation must display each wave completely. Interference could have helped to simplify (5.1), in which case the terms proportional to A_+ and A_-, respectively, are "waves" only by analogy, not in the sense of each satisfying (3.6) approximately. On the other hand, the choices of interest will make V and V' linearly independent functions and the terms proportional to them, respectively, must then each satisfy (3.6) approximately. Accordingly, we write (5.1) as

$$\Psi \sim \Psi_0 = e^{i\theta/\varepsilon}(\lambda_1 V - i\varepsilon^{1-\mu}\lambda_2 V'),$$

and substitute this into (3.6). In $\nabla\Psi_0$, V'' may be eliminated by (5.2) to regroup the result of the substitution in the form

$$e^{-i\theta/\varepsilon}\left[\nabla(G\nabla\Psi_0) + \frac{k^2}{\varepsilon^2}G\Psi_0\right] = \alpha V - i\varepsilon^{1-\mu}\beta V'.$$

To abbreviate the expressions for α and β, let

$$\lambda_1 \pm Q\lambda_2 = \lambda_\pm, \qquad \nabla\theta \pm Q\nabla v = \nabla S_\pm,$$

then straightforward use of (5.2) and regrouping gives

$$\alpha \pm Q\beta = \frac{i}{\varepsilon}M^\pm\nabla S_\pm + \nabla M^\pm \mp (PQ + Q')M^*\nabla v + \frac{k^2}{\varepsilon}G\lambda_\pm,$$

where

$$G^{-1}M^\pm = (i/\varepsilon)\lambda_\pm\nabla S_\pm + \nabla\lambda_\pm \mp (PQ + Q')\lambda_2\nabla v,$$

$$G^{-1}M^* = (i/\varepsilon)(\lambda_2\nabla\theta + \lambda_1\nabla v) + \nabla\lambda_2 - P\lambda_2\nabla v.$$

If this be regrouped by powers of ε,

$$\alpha \pm Q\beta = \varepsilon^{-2}G\lambda_\pm(k^2 - |\nabla S_\pm|^2) + \frac{i}{\varepsilon}\{\nabla(G\lambda_\pm\nabla S_\pm) + (\nabla S_\pm) \cdot G\nabla\lambda_\pm$$

$$\mp G(PQ + Q')[(\nabla S_\pm) \cdot \lambda_2\nabla v + (\nabla v) \cdot (\lambda_2\nabla\theta + \lambda_1\nabla v)]\} + \Sigma_{16}$$

$$(5.14)$$

results, with Σ_{16} bounded as $\varepsilon \to 0$, and since $(\nabla v) \cdot \nabla\theta = 0$ and $PQ + Q' = 2QR'/R$, the last brace may be rewritten as

$$\{\nabla(G\lambda_\pm\nabla S_\pm) + (G\nabla S_\pm) \cdot \nabla\lambda_\pm - 2(G/R)(\nabla R) \cdot \lambda_\pm\nabla S_\pm\}$$
$$= (R^2/\lambda_\pm)\nabla\{G(\lambda_\pm/R)^2\nabla S_\pm\}.$$

   Now, if Ψ_0 is to be a first approximation to the surface potential, then since V and V' are linearly independent, α and β must both vanish to the order of approximation warranted by (3.6). Equivalently $\alpha \pm Q\beta$ must both vanish (except for the term Σ_{16}, see Appendix IV,a), and as $\varepsilon \to 0$ this implies firstly the eiconal equation (4.1) for both the phase functions

$$S_\pm = \theta \pm \int_0^v Q(t)\, dt.$$

Second, it implies vanishing of the brace in (5.14), where $\lambda_\pm = \lambda_1 \pm Q\lambda_2$ are abbreviations for $\frac{1}{2}\cosh(kh)R(v)A_\pm$, respectively, in (5.1). But that amounts to just the conservation law (4.2) for $A_s = A_\pm \cosh(kh)$, respectively.

b. *Caustic reflection condition.* With $Q(v) = v^{1/2}$, (5.4) gives

$$S_\pm = \theta \pm \tfrac{2}{3}v^{3/2},$$

and for a smooth, strictly convex caustic C, Ludwig (1966) has shown the eiconal equation (4.1) for S_+ to have a solution with normal derivative $\partial v/\partial n = |\nabla v| \neq 0$. Near C, v can therefore be regarded as the normal distance from C. The expansion ratio $\Xi(\sigma)$ in (4.6) measures distance between neighboring rays as function of arclength σ on a ray, which may be measured from the point of contact with C. It is then seen from Fig. 12 that Ξ is directly proportional to σ, close to C, and therefore directly proportional to $v^{1/2}$, because the ray is tangent to C at the point of contact.

It is of course assumed that $h(x, y)$ is continuous and positive near C, and the same follows for $k(x, y)$ and $G(x, y)$ by (3.5) and (3.6). By (4.7) therefore,

$$\lim_{v \to 0} (v^{1/4} A_-)$$

exists because finite values of A_- have been specified at noncaustic points of the incident rays. The same follows for $\lim(v^{1/4}A_+)$ on any reflected ray on which A_+ exists anywhere. Conversely, if that holds, then the first term in the brace of (5.6) is bounded at the caustic C, and the condition (5.7) is necessary for this to extend also to the second term. This condition is also sufficient: Since

$$A_{s,\pm}(\Xi_\pm kG)^{1/2} = A_\pm^*$$

are invariant on the respective rays, their difference at a point P *close* to C is, according to (5.4), given by the change in $A_\pm^*$ along C between the respective points of contact with C of the rays through P (Fig. 12). Assuming a regular specification of A_- on the incident sheet, with continuous gradient, that change in $A_\pm^*$ is proportional to the arclength of C between the points of contact and hence, again to the value of $v^{1/2}$ at P. The condition (5.7) therefore makes $v^{-1/2}(A_+^* - A_-^*)$ independent of v, to first order, close to the caustic C. Since (5.7) also makes $A_\pm^* \propto v^{1/4}A_\pm$ finite away from C on the respective sheets, $v^{-1/4}(A_+ - A_-)$ in (5.6) remains bounded at C.

c. Effect of ray geometry on caustic amplitude enhancement. The influence of the ray pattern on the amplitude ratio (5.9) may be conveniently divided into three parts. First, the caustic position and the contact point Π, given the point P at which the incident wave is known, must be estimated, and this information has a major bearing on the value of (5.9). By comparison, the last factor in (5.9) is less significant, but also harder to pin down quantitatively, and a way of estimating it roughly may sometimes be helpful.

To this end, let r denote the radius of curvature of the caustic at the contact point Π and suppose that the linear variation of the expansion ratio $\Xi(\sigma)$ with arclength σ along the ray be an acceptable approximation for a distance l along the ray from Π. Let σ be measured from Π and normalize Ξ by $\Xi(l) = 1$, without loss of generality. Then Ξ_p measures ray convergence away from the caustic, from P to the normalization point.

The factor $\lim(v^{1/2}/\Xi)$, finally, represents a local geometrical effect at Π. Suppose first that the ray curvature at Π is insignificant by comparison with $1/r$. Then along a ray, arclength σ and normal distance n from the caustic C are related by $\lim(\sigma^2/n) = 2r$. In turn, $\lim(v/n) = (2k^2/r)^{1/3}$ (Lewis *et al.*, 1967). Since Ξ is linear in σ near C, $\Xi(\sigma) \sim \sigma/l$ and therefore,

$$\lim_{v=0} \left[v^{1/2}/\Xi(\sigma) \right] = (\tfrac{1}{2}k/r^2)^{1/3} l.$$

If ray curvature $1/r'$ at Π is not unimportant, the estimate is obtained by replacing $1/r$ by the relative curvature $(1/r) - 1/r'$ with r and r' understood to be of the same sign if ray and caustic are convex toward the same side.

d. Rays at a shore. To study the rays near a shore, we may normalize S_+, S_- without loss of generality by

$$S_+ = S_- = \theta \qquad \text{at the shore} \quad n = 0.$$

Then from (4.1), $|\nabla S_\pm| = k \sim h^{-1/2}$ as $h \to 0$ and therefore, $\partial S_\pm/\partial n \sim \pm h^{-1/2} = \pm(\alpha n)^{-1/2}$ and

$$2v = S_+ - S_- \sim 4(n/\alpha)^{1/2}.$$

The ray direction, ∇S, is therefore normal to the shoreline and the local rate of ray convergence is zero, in consonance with the general observation that the wave crests end up parallel to shore.

e. Mathematical comment. It may be pertinent to append a remark on the logical status of uniform ray theory (for real frequency) at the time of writing. While (3.3) is as "exact" as (2.10), the asymptotic analysis outlined in this article is entirely formal. The difference between the two ray theories is that simple ray theory cannot be a candidate for a proper asymptotic approximation to solutions of (3.3), on account of its singularities, but uniform ray theory can be such a candidate.

For the reduced wave equation, the uniform ray theory has, in fact, been proven to give a correct asymptotic approximation for real frequency (Buslaev, 1964; Grimshaw, 1966; Morawetz and Ludwig, 1968). It is natural to hope, therefore, that a general proof for surface waves is only a matter of time. The task is made more formidable, of course, by the problem of approximating the vertical structure of water waves.

VI. Shore Condition

It is the custom in the literature on water waves to mention the boundary condition at the shore as briefly as possible. Fairness to the reader, however, is better served by a frank admission that it is among the most troublesome problems of coastal oceanography. The problem is equivalent to that of wave reflection from shore, and it is disconcerting how little reliable knowledge has been established on it, despite many generations of use of beaches as wave absorbers in laboratories.

It should be observed promptly that a study of waves at the shore is entirely outside the scope of refraction theory. This may be seen readily from the change in the character of the refraction equation (3.7) as the water depth h decreases. From the dispersion relation (3.5), kh then also decreases and

$$\varepsilon^{-2}k^2 \sim \omega^2/(\varepsilon h),$$

but

$$G^{-1/2}\nabla^2 G^{1/2} \sim ch^{-2}$$

for some constant c, so that the shore singularity of G ultimately dominates the bracket in (3.7) and (3.6) becomes

$$(G\Psi)^{-1}\nabla(G\nabla\Psi) \to 0$$

independently of the small parameter ε of refraction theory. In other words, the refraction equation (3.6) itself predicts that the motion must lose any refraction character, close enough to shore. The issue here, therefore, is not what refraction theory can predict at the shore, but what it can draw from other sources in order to make definite predictions on wave motions involving a shore.

Some theoretical indications have been obtained for two-dimensional water motion on a plane beach. In dimensional notation, a limit on deep-water amplitude a is then defined in terms of beach slope ε and wave period $2\pi/\omega$ by

$$a\omega^2/g < (2\pi)^{-1/2}\varepsilon^{5/2}. \tag{6.1}$$

When this inequality holds, a consistent theoretical wave description from deep water to the very shore is available for small beach slope (Meyer and Taylor, 1972) in terms of a combination of the exact linear theory with the nonlinear long-wave theory. This solution is *bounded at the shore*. It also involves total energy reflection from shore, which is made somewhat more plausible by the very restrictive nature of the bound; on a beach of slope 1/100, waves of even 1 minute period must have a deep-water amplitude $a < 0.36$ cm to satisfy (6.1). Confirmed observations, moreover, show that the same bound marks a transition in the nature of wave shoaling (Galvin, 1972).

For amplitudes exceeding the bound (6.1), a similar theoretical description in terms of bounded solutions is not possible, and the manner of its break-down enhances doubts that such waves can reach the shore without breaking. For broken waves, the local nature of the shore singularity has been eluci-dated by Ho *et al.* (1963) and LeMehaute (1963) in terms of breaker collapse, run-up and backwash on the basis of inviscid nonlinear long-wave theory (Meyer, 1966). Some of the qualitative predictions of this theory have been confirmed by observation and experiment, but dissipation has been shown (Miller, 1968; LeMehaute, 1963; Meyer, 1970) to have a dominant quan-titative influence which increases with decreasing beach slope ε. Even on the nondissipative model, moreover, the wave transformation from early shoaling to incipient bore collapse remains unknown.

These indications recall the conjecture of Meyer and Taylor (1972) that the double limit of reflection as amplitude and beach slope tend to zero may be nonuniform. If the amplitude tends to zero before the beach slope does, then (6.1) is satisfied and perfect shore reflection may result, perhaps, by the process of underwater reflection. (Section IX). By contrast, if the beach slope tends to zero before the amplitude does, then (6.1) is violated and total wave absorption may result, perhaps, by the process of wave breaking and dissi-pation. In short, the conjecture is that

$$\lim_{\varepsilon \to 0} \lim_{\nu^{1/2} A_- \to 0} (A_+/A_-) = 1,$$

but

$$\lim_{\nu^{1/2} A_- \to 0} \lim_{\varepsilon \to 0} (A_+/A_-) = 0.$$

Of course, the transition from one extreme to the other could not be described by a simple inequality like (6.1). However, the relevance of some parameter like $a\omega^2/(g\varepsilon^{5/2})$ would not appear inconsistent with the experiments of Greslou and Mahe (1955), which are in urgent need of confirmation and amplification.

For three-dimensional wave motion, refraction turns the wave crests ultimately parallel to shore, and knowledge on two-dimensional shore reflection should therefore still be relevant, but to what degree, is not yet known. The theoretical description for very small amplitude has not yet been extended to edge waves. Their theory (Ursell, 1952) is based on solutions of (2.10) bounded at the shore and its predictions concerning frequency have been confirmed experimentally (Ursell, 1952), but less is known on amplitudes.

In sum, the shore condition of *boundedness* has significant support for progressive waves of the very small amplitudes admitted by (6.1) and for edge waves of perhaps larger amplitude. From (5.12), it is seen to imply *perfect energy reflection*, if the relevance of the long-wave approximation close to shore be accepted, as it must be (Friedrichs, 1948) for waves normally incident on a plane beach.

The paradoxical connection of plausible boundedness of the surface elevation with implausible total reflection of wave energy from the shore deserves emphasis. A precise, theoretical statement is available (Meyer and Taylor, 1972) for waves normally incident on a plane beach (two-dimensional water motion). Equations (2.10) of "exact" linear theory then have two solutions, say, B and Σ, the first bounded and the second, singular (Stoker, 1947). They are standing waves and hence, each by itself corresponds to perfect reflection. Any progressive wave and hence also any solution corresponding to imperfect reflection, is a linear combination of B and Σ. The shore itself is a singularity of (2.10), but for small beach slope ε, B and Σ approach the corresponding solutions of long-wave theory, closer to shore (Friedrichs, 1948). The *nonlinear* long-wave theory, moreover, has a regular solution bounded at the shore (Carrier and Greenspan, 1958) which approaches B with distance from shore, but it has no regular solution that approaches Σ or a combination of B and Σ. Boundedness within the linear framework therefore implies perfect reflection; conversely, imperfect reflection implies a solution unbounded within the linear framework and not regularly extensible to the shore by nonlinear long-wave theory.

Uniform ray theory (Section V) indicates that the same conclusions may apply to three-dimensional water waves, within the framework of the linear theory for small seabed slope. The bounded solution implies (5.12), hence perfect reflection. Imperfect reflection implies a linear combination of the bounded and unbounded solutions, with nonzero coefficient of the latter. The function of the shore condition in refraction theory is to determine the ratio of these coefficients.

Apart from the question mark for edge waves, perfect reflection is quite implausible for waves with amplitudes of practical concern. The condition

(5.12) must then be given up at the shore. It is important to note, moreover, that the coefficient ratio to be specified is a complex number, it determines both the amplitude ratio and the phase shift. More is therefore needed than a specification of percent energy absorbed.

It is a saving grace of refraction theory that the notion of rays or other mathematical techniques permit us to solve the differential equations independently of the boundary conditions. The ray pattern and conservation relations are the same for all boundary conditions. Virtually all the analytical or numerical work can therefore be completed independently of the shore condition and can later serve one condition as well as another. Only the final results, such as the maximal shore amplification or trapping frequencies, depend on the shore condition.

In some respects, conversely, any shore condition can serve as a stand-in for the unknown, correct condition for illustrative purposes. In this sense, the condition of boundedness will be used in the following where it would be worse to leave the discussion unillustrated.

VII. **Resonant Wave Trapping: Frequency**

It has been noted in Sections I and IV that refraction can trap waves in a bounded wave region even when the water surface is unbounded. The possibility of self-sustained standing waves analogous to those typical of waves in lakes and closed basins must then be suspected. Indeed, a trapped wave must, upon reflection at a boundary, start to sweep back over the wave region until it meets another boundary segment, whence it is again reflected to sweep back over the wave region. If the wave region is bounded, then after a number of such reflections, the wave must come to repeat earlier sweeps in sense and direction and when it does that, it must interfere with itself. The effect of such self-interference will depend on the phase relationship between the wave in its original sweep and the wave in its repeat sweep. If that phase relationship be just right, then resonant enhancement becomes a plausible possibility.

Determination of the phase matching condition characteristic of resonant self-interference amounts to formulation of an eigenvalue problem for the resonant standing wave modes, just as for lakes and basins. Keller and Rubinov (1960) have shown how simple ray theory (Section IV) together with the reflection conditions at shores and caustics (Sections IV, V) can be used to formulate and solve this eigenvalue problem with disarming directness and simplicity.

To explore the possible standing wave patterns, they exploit the decomposition (4.11) into individual waves which sweep over the water surface and generate others by reflection. The simultaneous sweep of several waves over the same water surface in different directions, however, calls for an accounting device to clarify the decomposition (4.11) and to keep track of the connections between the individual waves. To understand how the ray sheets (Section IV) serve this purpose, it helps to return to the simplest example (Section IV) of a semi-infinite laboratory channel with plane, parallel vertical sidewalls at $y = \pm b$, closed at the end $x = 0$ by a beach. The depth h is again taken to depend only on the distance x from shore and for definiteness, $h(x)$ will again be assumed strictly monotone increasing.

Regardless of the effect of the sidewalls, wave trapping between the shore $x = 0$ and a caustic $x = a > 0$ has then been seen in Section IV to occur whenever the initiation parameter c in (4.9) and (4.10) has a value in the range (k_∞, ∞) of the wave number function $k(h)$ given by the dispersion relation (3.5). It may be recalled, moreover, that $c \geq 0$, without loss of generality, and that the two signs in (4.9) and (4.10) may then be associated with distinct sheets of the wave region $0 < x < a$, each carrying just one ray congruence, and that these two sheets are connected at the shore and at the caustic by the reflection process. In the absence of sidewalls, that is, for a continental slope, these are the only connections between the sheets; in the y-direction, they extend indefinitely and the wave region $0 < x < a$ is therefore unbounded.

For the laboratory channel, however, reflection at the sidewalls is added, in which the y-component of the wave number vector changes sign, but not the x-component, as described explicitly by Snell's law (4.12). It is now relevant that, given the ray slope, the direction of increasing phase S on a ray can still take either sense. Given c, therefore, four oriented ray directions are possible at each point in the interior of the wave region. They are all physically distinct because the sense of S increasing is, by definition, the sense of wave propagation. Each of the four oriented wave directions therefore corresponds to a physically distinct wave and deserves a separate sheet of the water surface. When all four waves are present, then $N = 4$ in (4.11). Of the four sheets, one pair is connected at a sidewall to represent the physical connection between the corresponding waves by the reflection process there. A different pair is similarly connected at the shore, and so on.

The surface formed by the ray sheets *connected* in this manner by the reflection conditions (including that of caustic refraction) is called the *cover space* of the wave region, and the main point made by Keller and Rubinov (1960) is that its global, geometrical nature mirrors the structure of all the possible standing wave patterns.

To see that, it is best to begin with the case of progressive waves ($c < k_\infty$) where the caustic is absent and (4.10) predicts real ray slope throughout the whole, semi-infinite channel. Consider now a wave incident from the distant, deep end; its specification will fix c. Observed at some point far from shore, it will be seen to sweep obliquely towards one sidewall and then, after reflection from it, toward the other sidewall. That involves both ray directions (but both only in the shoreward sense of propagation), hence two ray sheets. They are connected at the sidewalls to form a surface like a (squashed) cylinder.

As we follow the wave, it approaches the shore in zigzag fashion and is finally reflected there (at least, partially). It then returns seaward in a similar manner, sweeping from sidewall to sidewall, which involves again both ray directions, but now only in the seaward sense of propagation. The return journey of our wave is therefore on a different pair of sheets, also connected at the sidewalls: our squashed cylinder is double-skinned, and the total number of sheets in (4.11) is indeed $N = 4$. The two skins, moreover, are sewn together at the shore to represent that particular reflection process, by which the shoreward skin generates the seaward skin. In sum, the cover space of progressive waves in the semi-infinite channel is a double-skinned (squashed) cylinder, similarly semi-infinite, with the skins connected at the shore end of the cylinder, but nowhere else. The surface is connected, but not bounded, which mirrors the physical fact that it represents a single wave propagation process, but one in which no sweep is ever repeated in *both* direction and sense. Topologically, on account of the shore connection between the skins, the surface is equivalent to a single-skinned, fully infinite cylinder.

Now consider the case of trapped waves ($c > k_\infty$) confined between the sidewalls at $y = \pm b$, shore at $x = 0$ and caustic at $x = a > 0$. For easier visualization, suppose first that a is rather large so that a wave incident from the deeper part of the channel will again sweep from sidewall before it is reflected from the shore. It is then apparent that $N = 4$ sheets are again needed. However, as the wave returns seaward, it comes to be refracted back at the caustic, so that the two skins of the (squashed) cylinder are also connected by the caustic reflection condition (4.13). The cover space is therefore a bounded surface. Of course, the magnitude of a has no influence on the number of sheets and the nature of their connections, and as we vary a continuously until it is rather small compared with b, another aspect of the surface becomes easier to visualize. The double-skinned cylinder becomes short and, with the shore and caustic connections between the skins, comes to look like the deflated inner tube of a tire. The cover space of trapped waves in the channel is therefore topologically equivalent to a torus. Since the ray direction and sense are single-valued on each sheet, a wave represented by this cover space must come to repeat earlier sweeps in both direction and sense.

For natural seabed topographies, the cover space can be quite complicated (Figs. 2, 3, 5), although it need not be (Fig. 1). In any case, the bounded wave region characteristic of trapping corresponds to a cover space which is a *closed* surface, if the number of sheets is finite, as is the norm. Clearly, for any reliable discussion of self-interference of waves and potential phase-matching, something equivalent to the topological genus of the cover space must be worked out first. Let us now see why "the rest is easy."

Equation (4.11) represents not only the velocity potential, but also the surface elevation $\zeta = iL\omega\phi(x, y, 0)$ which must be a single-valued function by its physical definition. This also applies, moreover, to the individual terms in the sum (4.11) because this sum is meant to decompose ϕ into its individual physical waves so that each term separately approximates a solution of the classical, linear water-wave equations (2.10). Each individual wave and the term in (4.11) representing it is associated with its own ray sheet of the wave region, and hence, each term must be single-valued on its sheet. However, the waves and associated terms are related to each other by the reflection conditions that join the sheets at the appropriate edges, and for this reason, it is the cover space, rather than the individual sheets, which really represents the complete physical wave. The precise expression of the physical condition is therefore that the surface elevation, and hence also *the potential, must be single-valued on the cover space*, and that is the famous eigencondition of Keller and Rubinov (1960).

Consider now the factors in the individual terms of the sum (4.11). Since k depends only on the local depth h, $\cosh[k(z + h)]$ is single-valued on the wave region. The rays are defined by (4.1), (4.4), and the reflection conditions (4.12) or (4.13), and they are single-valued on the sheets, by the definition of the sheets, inclusive of their orientation, and are therefore single-valued on the cover space. The amplitude function $A(x, y)$ is governed by the energy conservation law (4.7) along the rays and the reflection conditions (4.12) or (4.13), in which A^* is defined by (4.7) and $A_s = A\cosh(kh)$. The natural choice $A \geq 0$ is seen from them to involve no loss of generality. Since G depends only on h, by (3.5) and (3.6), the factors kG and $\cosh(kh)$ are single-valued already on the wave region, and the expansion ratio Ξ is single-valued on the cover space, like the rays which define it. If (4.13) holds at the shore as well as the caustic, it now follows from (4.7), (4.12), and (4.13) that A is single-valued on the cover space. (If (4.13) does not hold, then the frequency is complex and (4.11) is not tenable, see below and Section VIII.) Hence, the eigencondition of single-valuedness of ϕ on the cover space concerns only the factor $\exp(iS/\varepsilon)$ in (4.11), it has turned out to be indeed a pure phase-matching condition.

The phase $S(x, y)$, on the other hand, is not usually single-valued on a closed cover space. If we follow a ray on such a cover space from a phase line $S(x, y) = \text{const.} = c_1$ in the sense of propagation, then the ray continues from one sheet to the next by the reflection conditions, and as we keep thus following it, always in the sense of propagation, then if the number of sheets is finite as is the norm, the ray must in due course return to the sheet whence we had started. There it must intersect the phase line $S(x, y) = c_1$ again. However, since the sense of propagation is also that of S increasing and since $|\nabla S| = k > 0$, by (4.1), the phase S must have increased all along our ray path around the cover space, except for phase shifts $-\varepsilon\pi/2$ in S at each crossing of a shoreline or caustic on our ray path, by (4.13). Generally, therefore, our ray path around the cover space brings us back to the sheet and phase line whence we had started, but with a different value of the phase function S.

To quantify this change in S, let Γ denote any closed oriented path in the cover space and m_Γ, the number of shore and caustic crossings on Γ. Then by (4.12) and (4.13), the total change in S along Γ is

$$\int_\Gamma \nabla S \cdot d\mathbf{s} - \tfrac{1}{2}m_\Gamma\pi\varepsilon.$$

Furthermore, ∇S is single-valued on the cover space, by (4.5) and because the rays and $k(h)$ are single-valued. The last integral therefore has the same value on any two paths which are homologous (e.g., Meyer, 1971a, p. 19) in the cover space. For illustration, two cases had emerged in the channel example. For progressive waves, the cover space is a cylinder and there is just one class of irreducible closed paths in it, the belts of the cylinder. Up to orientation, all belts encircling the cylinder just once are homologous in it and the last integral can assume only one nonzero value (and integral multiples of it) for this cover space, regardless of the choice of Γ. For trapped waves, on the other hand, the cover space is a torus and there are two basic ways of going around it. Accordingly, the last integral can assume two nonzero values (and integral linear combinations of them) for the cover space, but no more. It thus begins to emerge gradually that the genus of the cover space mirrors fundamental features of spectral structure.

In any case, if the number of homology classes of irreducible closed paths in the cover space is N_Γ, then the integral can take just N_Γ nonzero values (and integral linear combinations of them). It follows that the factor $\exp(iS/\varepsilon)$ in (4.11) is also single-valued on the cover space if, and only if

$$\int_{\Gamma_q} \nabla S \cdot d\mathbf{s} - \tfrac{1}{2}m_q\pi\varepsilon = 2\pi n_q\varepsilon, \qquad q = 1, 2, \ldots, N_\Gamma, \qquad (7.1)$$

with integer n_q, for any one representative path Γ_q of each of the N_Γ homology classes; m_q denotes the net number of shore and caustic crossings on Γ_q. This is the condition of Shen et al. (1968) which is necessary and sufficient for the single-valuedness of the potential on the cover space. It does not have quite the expected form of a characteristic determinant for an eigenvalue problem. This is because reference to the homology classes has already permitted us to factor the N_Γ-by-N_Γ determinant.

The integer n_q represents the number of periods through which $\exp(iS/\varepsilon)$ goes when Γ_q is followed once around the cover space and therefore counts the number of wave periods that fit into Γ_q. It therefore also counts the number of wave crests and troughs (in the sense of Γ_q) of the standing wave mode characterized by (7.1), as will be illustrated presently. In this way, each of the N_Γ homology classes gives rise to a crest-counting parameter n_q which can run through the integers. Each N_Γ-tuple of integers n_q admitted by (7.1) thus characterizes a separate wave mode.

The evaluation of the integral in (7.1) is simplified by the arbitrariness of Γ_q within its class. If h depends only on distance x from the shore, for instance, ∇S is given by (4.9), and the obvious choices are paths Γ_1 at constant y and Γ_2 at constant x. Then (7.1) becomes for the case of trapping in a channel of width $2b$

$$2 \int_0^a |k^2 - c^2|^{1/2} \, dx - \pi\varepsilon = 2\pi n_1 \varepsilon, \qquad n_1 = 0, 1, 2, \ldots, \qquad (7.2)$$

$$2 \int_{-b}^b c \, dx = 4bc = 2\pi n_2 \varepsilon, \qquad n_2 = 1, 2, \ldots, \qquad (7.3)$$

because the closed path Γ_1 on the torus at constant y from shore to caustic and back must cross the shore and caustic once each (i.e., $m_1 = 2$), but the path Γ_2 at constant x from sidewall to sidewall and back does not cross either ($m_2 = 0$). These are the spectral conditions of Shen et al. (1968) for a semi-infinite channel. The value $n_2 = 0$ has been omitted because it is not compatible with trapping since it corresponds to $c = 0$, i.e., crests parallel to shore so that the motion is two-dimensional and can have no caustic. The parameter n_1 now counts the number of periods once around the cover space in the sense of Γ_1, so that it represents just the number of zeros of the standing wave mode counted at constant y from shore to caustic. Similarly, n_2 counts the number of zeros at constant x from sidewall to sidewall.

For the progressive wave case, by contrast, only the path Γ_2 leads around the cylinder, so that (7.3) is necessary and sufficient, while (7.2) does not apply. However, $n_2 = 0$ is now admissible.

For the evaluation of (7.2), (7.3), it should be recalled that $k(h)$ is defined by the dispersion relation (3.5), in which no generality is lost (for real frequency ω) by renormalizing ε so that $\eta \equiv \varepsilon\omega^2 = 1$. The distance a from shore to caustic is defined by $k(h(a)) = c$ in the case of trapping (Section IV,B). These definitions,

$$k \tanh(kh) = 1, \qquad k(h(a)) = c, \qquad (7.4)$$

together with (7.2) and (7.3) are four equations for a, c, ε, and $k(h)$, given $h(x)$, n_1, and n_2. Not all the values of the counting parameters n_1, n_2 need be compatible with Eqs. (7.2)–(7.4), but if $h(x)$ is monotone, then the strict monotonicity of $k(h)$ is readily seen to assure a discrete set of solutions for a and $\varepsilon = 1/\omega^2$. In practice, all we need is a root-finding routine for (7.4) and a quadrature routine for (7.2). Of course, there are special topographies, including the plane beach (Shen et al., 1968), for which Eqs. (7.2)–(7.4) can be solved explicitly, but they add little to the understanding. More details are found in Shen et al. (1968) and Meyer (1971b).

In a similar way, the eigencondition of Keller and Rubinov (1960) yields normally an N_Γ-parameter discrete spectrum of trapping frequencies when the cover space is closed and has N_Γ homology classes of irreducible closed paths. It shows, moreover, that no other frequencies can then correspond to solutions of (3.3) approximable by ray theory.

In the progressive wave case for a channel, by contrast, where (7.2) is inapplicable, the eigencondition is expressed by (7.3) and (7.4), which are only three equations for a, c, ε, and $k(h)$. They can therefore be solved for any c in the range $0 \le c \le k_\infty$ corresponding to progressive waves. Ray solutions of (3.3) are thus obtained for all frequencies $\omega = \varepsilon^{-1/2}$ exceeding

$$\omega_c = [\pi n_2/(2bk_\infty)]^{1/2}, \tag{7.5}$$

i.e., the frequency spectrum for progressive waves is continuous above the cutoff ω_c.

The nature of the frequency spectrum is illuminated further by consideration of the spectral "components" associated with fixed values of n_2. Then c/ε is fixed, and given $c > k_\infty$, the integral in (7.2) is bounded, and an upper bound on n_1 results. Each spectral component therefore has a finite set of discrete resonant frequencies below its cutoff. The cutoffs, on the other hand, depend on n_2, so that discrete frequencies of spectral components for high n_2 are embedded in the continuous spectrum of the spectral components for low n_2 . . . .

It is noteworthy that Eqs. (7.2)–(7.4) show the discrete eigenvalues ε to depend, not directly on local details of the topography, but only on a certain weighted average represented by the integral in (7.2)—as is quite typical of eigenvalues in mathematical physics. This feature simplifies the application of the eigencondition to natural topographies.

It should be repeated, finally, that the results for our illustrative example do depend markedly on the condition of boundedness at the shore (Section VI) which implies (4.13). If the amplitude reflection coefficient at the shore is $R_A \ne 1$, so that the energy fraction $1 - R_A^2$ is absorbed per reflection, then single-valuedness of ϕ is equivalent to single-valuedness of

$$A \exp(iS/\varepsilon) = \exp i(S/\varepsilon - i \log A),$$

i.e., the total change of $S/\varepsilon - i \log A$ along Γ_q must be an integer multiple of 2π. That changes (7.2) into

$$2 \int_0^a |k^2 - c^2|^{1/2} \, dx - \tfrac{1}{2}\pi\varepsilon + \delta_s - i\varepsilon \log R_A = 2\pi n_1 \varepsilon, \tag{7.6}$$

where δ_s denotes the phase change under shore reflection. On the other hand, (7.3) remains unchanged because Γ_2 crosses only sidewalls, at which (4.12) does hold. It is now seen that the new spectral relations have no real solutions for $R_A < 1$. In particular, $\varepsilon\omega^2$ must be complex, any trapped mode must decay in time (Section VIII).

The eigencondition can be applied with equal ease to a large variety of natural topographies. As an example, consider an underwater ridge or a long underwater bar. For a certain range of periods and, given the period, within a certain range of angles of incidence, wave rays will form a pair of caustics over such topographies so that the waves are trapped between the caustics (Shen *et al.*, 1968). To divide the physical difficulties, suppose first that the depth depends only on the coordinate normal to the bar, but not on that along the bar. The cover space is now an infinitely long, squashed cylinder of two sheets and the rays lead from one caustic to the second caustic on one sheet, and then back to the first caustic on the other sheet: they wind helically around the cover space. The cylinder admits just one class of irreducible closed paths which yield one spectral condition closely analogous to (7.2): the integral represents an intrinsic, normal (to the bar) wave distance between the caustics, and this must equal an odd half-integer multiple of $\pi\varepsilon$ for phase-matching.

Since the cover space is unbounded, we can anticipate a purely continuous spectrum, but it will have a finite number of components. It is continuous over an interval of wave number components along the bar, but given that component, then the spectral condition analogous to (7.2) together with the dispersion relation (3.5) and the two caustic conditions analogous to the second equation of (7.4) will have solutions for a finite set of odd-half integers. They determine the frequencies for which the bar acts as a waveguide, and such effects have been observed (Lozano, 1979).

More generally, the bar shape will not remain constant along the bar. Given a straight, incident wave train, however, ray tracing will tell whether two caustics are formed to trap the waves, and if so, then the waveguide mechanism is present. Any segment of the bar with near-parallel caustics will guide an associated set of waves. Another segment of the bar may be unable to guide these same waves; on the other hand, it may guide others. We may speculate on the modifications of the waveguide effect that could result, but these remain to be explored. Ray theory now offers the tools for this.

The channel example serves to illustrate the eigencondition and is helpful for experiment and for the discussion of edge waves. For many natural topographies, however, an essentially new element enters, which was discovered (Longuet-Higgins, 1967; Shen *et al.*, 1968) in theoretical work on axisymmetric topographies: *there can be no bounded wave regions!*

A simple kinematical explanation for this has been given by Longuet-Higgins (1967) for longwaves. As noted in the Introduction, the phase velocity of surface waves increases with the local depth, so that the parts of a wave crest lying over deeper water travel faster than those lying over shallower water, and this tends to turn all wave crests toward the shallows. That tendency, however, can win out over axisymmetric topographies only when the crests turn faster than the circular level contours of the seabed do already. For longwaves, the phase velocity is $(gH)^{1/2}$, the angular phase velocity about the center is therefore $(gH)^{1/2}/R$, if R denotes the radius. Refraction inward relative to the level contours of the seabed therefore requires that $H^{1/2}/R$ be an increasing function of R. However, a topography for which H increases faster than R^2 at large distances from the center is utterly unnatural. The argument can, of course, be extended to the general phase velocity of surface waves and a similar, even if more complicated, condition results (Shen et al., 1968). At sufficiently large distances from the center of an axisymmetric topography, refraction is therefore always unable to trap waves.

At an island shore, on the other hand, H always increases faster than R^2 locally, if the beach has nonzero slope. Local trapping near shore is therefore possible, but it must be accompanied by progressive waves far from shore. It will be shown in the next section how this implies radiation toward the open sea and any natural modes must exhibit time decay. Nonetheless, if there should be natural modes of very slow decay, then their frequency would be nearly real and an approximate treatment as if the frequency were real is helpful as a first step.

On this tentative basis, Shen et al. (1968) applied the eigencondition to trapping over axisymmetric topographies to obtain (real) frequency predictions with remarkable ease, as will now be sketched. The depth is then $\varepsilon L h(r)$ in polar coordinates $R = Lr$, θ; for simplicity, it will be assumed strictly monotone increasing. In the case of an island, the shore radius may be conveniently adopted as the unit of length L, and $h'(1)$ will then be assumed >0 to keep attention focused on what is typical.

By (3.5), $k = k(h)$, so that $\partial k/\partial\theta = 0$ and (4.4) shows

$$\frac{d}{d\sigma}\frac{\partial S}{\partial\theta} = \frac{\partial^2 S}{\partial r\,\partial\theta}\frac{1}{k}\frac{\partial S}{\partial r} + \frac{\partial^2 S}{\partial\theta^2}\frac{1}{kr^2}\frac{\partial S}{\partial\theta}$$

$$= \frac{1}{2k}\frac{\partial}{\partial\theta}\left[\left(\frac{\partial S}{\partial r}\right)^2 + \frac{1}{r^2}\left(\frac{\partial S}{\partial\theta}\right)^2\right] = 0,$$

by (4.1). Much as in (4.9), therefore,

$$\partial S/\partial\theta = \pm c = \text{const.} \qquad \partial S/\partial r = \pm(k^2 - c^2/r^2)^{1/2}, \qquad c \geq 0, \qquad (7.7)$$

by (4.1), and (4.4) now shows the ray slope to be

$$r\, d\theta/dr = \pm c(k^2 r^2 - c^2)^{-1/2}. \tag{7.8}$$

Given c, therefore, a plot of the "spectral function"

$$p_s(r) \equiv rk(h),$$

determined already by (3.5), indicates right away the wave regions where the ray slope is real and the caustics which bound them; by (7.8), the caustic radii are the roots of $p_s(r) = c$. Figure 13 shows the typical shape of $p_s(r)$ for reefs, where k decreases from a finite value at $r = 0$ to $k_\infty > 0$ as $r \to \infty$. Figure 14 shows the typical shape for islands, where $(1 - r)^{1/2} k$ tends to a limit at the shore $r = 1$. In either case, $p_s(r) > c$ for sufficiently large r, for any c, and the far field is always a wave region of real rays. Similarly, the region close to an island shore is always a wave region. The center of a reef, by contrast, is now seen to be always a region of evanescence or damping without real rays.

Spectral curves $p_s(r)$ with more than one extremum are possible (Shen *et al.*, 1968), if a reef has a very flat top or an island has a very pronounced shelf, but are too rare to be worth consideration here: we exclude all but "typical topographies," meaning those with a single extremum of $p_s(r)$, as in

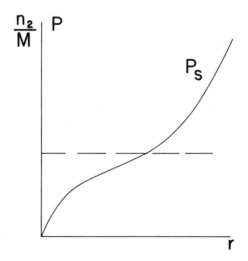

FIG. 13.   Typical shape of the spectral function $p_s(r) = rk(h)$ for a round reef. For any given value of c (denoted by n_2/M in the figure), there is just one caustic radius r_c at which $p_s(r_c) = c$. For $r > r_c$, (7.8) predicts real rays, but for $r < r_c$, it predicts no rays or waves. From Fig. 19, R. E. Meyer, Resonance of unbounded water bodies. Reprinted from "Mathematical Problems in the Geophysical Sciences," Lect. Appl. Math., Vol. 13, p. 224, by permission of the American Mathematical Society. © 1971 by the American Mathematical Society, Providence, R.I.

R. E. Meyer

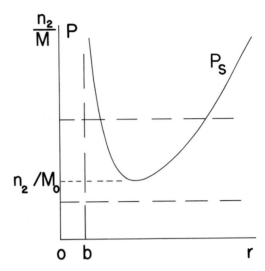

FIG. 14. Typical shape of the spectral function $p_s(r) = rk(h)$ for a round island of shore radius b. The minimum value c_m of the spectral function defines a cutoff. For any given c (denoted by n_2/M in the figure) below c_m, there is no caustic radius r_c at which $p_s(r_c) = c$. For any given $c > c_m$, on the other hand, there are two such caustic radii, and (7.8) predicts real rays and waves on either side of this pair, where $p_s(r) > c$. Between the two caustic radii, however, (7.8) predicts no rays or waves, the zone between them is a "ring of evanescence." From Fig. 17, R. E. Meyer, Resonance of unbounded water bodies. Reprinted from "Mathematical Problems in the Geophysical Sciences," Lect. Appl. Math., Vol. 13, p. 222, by permission of the American Mathematical Society. © 1971 by the American Mathematical Society, Providence, R.I.

Figs. 13 and 14. A "typical reef" is then seen incapable of trapping, all waves around it are progressive; the question of interest is where the caustic lies for given incident waves and how much the waves are enhanced there (Section V). Meyer (1971b) relates an instance where waves, apparently of this type, were observed to grow big enough to threaten even large offshore structures.

For a "typical island," by contrast, the minimum of $p_s(r)$ defines a cutoff value c_m such that values $c < c_m$ correspond to purely progressive waves, incident from the open sea and reflected at the shore. For $c > c_m$, however, a new phenomenon arises because (7.8) predicts a wave region adjacent to shore but separated from the far wave region by a ring of evanescence.

The inner wave region, for $c > c_m$, is bounded and its cover space is closed and certainly, ϕ must be single-valued on it. That implies again (7.1), with ∇S now given by (7.7). There are again four oriented ray directions at each point in the interior of this wave region, by (7.8), but there is no reflection mechanism by which clockwise propagation can generate anticlockwise propagation, or vice versa. For anticlockwise propagation, say, there are therefore $N = 2$ waves in this region, one sweeping from the shore out to the caustic, the other sweeping back shoreward. The two sheets are connected by the shore and caustic reflections to a cover space which is a torus. Hence, $N_{\Gamma} = 2$ again, and (7.7) indicates that the most convenient paths are at fixed θ and fixed r, respectively. The former crosses the shore and caustic once each on the cover space, the latter crosses neither and can conveniently be taken arbitrarily close to shore. For definiteness, take the boundedness condition at the shore (Section VI), then (4.13) applies to both caustic and shore reflection, and (7.1) yields

$$2 \int_{1}^{r_1} \left| k^2 - \frac{c^2}{r^2} \right|^{1/2} dr - \pi = 2\pi(n_1 - 1)\varepsilon, \qquad n_1 = 1, 2, \ldots,$$

$$\int_{0}^{2\pi} c \, d\theta = 2\pi c = 2\pi n_2 \varepsilon, \qquad n_2 = 1, 2, \ldots,$$

by (7.7). Here k is defined by the dispersion relation (3.5), in which $\varepsilon\omega^2$ may again be normalized to unity, r_1 is the smaller root of $p_s(r) - c$ (Fig. 14), and c may be eliminated conveniently to obtain

$$\int_{1}^{r_1} \left| k^2 - (n_2\varepsilon/r)^2 \right|^{1/2} dr = \pi(n_1 - \tfrac{1}{2})\varepsilon, \qquad n_1 = 1, 2, \ldots,$$

$$k \tanh(kh) = 1, \qquad r_1 k(h(r_1)) = n_2\varepsilon, \qquad n_2 = 1, 2, \ldots,$$

(7.9)

These relations are quite analogous to (7.2)–(7.4) and predict the same type of mixed spectrum: a countably infinite set of components, one for each value of n_2, each with a finite discrete spectrum of frequencies $\omega = \varepsilon^{-1/2}$ below the cutoff of a continuous spectrum.

More details are discussed by Shen *et al.* (1968), for instance, the lowest values of n_2 are shown to be frequently associated with an empty discrete spectrum. The accuracy of (7.9) can also be improved by use of the uniform ray approximation (Section V) in the place of (3.4) to evaluate the eigencondition (Shen and Keller, 1975; and Section VIII).

What meaning, however, can be attached to real frequency predictions? We may hope that they give a good approximation to the real part of ω when the imaginary part is small enough, and this will be borne out in Section VIII. But, the theory of the present section cannot provide more than a tentative distinction between those eigenvalues which have a small imaginary part and those which have not. The time decay represented by im ω arises from the leakage of energy through the ring of evanescence between the two caustics at $r = r_1$, r_2 (Fig. 14). The amplitude decay of evanescent waves is exponentially fast with distance from the caustic (Section V), and the leakage should therefore be small when the ring of evanescence is wide. The measure of width, however, should be an intrinsic one, related to the local wavelength, and in the ring of evanescence, the wavelength is is not real . . . .

A reliable measure of leakage, moreover, is needed because of its intuitively plausible connection with the *degree of resonance* of a mode. Even an accurate prediction of the real part of an eigenvalue ω^2 is of very limited use without an estimate of whether the eigenvalue is dangerously resonant or effectively harmless.

VIII. Resonant Wave Trapping: Response

In Sections IV to VII, refraction theory has been developed for real frequency, because it is much simpler. Two causes of time decay, however, have been identified for trapped waves; the first arises from shore absorption and the second, from energy leakage through evanescent zones. For complex frequency ω—to describe the time decay with a potential $\Phi = (L^3g)^{1/2}\phi(x, y, z)\exp(-i\omega t)$—the quantity $\varepsilon\omega^2 \equiv \eta$ in the dispersion relation (3.5) becomes complex, if ε is to serve its function as a scale of the seabed slope. Complex η, in turn implies complex wave number k and phase S, by (3.5) and (4.1), and as noted in Section III, any asymptotic scheme involving powers of ε then becomes useless. The form of the solution of (3.3) must be altogether different from (3.4). The uniform ray theory of Section V is similarly based on a comparison of powers of ε (Appendix V,a) and this rationale therefore collapses similarly.

This dictates a retreat all the way to "exact" linear theory (3.3), or at least, to the refraction equation (3.6), because the conjecture of a vertical structure of the motion dependent primarily on the local depth is not discredited by gradual time changes. Two mathematical lines of attack are then available. One is based on the method of characteristics for time-dependent problems (Section X) and is more suited to initial value problems than to the study of large-time solutions of most common interest in refraction. The other line of attack is available only for special topographies such as channels and axially symmetrical reefs and islands, but is then very effective. It separates the variables in (3.6) to assume, e.g., for waves around a reef or island,

$$\Psi \equiv \phi(x, y, 0) = e^{in_2\theta}r^{-1/2}w(r) \tag{8.1}$$

in polar coordinates $r = R/L, \theta$.

Is that legitimate? Two points may be made. First, ray theory (Sections IV–VII) does not assume such a form for the solutions, but finds no others, as is shown by (4.9) and (7.7), and uniform ray theory (Section V) preserves this structure. That encourages conjecture that no solutions are excluded by

(8.1). Second, the main task before us is to identify trapped modes of small leakage and much is gained by a reliable method for this, even if it is not known with certainty to be exhaustive.

Once (3.5), (3.6), and (8.1) are accepted, moreover, reliability of the further analysis can be assured because the refraction equation (3.6) is now reduced to an ordinary differential equation,

$$d^2w/dr^2 = [\varepsilon^{-2}f(r) + g''/g]w, \tag{8.2}$$

with

$$f(r) = (n_2\varepsilon/r)^2 - k^2, \tag{8.3}$$

$$g(r) = (rG)^{1/2} = \tfrac{1}{2}[2rh + (r/k)\sinh(2kh)]^{1/2}\operatorname{sech}(kh), \tag{8.4}$$

and $k(h)$ given by the dispersion relation (3.5)

$$k\tanh(kh) = \varepsilon\omega^2 \equiv \eta. \tag{8.5}$$

For small seabed slope ε, (8.2) has the form of a typical WKB problem of a rapid oscillator with slow modulation, and it is of interest to understand the physical reason why this appearance is partly misleading. To this end, let us begin by noting that the wave trapping problem for an island now reveals itself as a scattering problem. The main coefficient in (8.2) is closely related to the spectral function $p_s(r) = rk(h)$ of Section VII, and for real k, Fig. 14 allows us readily to plot $f(r)$ for a typical island; when $n_2\varepsilon$ is not too small, Fig. 15 results. It shows f to be negative at large r, in fact, $f \to -k_\infty^2 = -\eta^2$, by (8.3) and (8.5), and the solutions of (8.2) therefore tend to superpositions of incoming and outgoing waves of (radial) wave number η/ε and frequency ω as we approach the open sea. The question of leakage, or degree of trapping, now takes the following form: *if there is a motion without incoming wave, how large is the outgoing wave?*

To answer it, we need to trace the two waves to the shore, by the help of the differential equation (8.2), and see how the shore condition connects them. Now, the roots of $f(r)$ (Fig. 15) are "turning points" of (8.2), between which $f > 0$, so that the solutions have their exponential character, and the subdominant solution needs to be traced through this evanescent ring, however much it be hidden beneath the dominant solution. That is a typical WKB connection problem or "tunneling problem" which turning point theory is designed to solve. The new element arises from the practical importance of very small leakage.

The "tunneling" of energy through the evanescent ring can proceed in both directions and the reverse leakage provides a mechanism for exciting the motion in the inner, trapped wave ring by waves incident from the open ocean. This is a *first-order* mechanism of excitation completely absent in more classical, self-adjoint problems. For a channel topography, for instance,

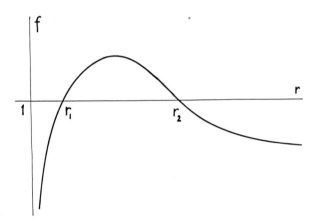

Fig. 15.   Typical shape of the main coefficient function $f(r)$ in (8.2) for a round island for a value of $n_2\varepsilon$ above the cutoff value $c_m = \min p_s(r)$ (cf. Fig. 14). The roots of $f(r)$ are the caustic radii for real frequency. From Fig. 2, R. E. Meyer, Resonant refraction by round islands. Reprinted from *Proc. Coastal Eng. Conf., 15th, 1976* p. 875, by permission of the American Society of Civil Engineers, © 1977 by the American Society of Civil Engineers, New York.

the mechanism is absent and edge waves were first excited by shaking the beach (Ursell, 1952), then subharmonically (Galvin, 1965). The latter mechanism is only of second order in the small wave amplitude. By contrast, reverse leakage around islands is an excitation mechanism of first order in the amplitude.

It is the more effective, moreover, the smaller the leakage rate. A standard measure of resonant excitation derives from the problem of *quasi-resonance*: *If a standing wave of given frequency be set up around an island, how large is the energy supply from the open sea needed to maintain it?* The ratio of the trapped wave amplitude in this situation to the amplitude of the wave coming from the open sea is called the *response coefficient* and will be seen below to be the reciprocal of the leakage rate (defined as the imaginary part of the eigenvalue ω^2). It is this response coefficient which provides a basic, practical measure of degree of resonance and directs attention to the very smallest leakage rates because they characterize the most resonant eigenvalues.

Now, WKB theory or conventional turning point theory is designed to determine the asymptotic expansions of the solutions of equations such as (8.2) for small ε. Those in turn permit us to connect the shore condition with the fundamental wave solutions in the far field to obtain the asymptotic expansion of the characteristic equation $\Delta(\varepsilon\omega^2) = 0$ of the eigenvalue problem: *is a nontrivial solution possible without incoming wave from the far field?* From the expansion of the characteristic equation, finally, we can try to extract the asymptotic expansion of the eigenvalues. For bounded solutions, however, the whole asymptotic expansion of the eigenvalues in powers of ε turns out to be real. It can give no information at all on leakage and resonant response!

Eigenvalues of extraordinarily large response are not implausible, moreover. For natural wave modes with a relatively wide ring of evanescence, the exponential wave decay with distance from the nearest caustic should become fully effective so that only an exponentially small fraction of the wave could penetrate through the ring to the other caustic. This is exemplified by explicit tunneling solutions for special potentials in quantum mechanics (Epstein, 1930). The very hugeness of such response, however, makes it entirely inaccessible to standard WKB or turning point theory because it is unobtainable from the asymptotic expansion of the characteristic equation. A basic extension of turning point theory (Lozano and Meyer, 1976) was therefore needed to show that typical, round islands do have such a huge resonant response at certain frequencies and to compute it.

The basic idea for this extension is that negligible signals may be filtered out of a large amount of noise, if they are qualitatively different. In particular, water waves of real frequency respect energy conservation, and this is intimately related to the fact that the differential equations of the exact theory have real coefficients for real frequency. Part of the strength of the refraction equation (3.6) lies in the preservation of this basic property, and the differential equation (8.2) inherits it, in turn. For real frequency, therefore, the solutions $w(r; \varepsilon, n_2, \eta)$ come in complex-conjugate pairs and this *complex symmetry* remains a relevant qualitative feature at slightly complex frequency, because the coefficients in (8.2) depend *analytically* on $\varepsilon\omega^2 = \eta$, by (8.5). From (8.3), for instance,

$$f(r; \varepsilon, n_2, \bar{\eta}) = \overline{f(r; \varepsilon, n_2, \eta)}$$

near the real η-axis, because $f = \bar{f}$ for real η. The solutions w inherit this complex symmetry in η, and so do any analytic functionals of the solutions. It is an exact property, moreover, on which we can rely regardless of our inability to pin w or its functionals down quantitatively with accuracy. By tracing it carefully through the elaborate process of turning point approximation, Lozano and Meyer (1976) were able to filter in the characteristic equation $\Delta(\eta) = 0$ a critical, exponentially small term out of the qualitatively different (and quantitatively all but inaccessible) asymptotic expansion of $\Delta(\eta)$.

To state their results, recall that ε is a small, real parameter representing the scale of the seabed slope: "asymptotic" refers to the first approximation for all sufficiently small, fixed ε, unless a different meaning is stated explicitly. The surface potentials admitted are of the form (8.1) with counting parameter

n_2 large enough that $n_2\varepsilon$ exceeds appreciably the minimum of the spectral function $p_s = rk(h(r))$ (Fig. 14) for real ω^2 near the eigenvalue under consideration. By definition of a "typical" island [or "simple island" in Lozano and Meyer (1976)], $p_s(r)$ has only one minimum, and simple trapping theory (Section VII) showed it to be a cutoff value for n_2 below which all waves are progressive and no trapping occurs. In (8.1), the distance r from the island center is measured in units of the island radius L, the dimensional undisturbed water depth is $H(rL) = \varepsilon L h(r)$, and $h(r)$ is assumed analytic on some (sectorial) neighborhood of the segment $[1, \infty)$ of the real r-axis, with simple root at the shore $r = 1$ and monotone increase along $[1, \infty)$.

Physics would not appear to give any definite guide on whether real, sedimentary seabed topographies are analytic. The justification for this assumption lies probably in the nature of the spectral relations, such as (7.2) or (7.9), which depend only on weighted averages of the topography. A nonanalytic topography then becomes equivalent to an analytic one with the same, relevant average. The averages in (7.2) and (7.9) will, in fact, reveal themselves as typical WKB integrals, and the results of Lozano and Meyer (1976) are also expressible in such familiar forms, even though they were not obtainable by the familiar methods.

For this general class of axially symmetric seabed topographies, the coefficient function of (8.2) is then defined by (8.3) to (8.5) and Fig. 15 shows the general form of $f(r)$ for real r and ω^2. Actually, there are no real eigenvalues and η must also be thought of as a complex variable on a neighborhood of an appropriate point of its real axis. The roots r_1, r_2 of $f(r)$—which mark the caustic radii of (3.6) or turning points of (8.2)—are then also complex and Fig. 16 shows them in the complex r-plane for small $\operatorname{Im}\eta < 0$. However,

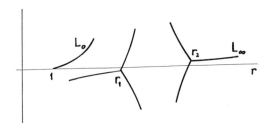

FIG. 16.   Roots r_1, r_2 of the main coefficient function $f(r)$ in (8.2), and Stokes lines L_i through them, in the plane of the complex radius r for small $\operatorname{Im}\eta < 0$. To each Stokes line corresponds a solution pair of (8.2) which are purely progressive waves (without growth or decay) along that Stokes line. The radiation condition takes its natural form in terms of the solution pair of L_∞; the shore reflection coefficient refers to the pair of L_0. From Fig. 3, R. E. Meyer, Resonant refraction by round islands. Reprinted from *Proc. Coastal Eng. Conf., 15th*, 1976 p. 876 by permission of the American Society of Civil Engineers. © 1977 by the American Society of Civil Engineers, New York.

the shore $r = 1$ is also a "transition point" of (8.2) because $f(r)$ has there a simple pole and g''/g, a double pole. In WKB parlance, our scattering problem is therefore a "three-turning-point problem."

In oceanographical parlance, the appearance of (8.2) could be simplified by the long-wave approximation (Section II) which replaces (8.5) by $k^2 h \approx \eta$ and (8.4), by $g \approx (rh)^{1/2}$, but that would raise doubts on the validity of the work without appreciable simplification of either analysis or results. A simple statement of the results depends rather on introduction of the familiar natural length variable of (8.2),

$$\xi = \int f^{1/2} \, dr = \int [n_2^2 \varepsilon^2 - k^2 r^2]^{1/2} r^{-1} \, dr,$$

often called WKB or Liouville–Green or Langer variable, which rescales (8.2) to local wave number ε^{-1}, and thus measures radial distance in (fixed multiples $(2\pi\varepsilon)^{-1}$ of) local, radial wavelengths. It is also an analytic function of r and η, once a definite lower limit of the integral and branch of $f^{1/2}$ have been chosen. A statement of the radiation condition, moreover, requires reference to the "Stokes lines" $\mathrm{Re}\,\xi = \mathrm{const.}$ through the transition points (Fig. 16). It is on those lines that (8.2) has solutions which are really pure waves, the "WKB solutions"

$$u_i^\pm \sim f_i(r)^{-1/4} \exp[\pm \xi_i(r_j, r)/\varepsilon]\{c_i + \mathcal{O}(\varepsilon)\}, \qquad (8.6)$$

where the subscript i refers to the Stokes Line L_i (Fig. 16) and j, to the transition point whence it issues $(r_0 = 1)$, and

$$\xi_i(r_j, r) = \int_{r_j}^r [f_i(s)]^{1/2} \, ds, \qquad (8.7)$$

with $f_i^{1/2}$ denoting, say, that branch of $f^{1/2}$ which makes u_i^+ a wave propagating out from r_j on L_i. Finally, c_i is a normalization constant and the brace in (8.6) represents the asymptotic expansion of $u_i^\pm$.

Conventional WKB theory tends to concentrate on the right-hand side of (8.6) and its meaning, but for Lozano and Meyer's (1976) extension, the right-hand side is mainly a housekeeping device for keeping track of the association of Stokes lines with branches of $f^{1/2}$ and ξ, while $u_i^\pm$ are thought of as a fundamental pair of *exact* solutions of (8.2) which have pure wave character on L_i. Then any solution w is

$$w(r) = \beta_\infty^+ u_\infty^+(r) + \beta_\infty^- u_\infty^-(r), \qquad (8.8)$$

with constants $\beta_\infty^\pm$, and the eigenvalue problem asks whether there are frequencies ω for which a nontrivial solution (8.8) is possible without incoming wave, i.e., can we have

$$\beta_\infty^- = 0, \qquad \beta_\infty^+ \neq 0 \qquad ? \qquad (8.9)$$

Actually, the physical question concerns $w(r)$ for large real r, rather than $w(r)$ on L_∞, but if $|\arg \omega|$ is not too large, the wave character of $u_\infty^\pm$ extends to the real r-axis (Lozano and Meyer, 1976), albeit with time decay.

The answer to the eigenquestion (8.9) must depend on the shore condition. There (8.2) has a singular point and standard arguments show it to have a fundamental solution pair gJ, gN characterized by boundedness of $J(r)$ and unboundedness of $N(r)$ at the shore $r = 1$, in fact, these are asymptotic to the Bessel functions familiar from long-wave theory and uniform ray theory (Section V). Thus

$$w(r) = \beta_b g(r) J(r) + \beta_s g(r) N(r), \tag{8.10}$$

with constants β_b, β_s. For illustrative purposes and lack of a definite alternative (Section VI), we begin with the shore condition of bounded surface elevation, i.e.,

$$\beta_s = 0. \tag{8.11}$$

Since (8.8) and (8.10) represent the same solution $w(r)$ of (8.2) exactly, the coefficients β in them must be linearily related, e.g.,

$$\beta_s = Z_+ \beta_\infty^+ + Z_- \beta_\infty^-,$$

with coefficients $Z_\pm(\eta)$ similarly independent of r, and by (8.9) and (8.11), the characteristic equation of the eigenvalue problem is

$$Z_+(\eta) = 0.$$

A more useful and informative equation is obtained by comparing (8.8) and (8.10) less directly: the solution $w(r)$ must also be a linear combination

$$w(r) = \beta_0^+ u_0^+(r) + \beta_0^- u_0^-(r) \tag{8.12}$$

of the fundamental pair associated with the Stokes line L_0 (Fig. 16). Standard WKB theory (Olver, 1974), or asymptotic approximation of the Bessel function asymptotic to J, now shows (8.11) to imply

$$\beta_0^+/\beta_0^- = -i + \mathcal{O}(\varepsilon). \tag{8.13}$$

From the wave character of $u_0^\pm$, the ratio (8.13) is seen to be just the (complex) shore reflection coefficient

$$R = \beta_0^+/\beta_0^-, \tag{8.14}$$

and indeed, (8.13) agrees with (4.13). Furthermore, since (8.8) and (8.12) represent the same solution, their coefficients must also be linearly related, e.g.,

$$\beta_\infty^- = t_+ \beta_0^+ + t_- \beta_0^-, \tag{8.15}$$

and by (8.9), an equivalent form of the characteristic equation is

$$\Delta(\eta) \equiv \beta_\infty^- / \beta_0^- = t_- + t_+ \beta_0^+ / \beta_0^- = 0. \tag{8.16}$$

By contrast to the shore reflection coefficient $R = \beta_0^+ / \beta_0^-$, $t_+(\eta)$ and $t_-(\eta)$ are unrelated to the shore condition and depend only on the differential equation (8.2). The characteristic equation (8.16) can therefore be used with any desired shore reflection condition and radiation condition, once the "transition coefficients" t_- and t_+ in (8.15) have been computed. In turn, the solution of the eigenvalue problem by determination of the roots of $\Delta(\eta)$ yields corollary answers to many physical questions. For this reason, the computation of the transition coefficients for all typical round islands on the basis of the general refraction approximation (Section III) may be regarded as the main result of Lozano and Meyer (1976).

The transition coefficients are one of the main objects of WKB or turning point theory, and Lozano and Meyer (1976) showed that it yields

$$t_- = \gamma_0 \exp[-2\xi_0(1, r_1)/\varepsilon]\{1 + \mathcal{O}(\varepsilon^2)\},$$
$$t_+ = \gamma_0\{i + \mathcal{O}(\varepsilon)\}, \tag{8.17}$$

and with the shore reflection condition (8.13), the characteristic equation (8.16) now yields $\exp[-2\xi_0(1, r_1)/\varepsilon] = -1 + \mathcal{O}(\varepsilon)$, which closer inspection reveals indeed as an implicit condition for ω^2. It is of a quite plausible form, since the natural distance $\xi_0(1, r_1)/\varepsilon$ between the shore $r = 1$ and inner turning point $r = r_1$, that is, the width of the trapped wave ring in units of $(2\pi)^{-1}$ radial wavelengths, is a typical WKB integral

$$\frac{1}{\varepsilon} \xi_0(1, r_1) = \frac{1}{\varepsilon} \int_1^{r_1} (n_2^2 \varepsilon^2 - k^2 r^2)^{1/2} r^{-1} \, dr.$$

Admittedly, it is complex, as are r_1 and the wavelength, but for near-real frequency, it is approximated by its value for real η,

$$\xi_0(1)/\varepsilon = \frac{i}{\varepsilon} \int_1^{r_1} |n^2 \varepsilon^2 - k^2 r^2|^{1/2} r^{-1} \, dr. \tag{8.18}$$

This is just the integral in (7.9), so that the first approximation to the characteristic equation is seen to yield the spectral condition of Shen *et al.* (1968) that twice the width of the trapped wave ring must be an odd-half integer number of radial wavelengths.

However, when (8.13) and (8.17) are substituted in the characteristic equation (8.16) to compute the imaginary part of the eigenvalues $\eta = \varepsilon \omega^2$, it is found to be hidden in the error terms. Those stand, of course, for the higher order terms in the asymptotic expansions of β_0^+ / β_0^-, t_- and t_+, which could, at least in principle, be calculated by conventional turning point theory. But, Lozano and Meyer (1976) show that it would not help because the imaginary part of the eigenvalues would always remain hidden in the error terms, however far that formidable process be carried!

It should be observed now that the characteristic equation (8.16) is exact because only the basic character of the fundamental system $u_i^\pm$, but not the approximations (8.6) and (8.13), were used to obtain (8.16). Similarly, (8.12) and (8.15) give exact, even if somewhat abstract, definitions of the terms in (8.16). Lozano and Meyer (1976) could therefore use them to trace the analytic dependence on η and the complex symmetry in the η-plane representing the property of energy conservation step by step through the long turning point calculations to show that (8.13) and (8.17) actually have the form

$$R = \beta_0^+/\beta_0^- = -i\exp[i\varepsilon\sigma_s(\eta)],$$
$$\gamma_0^{-1}t_- = \exp[-2\varepsilon^{-1}\xi_0(1,r_1) + i\varepsilon\sigma_-(\eta)],$$
$$\gamma_0^{-1}t_+ = i\exp[i\varepsilon\sigma_+(\eta)] - (1+i)\gamma^{-2}\{1 + \mathcal{O}(\varepsilon)\},$$
$$\gamma = \exp[-\xi_1(r_1,r_2)/\varepsilon],$$

(8.19)

where $\sigma_s, \sigma_+, \sigma_-$ are bounded on a suitable η-domain, and real for real η, but otherwise unknown: they contain the asymptotic expansions apart from the contribution in the common factor γ_0 which has magnitude

$$|\gamma_0| = \gamma\{1 + \mathcal{O}(\varepsilon)\}.$$

In (8.19), γ is recognized as another typical WKB expression; in

$$\xi_1(r_1,r_2) = \int_{r_1}^{r_2} f_1^{1/2}\,dr,$$

$f_1(r) > 0$ between r_1 and r_2 for real η (Fig. 15), but the appropriate branch $f_1^{1/2}$ is negative, so that the natural *width of the ring of evanescence* in (here imaginary) radial wavelengths is

$$(2\pi)^{-1}\log\gamma = -\xi_1/(2\pi\varepsilon) = \frac{1}{2\pi\varepsilon}\int_{r_1}^{r_2}|n_2^2\varepsilon^2 - k^2r^2|^{1/2}r^{-1}\,dr > 0 \quad (\eta\text{ real}) \quad (8.20)$$

and for near-real η, γ is therefore *exponentially large* whenever $n_2\varepsilon$ exceeds the minimum of $p_s(r) = rk(h(r))$ appreciably (Fig. 14).

That makes the second term of t_+ in (8.19) entirely negligible, by normal standards, in comparison with the unknown errors in the first term [which gives the same approximation to t_+ as (8.17)]. However the first term of t_+ in (8.19) has strictly unit magnitude and the second term can be distinguished as measuring the deviation of $|t_+|$ from 1, regardless of the much larger error in $\arg t_+$. It will now be observed that the first two quantities given by (8.19) are also strictly of unit magnitude, and this makes (8.19) an approximation of penetrating subtlety: The complex symmetry in the η-plane reveals a qualitative difference between the exponentially small contribution to t_+ and the asymptotic expansions.

An incidental corollary of (8.19) is that the shore condition (8.11) implies a reflection coefficient R of strictly unit magnitude $|R| = R_A$. This confirms perfect energy reflection to be rigorously implied by the boundedness condition (Section VI) for the refraction equation (3.6). Only the phase shift can have an asymptotic expansion.

Substitution of (8.19) into the exact characteristic equation (8.16) brings it into the form

$$\gamma_0^{-1}\Delta(\eta) = \Delta_0(\eta) + \Delta_1(\eta) = 0,$$

$$\Delta_0(\eta) = \exp[-2\xi_0(1, r_1)/\varepsilon + i\varepsilon\sigma_-] + \exp[i\varepsilon(\sigma_+ + \sigma_s)], \quad (8.21)$$

$$\Delta_1(\eta) = \gamma^{-2}(i - 1)\{1 + \mathcal{O}(\varepsilon)\},$$

in which again $|\Delta_1|$ is technically negligible in comparison with every term of the asymptotic expansion (contained in Δ_0) of the characteristic determinant Δ in powers of ε. However, the two terms of Δ_0 have strictly unit magnitude, and Δ_1 can be distinguished from them as the first approximation to the contribution of qualitatively different character to the characteristic determinant.

The great difference in the asymptotic sizes of Δ_0 and Δ_1 can be used to solve the characteristic equation (8.21) as follows. The scale ε of the seabed slope is not a precisely defined number and may again be normalized without loss of generality, this time, to make

$$\Delta_0(1) = 0,$$

i.e.,

$$-2\xi_0(1, r_1)/\varepsilon + i\varepsilon(\sigma_- - \sigma_+ - \sigma_s) = -(2n_1 + 1)\pi i,$$

with large integer n_1. Since $\xi_0(1, r_1)/\varepsilon$ is given by (8.18), for $\eta = 1$, this normalization amounts just to (7.9), except for the error term of σ's, and therefore determines just the sequence of values $\varepsilon(n_1, n_2)$ discussed in Section VII, except for a percentage error of order ε^2. That error remains otherwise unknown, but is of smaller order than might have been anticipated.

For large n_1, these ε-values are small and, with increasing $|\eta - 1|$, $\Delta_0(\eta)$ soon becomes much larger than $\Delta_1(\eta)$ in (8.21). The principle of the argument for analytic functions therefore assures a root of $\Delta(\eta)$ close to $\eta = 1$. To the first approximation, this eigenvalue is

$$\eta(n_1, n_2) \sim 1 - \Delta_1(1)/\Delta_0'(1) \sim 1 + \tfrac{1}{2}(i - 1)\varepsilon\gamma^{-2}/\xi_0'(1), \quad (8.22)$$

with γ given by (8.19), (8.20) and

$$\xi_0'(1) \equiv \frac{\partial}{\partial \eta} \xi_0(1, r_1)\Big|_{\eta=1} = i \int_1^r |n_2^2 \varepsilon^2 - k^2 r^2|^{-1/2} \frac{k^2 r \, dr}{1 + h(k^2 - 1)}, \quad (8.23)$$

by (8.18) and (3.5) (Lozano and Meyer, 1976).

Since γ is exponentially large in ε, $\eta(n_1, n_2)$ is seen from (8.22) to be exponentially small, the asymptotic expansion of the characteristic determinant makes no contribution to it. However, $\eta - 1$ contains the only nonreal contribution to the eigenfrequency

$$\omega(n_1, n_2) = [\eta(n_1, n_2)/\varepsilon(n_1, n_2)]^{1/2}.$$

As it should, Im $\eta(n_1, n_2)$ turns out negative, the time factor $\exp(-i\omega t)$ in the potential (Section III) describes a slow decay of the eigenfunctions with logarithmic decrement

$$|\mathrm{Im}\,\omega| \sim \tfrac{1}{4}\varepsilon^{1/2}\gamma^{-2}|\xi_0'(1)|^{-1} = \tfrac{1}{4}|\xi_0'(1)|^{-1}\omega(n_1, n_2)e^{-4\pi d},$$

if $d(n_1, n_2)$ denotes the width (8.20) of the ring of evanescence in (imaginary) radial wavelengths.

As was physically plausible, the leakage rate is thus seen to depend mainly on the intrinsic width (8.20) of the ring of evanescence and to decrease exponentially with increase of that width. Especially strong resonance would therefore be anticipated for trapping modes with very wide rings of evanescence. These occur for large values of $n_2\varepsilon$. The trapped wave region then becomes a narrow band around the shore of width $\eta/[h'(1)n_2^2\varepsilon^2]$, approximately, in units of the island radius (Lozano and Meyer, 1976). In the same units, the width of the ring of evanescence increases to $n_2\varepsilon/k(h(\infty))$, approximately (which here refers to asymptotics behavior as $n_2\varepsilon \to \infty$) and for deep water far from the island, $k(h(\infty)) \sim \eta \sim 1$, by (3.5). The intrinsic width becomes even larger $\xi_1/(2\pi\varepsilon) \sim (n_2/2\pi)\log(n_2\varepsilon) + \mathcal{O}(n_2)$, and (Lozano and Meyer, 1976)

$$\mathrm{Im}\,\eta \sim -\pi^{-1}h'(1)n_2^2\varepsilon^2 \exp\{-2n_2[\log(n_2\varepsilon) + \mathrm{const.}]\}.$$

On the other hand, since the trapped wave ring is so narrow, $k^{-2} \sim h \sim (r-1)h'(1)$ in it, and the integral in (7.9) becomes $\pi/(2n_2\varepsilon h'(1))$, approximately, where $\varepsilon h'(1) = \alpha$ is the actual beach angle. By (7.9) therefore, $n_2\varepsilon\alpha \sim (2n_1+1)^{-1}$ and

$$\omega^2 \sim \varepsilon^{-1} \sim (2n_1 + 1)n_2\alpha,$$

$$\mathrm{Im}\,\omega^2 \sim -\pi^{-1}n_2\alpha[(2n_1 + 1)\alpha]^{2n_2},$$

roughly.

While we have worked with the island radius as unit of length, for convenience, the physical scale of the wave pattern is really characterized by the width of the trapped wave ring. When this is taken as unit of length, the island radius becomes large in the limit $n_2\varepsilon \to \infty$ and the wave pattern is now seen to approach that of edge waves on a straight coast. The island results therefore help to bridge the gap between real trapped waves with their leakage and first-order resonant response and pure edge waves which have neither. Specifically, the last formulas indicate how leakage tends to zero (and resonant response becomes unbounded) as the ratio $n_2^2\varepsilon^2 h'(1)$ of the radius of curvature of the coast to the shore distance of the edgewave caustic grows large.

For the explanation of the reciprocal relation between leakage and resonant response, we return to the wave motion at real frequency (Section VII) on which the solution of the eigenvalue problem sheds much new light. Here the main physical question concerns the energy supply from the open sea needed to maintain such a motion when $n_2\varepsilon$ is large enough to cause partial trapping. The time-averaged energy flow outward across a cylinder of radius $R = rL$ is then (Stoker, 1957)

$$E = \frac{-\rho_0 L^3 g\omega}{2\pi} \int_0^{2\pi/\omega} dt \int_0^{2\pi} r\,d\theta \int_{-h}^0 dz \, \frac{\partial}{\partial t}(\mathrm{Re}\,\phi e^{-i\omega t}) \frac{\partial}{\partial r}(\mathrm{Re}\,\phi e^{-i\omega t}).$$

With the depth dependence approximated by $F_0(z)$ (Section III) and the surface potential $\phi(x, y, 0) = \Psi$, by (8.1), this becomes

$$E = \tfrac{1}{2}\rho_0 L^3 g i \pi \omega W(w, \bar{w})$$

in terms of the Wronskian $W(w, \bar{w}) = w\,d\bar{w}/dr - \bar{w}\,dw/dr$. For real ω and r, $\bar{w}(r)$ is a solution of (8.2) whenever $w(r)$ is, so that $W(w, \bar{w})$ is independent of r, the energy flux is conserved. The fundamental solution pair gJ, gN of (8.2) can now be taken real, and the boundary condition (8.11) then makes $w(r)$ and $\overline{w(r)}$ directly proportional, by (8.10), i.e., makes $W(w, \bar{w}) = 0$. Of course, that is just another way of deducing that the boundedness condition (8.11) at the shore implies perfect reflection: the energy flux E is zero at all r, the solution w is a standing wave.

For real frequency, (8.8) and (8.12) are therefore the familiar decompositions of linear, small amplitude theory (Section II) of the standing wave into progressive wave pairs of equal and opposite energy fluxes. It follows that $|\beta_0^- / \beta_\infty^-|^2$ is the ratio of the energy flux levels associated with the incoming waves in the inner and outer wave regions. By (8.15) and (8.16), therefore, the energy flux from the open sea associated with unit energy flux level in the trapped wave ring is just $|\Delta(\eta)|^{-2}$ for real η.

$$\rho \equiv |\beta_0^- / \beta_\infty^-| = |\Delta(\eta)|^{-1}$$

is therefore the natural measure of the amplitude *amplification* of the trapped part of the standing wave relative to the amplitude of the same wave mode in the open sea. In short, $\rho(\omega)$ is the response coefficient introduced physically in the first part of this section.

For real frequency, with the normalization adopted for ε, $\eta = 1$, and

$$\rho(\omega) = |\Delta(1)|^{-1},$$

which is now seen to be highly frequency dependent. By (8.16) and (8.19), it varies from an exponentially small minimum $\frac{1}{2}\gamma^{-1}\{1 + \mathcal{O}(\varepsilon)\}$ at a frequency midway between successive eigenfrequencies to an exponentially large maximum

$$\begin{aligned} \rho_m &= |\Delta'(1)\operatorname{Im}\eta(n_1, n_2)|^{-1}\{1 + \mathcal{O}(\varepsilon)\} \\ &= \gamma\{1 + \mathcal{O}(\varepsilon)\} \end{aligned}$$

at a frequency close to an eigenvalue $\eta(n_1, n_2)$, with γ given by (8.20). These extreme peaks of response are called "quasi-resonance" or "spectral concentration" in quantum mechanics.

Since these peaks of resonant response are due to the proximity of eigenvalues, the frequency bandwidth of the response is correspondingly narrow. Close to an eigenvalue, $1/\rho(\omega) = |\Delta(1)|$ is proportional to distance from the root of $\Delta(\eta)$, whence the half-width of the curve of ρ vs. ω^2 is found to be only

$$12^{1/2}|\operatorname{Im}\eta(n_1, n_2)| \sim 3^{1/2}(\omega\gamma)^{-2}|\xi_0'(1)|^{-1},$$

by (8.22). This striking feature of the quasi-resonance has defeated numerical refraction approaches (Lautenbacher, 1970).

It had been detected, however, in a remarkable study of long-wave trapping by shelf edges by Longuet-Higgins (1967). For a seabed of piecewise constant depth, the solution of the long-wave equations can be pieced together conveniently from well-known functions. Longuet-Higgins (1967) used this to treat waves around a flat-topped, submerged seamount dropping sharply to a seabed of constant depth. The characteristic equation is then expressible in terms of Bessel functions, and he succeeded by a strong computational effort to chart all its complex roots. The response curves (Longuet-Higgins, 1967) rise to peaks of remarkable height and narrowness as n_2 and n_1 begin to grow. His charts of eigenvalues, moreover, give an illuminating idea of the distribution of all the eigenvalues not covered by Lozano and Meyer (1976) because they do not have large response. These long-wave results were further extended by Summerfield (1972) to steep-sided circular islands with flat, circular shelf dropping sharply to the flat seabed.

In most practical circumstances, quasi-resonance overestimates the actual response either because the energy supply does not last long enough to establish an effectively standing wave or because it is distributed over a wider frequency band. Knowledge of the complex spectrum permits an estimate of the response to a supply of given frequency distribution, but the result depends considerably on that distribution which, in turn, varies with circumstances. Some estimates have been made by Longuet-Higgins (1967) and

Summerfield (1972). Basically, the usual expectation must be revised that the lowest modes are the most important. The narrow width of response only balances its extreme size to the extent of making the unsuspected, higher trapping modes of importance comparable to that of the strongest lower modes.

Viscous damping, by contrast, can suppress resonant trapping, if the scale is small, as was the case in the experiments of Pite (1973). On the full, natural scale, however, it is much less important (Longuet-Higgins, 1967). Deviations of the topography from axial symmetry lead to a splitting of eigenfrequencies and Coriolis effects also shift the eigenvalues slightly and cause wave beats (Longuet-Higgins, 1967).

We finally turn to the effect of energy absorption at the shore. Since it is accounted for explicitly in the characteristic equation (8.16) by the complex reflection coefficient $R = \beta_0^+/\beta_0^-$, its influence on the eigenvalues can be calculated in a general way (Meyer and Painter, 1979). If the phase shift arg R be denoted by δ_s and

$$\lambda = \log|R| \le 0, \qquad \delta = \delta_s + \pi/2,$$

so that δ denotes the part of the phase shift due to absorption, then the characteristic equation (8.16) becomes

$$\Delta(\eta) = t_- - it_+ \exp(\lambda + i\delta) = 0,$$

with t_-, t_+ still given by (8.19). There t_- is seen to depend on $\xi_0(1, r_1)$ which is (apart from its parametric dependence on the fixed number $n_2\varepsilon$) a function only of η, called briefly $\xi_0(\eta)$ in the rest of this section; the value of $\xi_0(1)$ is given explicitly by (8.18). Let ε be normalized as before to $\Delta_0(1) = 0$ so that

$$-2\xi_0(1)/\varepsilon + i\varepsilon(\sigma_- - \sigma_+) = -(2n_1 + 1)\pi i$$

(σ_s is now absorbed into δ) with large integer n_1, and recall that this determines just the sequence of values $\varepsilon(n_1, n_2)$ discussed in Section VII, apart from a small correction. Then the characteristic equation $\Delta(\eta) = 0$ is solved by

$$\xi_0(\eta) = \xi_0(1) - \tfrac{1}{2}\varepsilon[\lambda + i\delta - (1 - i)\gamma^{-2}\{1 + \mathcal{O}(\varepsilon)\}],$$

and it can again be shown (Meyer and Painter, 1979) by the principle of the argument that this determines a definite eigenvalue $\eta(n_1, n_2)$ close to 1 and that the eigenvalue is simple. To the first approximation in $\eta - 1$, it is

$$\eta(n_1, n_2) \sim 1 + \tfrac{1}{2}\varepsilon|\xi_0'(1)|^{-1}[i\lambda - \delta - (1 + i)\gamma^{-2}\{1 + \mathcal{O}(\varepsilon)\}], \qquad (8.24)$$

by (8.23).

If shore absorption is significant, then the term in (8.24) with exponentially small factor γ^{-2} is negligible and

$$\operatorname{Im} \eta(n_1, n_2) \sim \tfrac{1}{2}\varepsilon|\xi_0'(1)|^{-1} \log|R|,$$

where $|R| = R_A$ is the real amplitude reflection coefficient, and the logarithmic time-decrement of the potential is

$$|\text{Im } \omega| \sim [4\omega|\xi_0'(1)|]^{-1}|\log|R||. \qquad (8.25)$$

The phase shift due to absorption is seen from (8.24) to cause a further correction to the real part of the eigenfrequency. These results could have been obtained from the cruder approximation (8.17) of standard WKB theory. The more precise theory, however, shows that the asymptotic expansion in powers of ε of the characteristic determinant has no influence on the time decay of the eigenfunctions even in the presence of shore absorption. For very small amplitude, moreover, the possibility arises (Section VI) that shore reflection could approach perfection, and the very precise approximation (8.24) may then offer a quantitative comparison of wave decay due to shore absorption and radiation damping.

Similar results hold for two-dimensional channel topographies (Sections IV, VII), in fact, (8.25) remains valid with the appropriate interpretation of ω and $\xi_0'(1)$ (Meyer and Painter, 1979). It is noteworthy that the time decrement Im ω is inversely proportional to ω and hence, is quite small even for substantial shore absorption. Accordingly, although the peaks of resonant response are much reduced by this absorption, they remain high, unless the absorption is near-complete. Since the bandwidth of response is correspondingly broadened, moreover, these spectral concentrations at high frequency may still be anticipated to make a prominent contribution to the overall response in many circumstances.

IX. Underwater Reflection

When waves travel across a slope of the seabed, they should plausibly be reflected to some degree, but this effect has remained puzzling for a long time (Carrier, 1966; Mahony, 1967). For natural topographies, this is a refraction effect, for instance, in the propagation of tsunamis over the ocean, the amplitude is extremely small, so that linear theory is entirely plausible. At the same time, the wavelength greatly exceeds the depth, so that the major topographic features of the ocean bed must have a marked effect on wave propagation. On the other hand, the width of large submarine ridges and valleys is much greater than the wavelength, so that an approach suited primarily to the case of major depth changes over a wavelength (Kreisel, 1948) is inappropriate (except right at the shore). Rather, if h is the ocean depth and λ, the wavelength, a physically rational approach should be based (Carrier, 1966) on the smallness of $(\lambda/h)\nabla h$, which is the notion characterizing refraction theory (Section III).

Another instance arises in wave shoaling on natural, gentle beaches, where it is physically plausible that some reflection of wave energy should take place before the wave reaches the vicinity of the shore and surely, before it breaks. Indeed, even if much of shoaling be dominated by nonlinear processes, a small degree of reflection must be anticipated to occur already before nonlinearity becomes important. Since shoaling models have their own difficulties in pinning down reflection, it may be helpful to know what linear theory can predict about it. Moreover, even if such reflection be a very small effect, it must be anticipated to accumulate during the long journey of waves over continental shelves.

The difficulties encountered (Carrier, 1966) in determining wave reflection arose (Mahony, 1967) from a WKB paradox of mathematical asymptotics. Standard theory based on smallness of a typical value ε of $(\lambda/h)\nabla h$ can cope quite simply (Mahony, 1967) with the case where the seabed slope or curvature has a (finite) discontinuity. The reflection coefficient, however, is then found to depend directly and exclusively on that discontinuity, which is not plausible. Thus if the discontinuity be smoothed out locally, for realism, then reflection seems to disappear. Results of this nature, reported by many authors, may be suspected (Mahony, 1967) to represent an accident of mathematical procedure, rather than physical reality.

This paradox has since been broken (Meyer, 1975), and reflection predictions for realistically smooth seabed topographies have become available (Meyer, 1979), at least for long slopes, ridges and valleys that can be adequately represented by a depth distribution $h(x)$ dependent on only one coordinate. This carries over to the wave number k, which the dispersion relation (3.5) shows to depend only on h, and thence to the wave depth function G of the refraction equation (3.6) for the surface potential Ψ. Fourier transform with respect to the other coordinate then leads to modes

$$\Psi(x, y) = F(x)\exp(i\beta y)$$

with constant β, which (3.6) shows to be governed by

$$d(G\, dF/dx)/dx + (k^2 - \beta^2)GF = 0. \tag{9.1}$$

The coefficient

$$\alpha = (k^2 - \beta^2)^{1/2} \tag{9.2}$$

in (9.1) is the x-component of the local wave number k, and it is natural to refer the development to an intrinsic distance

$$\xi = \int_0^x \alpha(x')\, dx' \tag{9.3}$$

based more on local wavelength than on map distance x. This transforms (9.1) to

$$f'' + 2\phi f' + \varepsilon^{-2} f = 0,$$
$$\phi(\xi) = (2\alpha^2 G)^{-1}\, d(\alpha G)/dx \tag{9.4}$$

for the wave function $F(x) = f(\xi)$. The modulation of the wave by the influence of the seabed is therefore entirely characterized by the function $\phi(\xi)$ and reflection must similarly depend only on this modulation function. To help with a "physical" feel for its nature, the long-wave limit $kh \to 0$ may be noted, in which

$$k \sim \omega(\varepsilon/h)^{1/2}, \qquad \alpha \sim (\varepsilon h^{-1}\omega^2 - \beta^2)^{1/2}, \qquad G \sim h. \tag{9.5}$$

If $\beta = 0$, for simplicity, then $2\phi \sim (\varepsilon\omega^2)^{-1/2}\, dh^{1/2}/dx \sim dk^{-1}/dx$.

To concentrate on the effect of gradual reflection, shore lines (Section VI), caustic reflection (Sections IV, V), and wave trapping (Sections VII, VIII) will be excluded in the present section by the assumption that $\alpha(x)$ and $G(x)$ have positive lower bounds (and of course, upper bounds also). For concrete reflection results, it is also helpful to refer to clear-cut, classical wave states as $x \to \pm\infty$, and to this end, it will be assumed that the depth tends to limits

$$h(x) \to h_+ \quad \text{as } x \to \infty, \qquad h(x) \to h_- \quad \text{as } x \to -\infty$$

(which are positive because of the bounds for G) and that these limits are approached in a natural way so that $|dh/dx|$ is integrable. That is sufficient to assure for the refraction equation existence of solutions with the asymptotic character of purely progressive waves. They are described by $f \sim \exp(i\xi/\varepsilon)$ for waves propagating in the sense of ξ increasing, if $\omega > 0$ in the time factor $\exp(-i\omega t)$ in (3.2), (3.4), and by $f \sim \exp(-i\xi/\varepsilon)$, for waves traveling in the opposite direction. The reflection coefficient R and transmission coefficient T are then defined uniquely (Kreisel, 1948) by the radiation condition

$$e^{-i\xi/\varepsilon} f(\xi) \to T \qquad \text{as } \xi \to \infty$$
$$e^{i\xi/\varepsilon}(f(\xi) - e^{i\xi/\varepsilon}) \to R \qquad \text{as } \xi \to -\infty \tag{9.6}$$

for (9.4), which specifies a unit wave incident from $\xi = -\infty$, but none incident from $\xi = \infty$. The complex constants R and T contain information on the wave phase that is usually of secondary interest, at most, and only the amplitude reflection coefficient $|R|$ will be considered in this section. It is independent of the sense of wave travel (Kreisel, 1948) which is determined by the choice of signs for α and ω.

Since $|R|$ is just a number, it must be a functional of $\phi(\xi)$ and an integral representation for it more useful than (9.6) is obtainable, e.g., by the multiple scale approach based on recognition of the difference between the topographical scale on which ξ is based and the wavelength scale on which ξ/ε

is measured. If (9.4) had solutions

$$y = (\alpha G)^{-1/2}\{Ae^{i\xi/\varepsilon} + Be^{-i\xi/\varepsilon}\}$$

consisting of local progressive waves with "slowly varying" amplitudes $A(\xi)$, $B(\xi)$, then to first approximation for small ε,

$$dy/d\xi = i\varepsilon^{-1}(\alpha G)^{-1/2}\{Ae^{i\xi/\varepsilon} - Be^{-i\xi/\varepsilon}\}.$$

Unfortunately, (9.4) does not have such solutions, but to come as close as possible to such a wave decomposition, we may define unknown functions $A(\xi;\varepsilon)$, $B(\xi;\varepsilon)$ by the last two equations. Substitution in (9.4) then yields

$$dA/d\xi = \phi Be^{-2i\xi/\varepsilon}, \qquad dB/d\xi = \phi Ae^{2i\xi/\varepsilon},$$

and $a = -(A/B)\exp(2i\xi/\varepsilon)$ satisfies

$$da/d\xi = 2ia/\varepsilon + (a^2 - 1)\phi, \qquad (9.7)$$

a Riccatic equation associated with (9.4). Conversely, any solution

$$a = (i\varepsilon y' - y)/(i\varepsilon y' + y)$$

of (9.7) will furnish solutions y of (9.4) by a quadrature. As an initial condition,

$$a(-\infty) = 0 \qquad (9.8)$$

turns out convenient, and direct integration turns (9.7) and (9.8) into an integral equation

$$a(\xi)e^{-2i\xi/\varepsilon} = \int_{-\infty}^{\xi} ([a(s)]^2 - 1)e^{-2is/\varepsilon}\phi(s)\,ds. \qquad (9.9)$$

Since $\phi(\xi)$ is real, any solutions y of (9.4) come in complex conjugate pairs y, $\bar{y}$, which are found to be linearly independent by (9.8). A solution f of (9.4) satisfying the radiation condition (9.6) must therefore be a linear combination of y and $\bar{y}$, and substitution in (9.6) then yields by a straightforward computation (Meyer, 1975)

$$|R| = \lim_{\xi \to \infty} |ae^{-2i\xi/\varepsilon}| \qquad (9.10)$$

$$= \left| \int_{-\infty}^{\infty} (a^2 - 1)e^{-2i\xi/\varepsilon}\phi\,d\xi \right|,$$

by (9.9). This integral representation of the reflection coefficient resembles a Fourier integral, but a^2 is unknown and is actually a rapidly oscillating function that depends strongly on ε.

It is tempting to interpret $a(\xi;\varepsilon)$ in terms of a "local reflection coefficient" (Kajiura, 1961) and solve (9.7) or (9.9) by successive approximation based on the fact that $a = \mathcal{O}(\varepsilon)$. However, this is liable to lead to wrong results for naturally smooth seabed topographies because the integral in (9.10) is then very small indeed in the technical sense of asymptotics. As usually in Fourier integrals with large parameter ε^{-1}, but even much more here, the integrand in (9.10) oscillates so rapidly that all contributions to the integral cancel each other almost completely; the integral turns out to be smaller than any power of ε, which lies at the root of the WKB paradox. The embarrassment of cancellation, however, can be turned to advantage by complex embedding: in the lower half-plane of the complex wave distance ξ, the offending factor $\exp(-2i\xi/\varepsilon)$ in (9.10) becomes very small.

To turn (9.10) into a really efficient representation for the reflection coefficient, we therefore assume that $\phi(\xi)$ is analytic on a neighborhood N of the real ξ-axis. This is equivalent to an analogous assumption on the water depth $h(x)$; more general classes of smooth functions can be treated as limits of analytic functions, but the analysis and results are liable to be more complicated (Meyer and Guay, 1974; Stengle, 1977). The radiation condition (9.6) then also needs interpretation in the complex ξ-plane and the following hypothesis has been proposed (Meyer, 1975, 1979). (H) If $2m$ denotes the *minimum width of the analytic strip* N of $\phi(\xi)$, then $\phi(\xi) \to 0$ as $|\mathrm{re}\,\xi| \to \infty$ uniformly in $|\mathrm{im}\,\xi| < m$ and for $\mathrm{im}\,\xi = -m$, $\phi(\xi)\,\varepsilon\,L(-\infty, -M) \cap L(M, \infty)$ with respect to $\mathrm{re}\,\xi$ for some M independent of ε. The corresponding hypothesis on dh/dx is very similar.

These assumptions permit us to shift the path of integration in the functional

$$a_+ = \int_{-\infty}^{\infty} (a^2 - 1)e^{-2i\xi/\varepsilon}\phi\,d\xi \qquad (9.11)$$

to a line $\mathrm{im}\,\xi = \mathrm{const.} = -c > -m$ parallel to the real ξ-axis. For $c > 0$, a very small factor $|\exp(-2i\xi/\varepsilon)| = \exp(-2c/\varepsilon)$ can then be extracted from the integral, and to this degree, the cancellation sickness has been cured. Clearly, c should be increased as far as possible so that the cure becomes complete. Since (9.7) has analytic coefficients, $a(\xi;\varepsilon)$ is also analytic on the

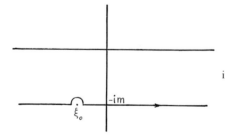

FIG. 17.   Path of integration for (9.11) with indentation of radius δ at the critical point ξ_0.

strip N of $\phi(\xi)$ and the optimal path is along im $\xi = -m$ with indentations avoiding the singular points ξ_c of $\phi(\xi)$ with im $\xi_c = -m$ (Fig. 17). An organic connection has thus emerged between wave reflection and the breakdown of analyticity of the modulation function $\phi(\xi)$ with increasing distance from the real axis of wave distance ξ.

To exploit it, begin with the case where only one singular point $\xi_c = \xi_0$ is present. The hypothesis (H) then ensures that the analytic half-width $m > 0$. The path of integration (Fig. 17) may be split into a critical interval $(\xi_0 - \delta, \xi_0 + \delta)$, with any $\delta > 0$ independent of ε, and the semi-infinite segments to its left and right. The contributions from re$(\xi - \xi_0) = \pm\infty$ are limited by the hypothesis (H). A contraction argument on the integral equation (9.9) along the left segment, with appeal to the Riemann–Lebesgue lemma (Meyer, 1975), shows $a(\xi;\varepsilon)$ to be analytic in ξ and $|a| \to 0$ as $\varepsilon \to 0$ uniformly on that segment. A similar contraction argument (Meyer, 1975) for the right segment shows $a(\xi;\varepsilon)$ to be analytic there also, and bounded, if bounded at $\xi = \xi_0 + \delta$. In that case, it follows from the Riemann–Lebesgue lemma that all these contributions are $o(\exp - 2m/\varepsilon)$, leaving

$$a_+ = \int_{\xi_0 - \delta}^{\xi_0 + \delta} (a^2 - 1)e^{-2i\xi/\varepsilon}\phi\, d\xi + o(e^{-2m/\varepsilon})$$

$$= \left[a(\xi;\varepsilon)e^{-2i\xi/\varepsilon} \right]_{\xi_0 - \delta}^{\xi_0 + \delta} + o(e^{-2m/\varepsilon}),$$

(9.12)

by (9.9). If the jump of a across ξ_0 is bounded and nonzero, the integral (9.10) therefore admits a stationary phase principle and (9.12) gives the first approximation to the reflection functional a_+. It is then revealed as a local property of the dominant singularity of the modulation function. Exponential precision, moreover, has then been achieved because a factor $\exp(-2m/\varepsilon)$ is evident in the bracket of (9.12), and the remaining factor needs determination only to the first approximation.

The forms of breakdown of the analyticity of $\phi(\xi)$ covered by available mathematical theory may be described by the following hypothesis. (C) The modulation function $\phi(\xi)$ in (9.4) is analytic for $|\text{im } \xi| \leq m$ except for a set (without limit point) of singular points ξ_c with $|\text{im } \xi_c| = m > 0$ arising from "transition points" x_c of the wave number component α or wave depth G of the type

$$\alpha(x) = (x - x_c)^{\nu/2}\alpha_1(x), \qquad \nu > -2,$$

or
$$G(x) = (x - x_c)^\lambda G_1(x), \qquad \lambda \leq 1,$$

(9.13)

with functions $\alpha_1(x)$, $G_1(x)$ analytic and nonzero at x_c. This covers a large class of smooth seabed topographies. The singularities admitted by (C) are turning points of fractional order of a related equation (Meyer, 1979), and approximations in terms of Bessel functions are known, which show the jump of $a(\xi;\varepsilon)$ across the critical point to be bounded (Meyer, 1975). The contribution of a single critical point to the reflection coefficient is thus found (Meyer, 1976) from (9.12) to be

$$|R| = |a_+| \sim 2\left|\cos\frac{\pi}{\nu+2}\right|e^{-2m/\varepsilon}, \tag{9.14}$$

to the first approximation, with the appropriate value of the exponent ν in (9.13); if the transition point arises from the wave depth G, then $\nu = \lambda/(1-\lambda)$ for $0 < \lambda < 1$, but $\nu = 2\lambda/(1-\lambda)$ for $\lambda < 0$ and $\beta \neq 0$, and $(2+\nu)^{-1} = 0$ for $\lambda = 1$ (Meyer, 1979). If several critical points share the minimum m of $|\text{im } \xi_i|$, then their contributions (9.14) are additive (Meyer, 1975).

Wave reflection is thus seen to depend primarily and strongly on the analytic halfwidth m of the modulation function $\phi(\xi)$ in (9.4), which is a topographical scale of more local than global nature, characteristic neither of the total depth change nor of the maximal seabed slope. Its most direct interpretation is as the radius of convergence of the Taylor series of $\phi(\xi)$ at the real value of wave distance ξ where this radius is smallest, and it has been called the maximal local intensity of modulation. It is noteworthy that, if $\phi(\xi)$ be the sum of two analytic functions of significantly different analytic half-widths, then the one with the smaller half-width will dominate reflection even though the other be responsible for most of the depth change or seabed steepness.

For the long-wave approximation, the relation between topography and reflection is slightly simplified by (9.5) and some illustrative examples have been worked out (Meyer, 1979): For waves normally incident on a submerged ridge represented by

$$h(x) = 2 - \exp(-x^2),$$

the reflection coefficient is

$$|R| = 2\exp[-2.7\varepsilon^{-1/2}\omega].$$

For waves normally incident on a shelf slope represented by

$$h = 1 + \tfrac{1}{2}\tanh x,$$

it is

$$|R| = \exp[-(2/3\varepsilon)^{-1/2}\pi\omega],$$

but reflection increases rapidly with obliqueness of incidence so that, e.g., for the same slope with $\beta = \omega(\varepsilon/2)^{1/2}$ in (9.1), it becomes

$$|R| = \exp[-(6\varepsilon)^{-1/2}\pi\omega].$$

The dependence on $\omega/\varepsilon^{1/2}$ is not unexpected since $\Lambda \sim 2\pi\omega^{-1}(h/\varepsilon)^{1/2}$ is the wavelength scaled in vertical units (which is large in the long-wave limit) and $\varepsilon\Lambda \propto \varepsilon^{1/2}/\omega$ is therefore the wavelength on the topographical scale of x; it must be small for refraction theory (Section III), and reflection calculations for observed topographies require a smoothing of the data to remove depth variations on any scale smaller than that of x.

Reflection, like radiation damping in Section VIII, is a very subtle, asymptotic effect for naturally gentle seabed topographies, and a question arises whether it can be discussed at all on the basis of refraction theory which concerns only the first approximation to the wave motion under such circumstances (Section III and Appendix III). For such reasons, reflection studies on the basis of the more exact, linear equations (2.10) have been made for normal wave incidence by Kreisel (1948), Fitz-Gerald (1976), and Harband (1977). The first two give iteration procedures converging to the potential Φ of the motion under circumstances applicable to engineering works, but not to natural topographies, and Fitz-Gerald gives another such procedure convergent generally for waves on deep water; Harband gives an integral representation approximate in some respect, but exact in regard to refraction and diffraction (Appendix III). In all cases, the reflection coefficient emerges as an integral of the form (9.10) and what matters, therefore, is not the degree of approximation of the potential, but the analytic width of the integrand. In the more exact theory, this depends only on a certain conformal mapping function, but a direct comparison of its width with that of the modulation function $\phi(\xi)$ in (9.4) has not yet been made.

A direct integral representation for the solution of (2.10) is known for normal incidence on a very special shelf slope (Roseau, 1952), and when the slope is gentle, the reflection coefficient (Kajiura, 1961) is again given by (9.14) with $v = 1$ and $m = kD$, where k and D are respectively the wave number and depth over the shelf. Reflection is then dominated by the top of the slope, where a certain mapping function has its minimal half-width m; the toe of the slope has little influence even when the total depth change is small.

X. **Developing Wave Patterns**

The preceding sections have all been devoted to the discussion of surface waves of fixed frequency and of methods suited in particular to the analysis of more-or-less permanent wave patterns, in view of their dominant importance in contemporary oceanography. Unsteady wave patterns may be Fourier-analyzed in terms of the modes that have been discussed. However, a direct refraction approach to the time development of wave patterns has also been developed by Hector *et al.* (1972) and by Shen and Keller (1975). These papers are rather demanding because the authors deal right away with a multitude of physical effects, including internal and acoustic waves and Coriolis effects, and moreover, Shen and Keller (1975) jump right to the formulation of a uniform refraction approximation generalizing that of Section V in all those directions. It may be worthwhile to offer here an introduction to those papers in the form of a direct explanation merely of unsteady, simple ray theory for surface waves of the classical type set out in Section II.

To specialize from the general, linear surface-wave equations (2.10) to the form appropriate for refraction, we introduce again the scaling (3.1),

$$X = Lx, \qquad Y = Ly, \qquad T = L^{1/2}(g\varepsilon)^{-1/2}t,$$

$$Z = \varepsilon Lz, \qquad H(X, Y) = \varepsilon Lh(x, y),$$

where the scaling now proposed for the time reflects the experience of the earlier sections that we risk locking ourselves into an analysis merely of degenerate limit cases unless frequencies of order $\varepsilon^{-1/2}$ be admitted. The classical, linear equations (2.10) for the potential $\Phi(X, Y, Z, T) = (L^3 g)^{1/2} \phi(x, y, z, t)$ then become

$$\partial^2\phi/\partial x^2 + \partial^2\phi/\partial y^2 + \varepsilon^{-2}\partial^2\phi/\partial z^2 = 0 \qquad \text{for } 0 > z > -h(x, y),$$

$$\partial^2\phi/\partial t^2 + \varepsilon^{-2}\,\partial\phi/\partial z = 0 \qquad \text{at } z = 0, \qquad (10.1)$$

$$\nabla\phi \cdot \nabla h + \varepsilon^{-2}\,\partial\phi/\partial z = 0 \qquad \text{at } z = -h,$$

where ∇ denotes again the gradient operator with respect to x, y. For a simple ray theory, we propose an asymptotic "ansatz"

$$\phi \sim e^{is/\varepsilon} \sum_0^\infty (-i\varepsilon)^m a_m(x, y, z, t)$$

analogous to (3.4) with phase function $s(x, y, t)$ independent of z and ε, or at least, an ansatz

$$\phi \sim e^{is/\varepsilon}(a - i\varepsilon b + \cdots),$$

since it is clearly premature to worry about satisfying (10.1) to more than the dominant two powers of ε. Then

$$\nabla\phi = \left(\frac{i}{\varepsilon}a\,\nabla s + \nabla a + b\,\nabla s\right)e^{is/\varepsilon} + \cdots,$$

$$\partial\phi/\partial t = \left(\frac{i}{\varepsilon}a\,\partial s/\partial t + \partial a/\partial t + b\,\partial s/\partial t\right)e^{is/\varepsilon} + \cdots,$$

where the dots indicate terms in positive powers of ε, which we ignore for our present purposes, and the first of (10.1) yields

$$e^{-is/\varepsilon} \nabla^2 \phi = -\varepsilon^{-2} a |\nabla s|^2 + i\varepsilon^{-1} \{\nabla s \cdot (\nabla a + b \nabla s) + \nabla(a \nabla s)\}$$
$$= -\varepsilon^{-2} \partial^2 a/\partial z^2 + i\varepsilon^{-1} \partial^2 b/\partial z^2,$$

whence any asymptotic approximation scheme starting as proposed needs both

$$\frac{\partial^2 a}{\partial z^2} = a |\nabla s|^2 \quad \text{and} \quad \frac{\partial^2 b}{\partial z^2} = b |\nabla s|^2 + \nabla(a \nabla s) + \nabla s \cdot \nabla a.$$

Similarly, the boundary conditions in (10.1) are seen to require both

$$\frac{\partial a}{\partial z} = a \left(\frac{\partial s}{\partial t}\right)^2, \quad \frac{\partial b}{\partial z} = \frac{\partial}{\partial t}\left(a \frac{\partial s}{\partial t}\right) + \frac{\partial s}{\partial t}\left(b \frac{\partial s}{\partial t} + \frac{\partial a}{\partial t}\right) \quad \text{at } z = 0,$$

and

$$\frac{\partial a}{\partial z} = 0, \quad \frac{\partial b}{\partial z} = a \nabla s \cdot \nabla h \quad \text{at } z = -h.$$

Since s is to be independent of z, the left-hand equations in this scheme set an eigenvalue problem for the vertical structure of the water motion, at each fixed x, y, t. The differential equation demands a solution of form $a = e^{kh}(Ae^{kz} + Be^{-kz})$ with z-independent coefficients A, B, and the boundary conditions show the characteristic condition for a nontrivial solution to be

$$|\nabla s| \tanh(h|\nabla s|) = (\partial s/\partial t)^2. \tag{10.2}$$

Thus reemerges our trusty dispersion relation (3.5) but now as a first-order differential equation for the phase $s(x, y, t)$, indeed, the classical Hamilton–Jacobi equation of analytical mechanics and optics. It is convenient to reintroduce a local frequency $\omega(x, y, t)$ and wave number vector $k(x, y, t)$ by

$$\partial s/\partial t = -\varepsilon^{1/2}\omega, \quad \nabla s = \mathbf{k}, \tag{10.3}$$

consistently with (3.5) and (4.1) so that (10.2) reads like (3.5),

$$k \tanh(kh) = \varepsilon\omega^2, \quad k = |\mathbf{k}|.$$

Once this is satisfied, the eigensolution for a is readily found to be

$$a = A(x, y, t) \cosh[k(z + h)]. \tag{10.4}$$

The "second-order" equations for b, on the right hand of the scheme developed above, turn out to imply a "transport equation" for the determination of $A(x, y, t)$, much as in Section IV. To see this, we may combine the left and right hand parts of the scheme into

$$\frac{\partial}{\partial z}\left(a\frac{\partial b}{\partial z} - b\frac{\partial a}{\partial z}\right) = \nabla(a^2\nabla s),$$

$$a\frac{\partial b}{\partial z} - b\frac{\partial a}{\partial z} = \frac{\partial}{\partial t}\left(a^2\frac{\partial s}{\partial t}\right) \qquad \text{at } z = 0,$$

but

$$= a^2\nabla s \cdot \nabla h \qquad \text{at } z = -h.$$

Therefore,

$$a\frac{\partial b}{\partial z} - b\frac{\partial a}{\partial z} = \frac{\partial}{\partial t}\left(a^2\left.\frac{\partial s}{\partial t}\right|_{z=0}\right) + \int_0^z \nabla(a^2\nabla s)\,dz$$

and

$$a^2\bigg|_{z=-h}\nabla s \cdot \nabla h = \frac{\partial}{\partial t}\left(a^2\left.\frac{\partial s}{\partial t}\right|_{z=0}\right) + \int_0^{-h} \nabla(a^2\nabla s)\,dz,$$

or

$$\frac{\partial}{\partial t}\left(a^2\left.\frac{\partial s}{\partial t}\right|_{z=0}\right) + \nabla\int_0^{-h} a^2\nabla s\,dz = 0,$$

an equation for a of typical conservation form. When account is taken of (10.3) and (10.4), it becomes the transport equation

$$\varepsilon^{1/2}\frac{\partial}{\partial t}(A_s^2\omega) + \nabla(A_s^2 G\mathbf{k}) = 0, \tag{10.5}$$

with the wavedepth function G defined in (3.6), where

$$A_s = A\cosh(kh) = a(x, y, 0, t)$$

is again the surface amplitude, and in the time-independent case, (4.2) is recovered.

The Hamilton–Jacobi equation (10.2) is traditionally written

$$\partial s/\partial t + \mathcal{H}(\mathbf{k}, \mathbf{x}) = 0, \tag{10.6}$$

where $\mathbf{x} = (x, y)$ and $\mathcal{H}$ is called the Hamiltonian. As a first-order partial differential equation for $s(\mathbf{x}, t)$, it can be solved by the method of characteristics or rays, which are just the "orbits" $\mathcal{H} = $ const. The ray equations, therefore, are just the canonical equations of the Hamiltonian,

$$d\mathbf{x}/dt = \mathbf{g}, \qquad d\mathbf{k}/dt = -\nabla\mathcal{H}, \tag{10.7}$$

where the time t plays the role of parameter along the orbit or ray and

$$\mathbf{g} = \partial\mathcal{H}/\partial k_1, \partial\mathcal{H}/\partial k_2$$

is the *group velocity* vector. In the present case, where the Hamiltonian depends only on $k = |\mathbf{k}|$, it follows that $\mathbf{g} = k^{-1}\mathbf{k}\partial\mathscr{H}/\partial k$, and with the Hamiltonian given by (10.2) and (10.3), a short calculation gives

$$\mathbf{g} = \varepsilon^{-1/2}\omega^{-1}G\mathbf{k}, \tag{10.8}$$

which is, of course, precisely the classical group velocity (Stoker, 1957).

The attraction of the ray equations, as in Section IV, is that the distribution of phase s and amplitude A along a ray can be written down explicitly. By (10.7), the time rate of change along a ray is

$$d/dt = \partial/\partial t + \mathbf{g} \cdot \nabla, \tag{10.9}$$

and from (10.3), $ds/dt = -\varepsilon^{1/2}\omega + \mathbf{k} \cdot \mathbf{g}$. Since the rays are lines of constant Hamiltonian, it follows from (10.6) that ω is also constant on them, whence

$$s(t) = s(t_0) - \omega(t - t_0) + \int_{t_0}^{t} \mathbf{k} \cdot \mathbf{g}\, dt \qquad \text{along a ray.} \tag{10.10}$$

In the transport equation (10.5), $G\mathbf{k} = \varepsilon^{1/2}\omega\mathbf{g}$ by (10.8), so that (10.5) may also be written

$$\frac{\partial}{\partial t}(A_s^2\omega) + \nabla \cdot (A_s^2\omega\mathbf{g}) = 0.$$

In this form, the transport equation bears a striking resemblance to the familiar mass-conservation equation (2.1) of general fluid motion, $\partial\rho/\partial t + \nabla \cdot (\rho\mathbf{v}) = 0$. In wave refraction, the analog of the mass density ρ is seen to be an energy density $A_s^2\omega$, and the analog of the fluid velocity $\mathbf{v}$, the group velocity $\mathbf{g}$. The orbits or rays are seen by (10.7) to be the analog of the "particle paths" (not streamlines, since the rays are not fixed curves in x-,y-space). We may conclude immediately that energy is convected with the group motion, just like mass is convected with the fluid motion, so that

$$\int_{D(t)} A_s^2\omega\, dx\, dy = \text{const.} \tag{10.11}$$

for any domain D that moves with the local group velocity. For a more local statement of this result, envisage a fairly small domain $D(t_0)$ bounded by a simple, smooth, closed curve $C(t_0)$; let its area be denoted by $\Sigma(t_0)$. As it is convected with the group velocity, each boundary point moves with speed $|\mathbf{g}|$ along the ray that starts from the position of the boundary point at time t_0. At time t, if that does not differ too much from t_0, the convected domain $D(t)$ will have a boundary curve $C(t)$ that is still smooth, closed and simple, and will have area $\Sigma(t)$. Then a first ray-expansion ratio may be defined by

$$\sigma(t) = \lim_{d(t_0) \to 0} [\Sigma(t)/\Sigma(t_0)],$$

where $d(t_0)$ is the diameter of $D(t_0)$, and since $\omega = \varepsilon^{-1/2} \mathscr{H}$ is constant on rays, (10.11) implies

$$A_s^2 \sigma = \text{const.} \qquad \text{along a ray.} \qquad (10.12)$$

This is not quite analogous to (4.7) because σ arises from measuring the area enclosed by the ray tube through $C(t_0)$ at fixed time, rather than measuring the cross-sectional area of the ray tube in the plane normal to the local ray direction; the latter area is $(1 + |\mathbf{g}|^2)^{-1/2} \Sigma$, in the limit $d(t_0) \to 0$. If a second expansion ratio δ based on cross-sectional area is used (Shen and Keller, 1975), the local interpretation of the transport equation is therefore

$$A_s^2 \, \delta (1 + |\mathbf{g}|^2)^{1/2} = \text{const.} \qquad \text{along a ray.}$$

In a sense, (10.10) and (10.12) reduce the refraction problem to the integration of the ray equations (10.7). Since they are ordinary differential equations, a numerical approach is possible and is well suited to the study of wave pattern development over relatively limited time intervals. In cases of special symmetry, as in the earlier sections, a general analytical solution free of such limitations may be available. For large-time development and approach to a permanent state, the traditional approach is by Fourier analysis of (10.1) and asymptotic approximation of inversion integrals.

In any case, it should be borne in mind that the approximation obtainable from (10.7), (10.10) and (10.12), must generally be expected to fail locally where the expansion ratio vanishes. A uniform asymptotic approximation (Shen and Keller, 1975) is then needed and the asymptotic notions of group velocity and energy convection also need careful reinterpretation in its light.

ACKNOWLEDGMENT

A major part of the research which, over the last two decades, produced all these advances in the theory of water-wave refraction was supported by the National Science Foundation. The work for this article, in particular, was supported by it in part under Grant MPS 77-00097-A01, 2.

REFERENCES

Berkhoff, J. C. W. (1973). Computation of combined refraction-diffraction, *Proc. Int. Conf. Coastal Eng., 13th, 1972* pp. 471–490.

Buslaev, V. S. (1964). On formulas of short wave asymptotics in the problem of diffraction by convex bodies. *Tr. Mat. Inst., Akad. Nauk SSSR* **73**, 14–117.

Carrier, G. F. (1966). Gravity waves on water of variable depth. *J. Fluid Mech.* **24**, 641–659.

Carrier, G. F., and Greenspan, H. P. (1958). Water waves of finite amplitude on a sloping beach. *J. Fluid Mech.* **4**, 97–109.

Chao, Y.-Y. (1974). "Wave Refraction Phenomena Over the Continental Shelf Near the Chesapeake Bay Entrance," Tech. Memo. TM-47. Dept. Meterol. Oceanogr. New York University, New York, Doc. AD/A-002056, Natl. Tech. Inf. Serv., U.S. Dept. of Commerce, Springfield, Virginia.

CHAO, Y.-Y., and PIERSON, W. J. (1972). Experimental studies of the refraction of uniform wave trains and transient wave groups near a straight caustic, *J. Geophys. Res.* **77**, 4545–4554.

CHRISTIANSEN, P.-L. (1976). Diffraction of gravity waves by ray methods. *In* "Waves on Water of Variable Depth" (D. G. Provis and R. Radok, eds.), Lect. Notes Phys. 64, pp. 28–38. Springer-Verlag, Berlin and New York.

COURANT, R., and HILBERT, D. (1962). "Methods of Mathematical Physics," Vol. II. Wiley (Interscience), New York.

EPSTEIN, P. S. (1930). Reflection of waves in an inhomogeneous absorbing medium. *Proc. Natl. Acad. Sci. U.S.A.* **16**, 627–637.

FITZ-GERALD, G. F. (1976). The reflexion of plane gravity waves travelling in water of variable depth. *Philos. Trans. R. Soc. London, Ser. A* **284**, 49–89.

FRIEDRICHS, K. O. (1948). Water waves on a shallow sloping beach. *Commun. Pure Appl. Math.* **1**, 109–134.

GALVIN, C. J. (1965). Resonant edge waves on laboratory beaches. *Trans. Am. Geophys. Union* **46**, 112.

GALVIN, C. J. (1972). Wave breaking in shallow water. *In* "Waves on Beaches" (R. E. Meyer, ed.), pp. 413–456, Academic Press, New York.

GRESLOU, L., and MAHE, Y. (1955). Etude du coefficient de reflexion d'une houle sur un obstacle constitue par un plan incline. *Proc. Int. Conf. Coastal Eng., 5th, 1954* pp. 68–81.

GRIMSHAW, R. (1966). High-frequency scattering by finite convex regions. *Commun. Pure Appl. Math.* **19**, 167–198.

HARBAND, J. B. (1977). Propagation of long waves over water of slowly varying depth. *J. Eng. Math.* **11**, 97–119.

HECTOR, D., COHEN, J., and BLEISTEIN, N. (1972). Ray method expansions for surface and internal waves in inhomogeneous oceans of variable depth, *Stud. Appl. Math.* **51**, 121–138.

HO, D. V., MEYER, R. E., and SHEN, M. C. (1963). Long surf. *J. Mar. Res.* **21**, 219–232.

HOMMA, S. (1950). On the behaviour of seismic sea waves around circular island, *Geophys. Mag.* **21**, 199–208.

JONSSON, I. G., and SKOVGAARD, O. (1978). A mild-slope wave equation and its application to tsunami calculations. *Mar. Geodesy* **2**, 42.

JONSSON, I. G., SKOVGAARD, O., and BRINK-KJAER, O. (1976). Diffraction and refraction calculations for waves incident on an island, *J. Mar. Res.* **34**, 469–496.

KAJIURA, K. (1961). On the partial reflection of water waves passing over a bottom of variable depth. *I.U.G.G. Monogr.* **24**, 205–230.

KELLER, J. B. (1958). Surface waves on water of non-uniform depth. *J. Fluid Mech.* **4**, 607–614.

KELLER, J. B., and RUBINOV, S. I. (1960). Asymptotic solution of eigenvalue problems. *Ann. Phys. (N.Y.)* **9**, 24–75.

KRAVTSOV, Y. A. (1964). A modification of the geometrical optics method. *Radiofizika* **7**, 664–673.

KREISEL, G. (1948). Surface waves. *Q. Appl. Math.* **7**, 21–44.

KRIEGSMANN, G. A. (1979). An illustrative model describing the refraction of long water waves by a circular island. *J. Phys. Oceanogr.* **9**, 607–611.

LAMB, SIR H. (1932). "Hydrodynamics," 6 ed. Dover, New York.

LAUTENBACHER, C. C. (1970). Gravity wave refraction by islands. *J. Fluid Mech.* **41**, 655–672.

LEMEHAUTE, B. (1963). On non-saturated breakers and the wave run-up. *Proc. Int. Conf. Coastal Eng., 8th, 1962* pp. 77–92.

LEWIS, R. M., and KELLER, J. B. (1964). "Asymptotic Methods for Partial Differential Equations: The Reduced Wave Equation and Maxwell's Equations," Res. Rep. No. EM-194. Courant Inst. Math Sci., New York University, New York.

LEWIS, R. M., BLEISTEIN, N., and LUDWIG, D. (1967). Uniform asymptotic theory of creeping waves. *Commun. Pure Appl. Math.* **20**, 295–328.

LONGUET-HIGGINS, M. S. (1953). On the decrease of velocity with depth in an irrotational water wave. *Proc. Cambridge Philos. Soc.* **49**, 552–560.

LONGUET-HIGGINS, M. S. (1967). On the trapping of wave energy around islands. *J. Fluid Mech.* **29**, 781–821.

LOZANO, C. J. (1979). Gravity surface waves on water of variable depth. *Phys. Fluids* **22** (in press).

LOZANO, C., and MEYER, R. E. (1976). Leakage and response of waves trapped by round islands. *Phys. Fluids* **19**, 1075–1088.

LUDWIG, D. (1966). Uniform asymptotic expansions at a caustic. *Commun. Pure Appl. Math.* **19**, 215–250.

MAHONY, J. J. (1967). The reflection of short waves in a variable medium. *Q. Appl. Math.* **25**, 313–316.

MEYER, R. E. (1966). An asymptotic method for a singular hyperbolic equation. *Arch. Rat. Mech. Anal.* **22**, 185–200.

MEYER, R. E. (1970). Note on wave run-up. *J. Geophys. Res.* **75**, 687–690.

MEYER, R. E. (1971a). "Introduction to Mathematical Fluid Dynamics." Wiley (Interscience), New York.

MEYER, R. E. (1971b). Resonance of unbounded water bodies. In "Mathematical Problems in the Geophysical Sciences" (W. H. Reid, ed.), Lect. Appl. Math., Vol. 13, pp. 189–227. Am. Math. Soc., Providence, Rhode Island.

MEYER, R. E. (1975). Gradual reflection of short waves. *SIAM J. Appl. Math.* **29**, 481–492.

MEYER, R. E. (1976). Quasiclassical scattering above barriers in one dimension. *J. Math. Phys.* **17**, 1039–1041.

MEYER, R. E. (1979). Surface wave reflection by underwater ridges. *J. Phys. Oceanogr.* **9**, 150–157.

MEYER, R. E., and GUAY, E. J. (1974). Adiabatic variation. Part III. A deep mirror model, *Z. Angew. Math. Phys.* **25**, 643–650.

MEYER, R. E., and PAINTER, J. F. (1979). Wave trapping with shore absorption. *J. Eng. Math.* **13**, 33–45.

MEYER, R. E., and TAYLOR, A. D. (1963). On the equations of surf. *J. Geophys. Res.* **68**, 6443–6445.

MEYER, R. E., and TAYLOR, A. D. (1972). Run-up on beaches. In "Waves on Beaches" (R. E. Meyer, ed.), pp. 357–411. Academic Press, New York.

MILLER, R. L. (1968). Experimental determination of run-up of undular and fully developed bores. *J. Geophys. Res.* **73**, 4497–4510.

MORAWETZ, C. S., and LUDWIG, D. (1968). An inequality for the reduced wave operator and the justification of geometrical optics. *Commun. Pure Appl. Math.* **21**, 187–203.

MUNK, W. H., and TRAYLOR, M. A. (1947). Refraction of ocean waves. *J. Geol.* **55**, 1–26.

OLVER, F. W. J. (1974). "Asymptotics and Special Functions." Academic Press, New York.

PEREGRINE, D. H. (1972). Equations for water waves and the approximations behind them. In "Waves on Beaches" (R. E. Meyer, ed.), pp. 95–121. Academic Press, New York.

PEREGRINE, D. H. (1976). Interaction of water waves and currents. *Adv. Appl. Mech.* **16**, 10–117.

PETERS, A. S. (1952). Water waves over sloping beaches and the solution of a mixed boundary value problem for $\Delta\varphi - k^2\varphi = 0$ in a sector. *Commun. Pure Appl. Math.* **6**, 87–108.

PIERSON, W. J. (1972a). The loss of two British trawlers—a study in wave refraction, *J. Navig.* **25**, 291–304.

PIERSON, W. J. (1972b). Wave behavior near caustics in models and in nature. In "Waves on Beaches" (R. E. Meyer, ed.), pp. 163–179. Academic Press, New York.

PIERSON, W. J., NEUMANN, G., and JAMES, R. W. (1955). "Practical Methods for Observing and Forecasting Waves by Means of Wave Spectra and Statistics," Publ. No. 603. U.S. Navy Oceanogr. Office, Washington, D.C.

PITE, H. D. (1973). Studies in frictionally damped waves. Ph.D. Thesis Sch. Civ. Eng. University of New South Wales, Australia.

ROSEAU, M. (1952). "Contributions à la théorie des ondes liquides de gravité en profondeur variable, "Publ. Sci. Tech. No. 275. Ministère de l'Air, Paris.

SHEN, M. C., and KELLER, J. B. (1975). Uniform ray theory of surface, internal and acoustic wave propagation in a rotating ocean or atmosphere. *SIAM J. Appl. Math.* **28**, 857–875.

SHEN, M. C., MEYER, R. E., and KELLER, J. B. (1968). Spectra of water waves in channels and around islands. *Phys. Fluids* **11**, 2289–2304.

SHEPARD, F. P. (1963). "Submarine Geology." Harper, New York.

SKOVGAARD, O., JONSSON, I. G., and Bertelsen, J. A. (1976). Computation of wave heights due to refraction and friction. *J. Waterw. Harbors, Coastal Eng. Div., Am. Soc. Civ. Eng.* **102**, 100–105.

STENGLE, G. (1977). Asymptotic estimates for the adiabatic invariance of a simple oscillator, *SIAM J. Math. Anal.* **8**, 640–654.

STOKER, J. J. (1947). Surface waves in water of variable depth. *Q. Appl. Math.* **5**, 1–54.

STOKER, J. J. (1957). "Water Waves." Wiley (Interscience), New York.

SUMMERFIELD, W. (1972). Circular islands as resonators of long-wave energy. *Philos. Trans. R. Soc. London, Ser. A* **272**, 361–402.

URSELL, F. (1952). Edge waves on a sloping beach. *Proc. R. Soc. London, Ser. A* **214**, 79–97.

VAN DORN, W. G. (1961). Some characteristics of surface gravity waves in the sea produced by nuclear explosion. *J. Geophys. Res.* **66**, 3845–3862.

WILSON, B. W., HENDRICKSON, J. A., and KILMER, R. E. (1965). "Feasibility Study of a Surge-action Model of Monterey Harbor, California," Contract Rep. No. 2-136. U.S. Army Eng. Waterw. Exp. Stn., Corps Eng., Vicksburg, Mississippi.

Polymer Fluid Mechanics*

J. D. GODDARD

Department of Chemical Engineering
University of Southern California
University Park
Los Angeles, California

I. General Survey and Review of Rheological Models

A. INTRODUCTION

Polymeric liquids, consisting of high-molecular weight polymers or "macromolecules" in the form of polymer solutions or polymer melts, provide one of the most ubiquitous examples of the viscoelastic, or non-Newtonian liquid. The growth 'of the synthetic-polymer industry and the related efforts to understand and represent mathematically the mechanical behavior of such fluids in nature and technology has provided an inestimable contribution to the overall progress of continuum mechanics in the past 30 years or so. The present work is a review, and a synthesis, of certain

* With Appendix by Dr. Chester Miller, E. I. DuPont de Nemours and Co., Wilmington, Delaware.

developments in the mechanics of viscoelastic fluids, as typified by polymeric liquids. Here, the term "mechanics" embodies both the underlying rheology, in the form of mechanical constitutive theories or models, and dynamics, as represented by the phenomena of fluid motion and the governing field equations of continuum mechanics.

This review is concerned mainly with developments, of the past decade approximately, in the continuum and microstructural theory of polymer-fluid rheology, and, to a lesser extent, in the fluid dynamics. The emphasis is decidedly on continuum mechanics, and, except insofar as they relate grossly to the characteristic departures from classical (Newtonian) fluid behavior, there is little said about the chemical structure and molecular physics of polymers.

The objectives here are to trace the progress in selected avenues of research, and to summarize and evaluate certain ideas and methods, which appear to be most in need of refinement or most promising for further development. At the same time several extensions of existing methods and some original results are presented.

A number of existing textbooks, including two very recent and comprehensive works by Bird *et al.* (1977a,b) and Schowalter (1978), provide a thorough discussion of the molecular or microstructural origins of viscoelasticity in polymer fluids and related systems, as well as a treatment of fluid mechanics. However, it is worthwhile to touch briefly here upon certain salient points. In particular, we recall that the typically large molecular dimension suggested by the term "macromolecule," and the correspondingly long molecular time scales, that is, the "relaxation times" associated with gross changes in molecular configuration, account largely for the origins of viscoelasticity in polymer fluids. Typically, a polymer molecule is built up from thousands or tens of thousands of ordinary atomic or low-molecular weight subunits or "monomers" and may have molecular weights ranging from ten-thousands to millions. A useful conceptual model is that of the flexible long-chain molecule depicted in Fig. 1, which is representative of a more or less "randomly-coiled" configuration of the molecule in a low-molecular weight (Newtonian) solvent or else as a segment of the molecule surrounded by similar neighbors in a polymer melt. Figure 1a is intended to represent a spherically diffuse coil in the fluid at rest, in an equilibrium thermodynamic state, whereas Fig. 1b suggests the deformation of the molecular coil induced by flow processes. As discussed in Section I,C below, the relevant molecular time scales, or relaxation rates, are those associated with the continual tendency of the coil to return elastically to its underformed state, under the competing action of thermal "vibrations" or Brownian motion and of frictional drag or momentum exchange with its immediate environment. Much of the classical literature and a good deal of the polymer science literature is concerned mainly with near-equilibrium states, linear relaxation

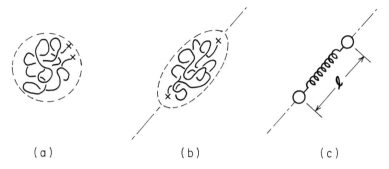

(a) (b) (c)

FIG. 1. Schematic representation of a long-chain flexible linear macromolecule in solution or melt, with + denoting terminal or entanglement point. (c) Bead-spring model.

processes and, hence, the linear viscoelasticity of polymeric materials. Much of the focus of that literature is on the relatively fine detail of the various internal modes of relaxation of polymer-chain segments and the corresponding relaxation spectrum. While this is relevant to many applications, particularly as regards the mechanical response of polymeric solids subject to small deformations, it is now generally recognized by polymer-fluid rheologists that linear viscoelasticity is inadequate to describe the strongly nonlinear processes resulting from flow-induced macromolecular deformations and orientation in fluids. This fact alone justifies the present-day development of nonlinear viscoelasticity. It also explains much of the recent concern with the distinction between "weak" and "strong" flows, according to the degree of macromolecular deformation and associated nonlinearity which is likely to occur, as illustrated in Fig. 2.

As for the general nonlinear viscoelasticity of polymers, it is now widely accepted that the general mathematical continuum theories of materials with memory provide an altogether adequate framework for the formulation of continuum rheological models. Moreover, with the exception of certain liquid-crystal systems, the *simple fluid* theory of Noll (1958) appears to provide a widely accepted foundation for the description of polymer-fluid rheology.

It is also worth recalling one other salient aspect of polymer-fluid rheology as it relates to the fluid dynamics for most applications. With the exception of very dilute polymer solutions, which are important for "turbulent drag reduction," most of the polymer fluids involved in technological processes are characterized by (effective) viscosities which are many orders of magnitude greater $[O(10^5)$ and larger$]$ than those of ordinary Newtonian liquids. Thus, for many applications, the low "Reynolds-number" or creeping motion régime provides an adequate description of dynamics. Although we may

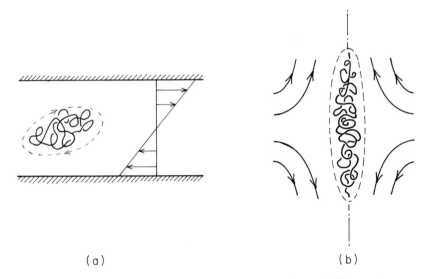

FIG. 2. Deformation of a macromolecular chain in (a) shear and (b) extension.

often dispense, then, with the *inertial* nonlinearities that are of paramount importance to Newtonian fluid mechanics, nature has unkindly replaced them with history-dependent rheological, nonlinearities.

With these introductory remarks, the balance of this chapter is concerned with recent progress in the development of rheological constitutive equations for viscoelastic fluids. Section I,B is devoted to the phenomenological and empirical approach, while Section I,C provides a brief survey of developments in molecular or microstructural models. Essentially all this discussion is focused on isothermal fluids.

The past decade has witnessed a sustained development of experimental rheology and dynamics and the associated testing and application of rheological models, a trend which is likely to continue, given the inadequacies of most existing fluid models. Hence, there is some merit in considering the form of rheological response to be expected of simple fluid models in various flows and rheological tests of increasing kinematical complexity. This provides some of the motivation for Section II which is something of an original contribution on the reduction of constitutive equations for a general class of deformations called "commutative motions" that appear to provide a convenient summary of most existing literature on special classes of deformation.

Section III is concerned with fluid dynamics and summarizes only a small body of the vast literature on the subject, relating mostly to hydrodynamic stability and other applications involving perturbations on elementary flows. The Appendix, authored by C. Miller, contains a previously un-published analysis of the Taylor–Couette stability in simple fluids (1967). The discussion of fluid dynamics does not directly address itself to the important field of application in industrial polymer processing, which would be well beyond the scope and intent.

The overall review is aimed at fellow workers in mechanics and rheology who have a modicum of familiarity with the ideas and language of continuum mechanics, as it has developed in the past two decades. Also, it is assumed that the reader is familiar with a number of established rheological ideas, such as viscometric flow and the related material functions.

The burden of reviewing, indeed comprehending even a small body of the recent literature is considerably lightened by the existence of a number of excellent and timely review articles and textbooks which have appeared over the past decade. Many of these are acknowledged by liberal citation throughout this review. Since many of the important research contributions of recent years have been adequately cited in recent textbooks and review articles, the author has by no means attempted an exhaustive list of all the works which touch upon the ideas discussed here.

B. Phenomenological Constitutive Models

In the case of isothermal motions of incompressible fluids, we recall that simple-fluid theory takes the well-known general form

$$\mathbf{T}(t) + p\mathbf{1} = \underset{s=0}{\overset{\infty}{\mathscr{F}}} \{\mathbf{G}(s)\} \tag{1.1}$$

for a specified material particle, where $\mathbf{T}(t)$ is the Cauchy stress tensor, s the (past) time lapse, and

$$\mathbf{G}(s) = \mathbf{C}_t(t') - \mathbf{1}, \qquad t' = t - s, \quad s \geq 0 \tag{1.2}$$

with, as usual, $\mathbf{C}_t(t')$ denoting the relative Cauchy (or Cauchy–Green) tensor. Also, $\mathscr{F}$ is an isotropic (symmetric) tensor valued functional, defined on the history $\mathbf{G}(s)$, $s \geq 0$.

In line with the introductory remarks, this review is based on the premise that essentially all differential ("rate-type") models and integral models that have been proposed as particular constitutive equations for simple fluids represent various special or limited forms of the above general theory, in line with Noll's earliest observations on the subject (Noll, 1958). Starting from that premise, the intent is to review here some of the various, specific forms of (1.2) which have been proposed as mainly empirical representations of polymer fluid behavior.

1. *Differential models and Time Derivatives*

Differential models, relating stress to kinematics through differential equations, have a venerable history in rheology. However, little of the recent work on the formulation of constitutive equations has concerned itself with elaboration on such models per se, no doubt, because of their special and sometimes singular nature (i.e., lack of fading memory) and the fact that differential models can, by a rational approximation scheme, often be made to correspond to integral representations (see, e.g., Walters, 1975). Accordingly, the author will not dwell on such models here, except to establish notation and to make, but hopefully not to belabor, a point concerning the associated time derivatives. The author excludes from the present remarks certain "structure" theories that relate stress to kinematics indirectly by means of differential equations for structural parameters, a subject which is taken up below, and elsewhere, in this review.

It will be recalled that rate-type* theories take the form

$$\mathbf{f}(\mathbf{T}, \mathbf{T}_1, \ldots, \mathbf{T}_n, \mathbf{E}_1, \mathbf{E}_2, \ldots, \mathbf{E}_m) = \mathbf{0}, \tag{1.3}$$

where $\mathbf{f}$ is an isotropic tensor function of all its arguments, $\mathbf{T}(t)$ is the present stress, $\mathbf{T}_k(t)$, $k = 1, 2, \ldots, n$, is a set of objective or frame-indifferent time derivatives of $\mathbf{T}(t)$, and the $\mathbf{E}_k$, $k = 1, 2, \ldots, m$, a set of objective kinematical tensors, which can also be regarded as time derivatives. One possible choice is

$$\mathbf{E}_k = \mathbf{A}_k, \qquad k = 1, 2, \ldots, \tag{1.4}$$

where $\mathbf{A}_k$ is the kth Rivlin–Ericksen tensor:

$$\mathbf{A}_k(t) = \left[\left(\frac{d}{dt'} \right)^k \mathbf{C}_t(t') \right]_{t=t'} \tag{1.5}$$

$$= \dot{\mathbf{A}}_{k-1} + \mathbf{A}_{k-1} \cdot \mathbf{L} + \mathbf{L}^T \cdot \mathbf{A}_{k-1}$$

for $k = 1, 2, \ldots$, with

$$\mathbf{A}_0 = \mathbf{1}$$

$$\left. \begin{array}{c} \\ \\ \\ \end{array} \right\} \tag{1.6}$$

and $\mathbf{L}(t)$ denoting the velocity gradient.[†] Here, d/dt and the superposed dot represent the usual material derivative. The second equality, (1.6), defines a well-known type of convected or "embedded" time derivative, which together with certain related derivatives, has been discussed at great length in several textbooks on the subject (e.g., Bird *et al.*, 1977a; Astarita and

---

* Rate type is used here in the sense of Noll (1958) and not of Huilgol (1978).

[†] We call $\mathbf{L} = (\mathbf{\nabla v})^T$, with $L_{ij} = v_{i,j}$, the *velocity gradient* which is to be distinguished, by the order of terminology and symbols, from the *gradient of velocity* $\mathbf{L}^T = \mathbf{\nabla v}$.

Marrucci, 1974). A formula similar to (1.5) with $\mathbf{A}$ replaced by $\mathbf{T}$ and $\mathbf{T}_0 \equiv \mathbf{T}$, defines a possible choice for the stress derivatives $\mathbf{T}_k$ in (1.3). However, an equally suitable choice for the arguments in (1.3) (implying, of course, a *different* function $\mathbf{f}$), can be based on the *Jaumann* or *corotational time derivative*. In particular, we may define an (objective) strain tensor, the *corotational strain*:

$$\mathbf{E}_t(t') = \int_t^{t'} \mathbf{Q}_t^T(\tau) \cdot \mathbf{D}(\tau) \cdot \mathbf{Q}_t(\tau)\, d\tau, \tag{1.7}$$

where, with

$$\mathbf{D} = \tfrac{1}{2}(\mathbf{L} + \mathbf{L}^T) \qquad \text{and} \qquad \mathbf{\Omega} = \tfrac{1}{2}(\mathbf{L} - \mathbf{L}^T), \tag{1.8}$$

representing the symmetric rate-of-deformation and antisymmetric vorticity tensors, $\mathbf{Q}_t(t')$ denotes the *mean rotation* tensor, defined by

$$\dot{\mathbf{Q}}_t(\tau) \equiv \frac{d}{d\tau}\mathbf{Q}_t(\tau) = \mathbf{\Omega}(\tau) \cdot \mathbf{Q}_t(\tau),$$

with

$$\mathbf{Q}_t(t) = \mathbf{1}.$$

As opposed to $\mathbf{C}_t(t')$ and related strain measures that are directly derivable from the deformation gradient $\mathbf{F}(t)$, the tensor $\mathbf{E}_t(t')$, which will be called here a *nonholonomic* or *generalized* strain measure (see below), is a *functional* defined on $\mathbf{F}(\tau)$, for $t' \leq \tau \leq t$, a property which has already been demonstrated for the associated "corotational" strain rate (Goddard, 1967),

$$\dot{\mathbf{E}}_t(t') = \frac{d}{dt'}\mathbf{E}_t(t'). \tag{1.9}$$

Thus, analogously to (1.5) and (1.6), one has the derived kinematic tensors

$$\mathbf{E}_k(t) = \left[\left(\frac{d}{dt'}\right)^k \mathbf{E}_t(t')\right]_{t=t'} \tag{1.10}$$

$$= \frac{\mathscr{D}}{\mathscr{D}t}\mathbf{E}_{k-1}(t) = \dot{\mathbf{E}}_{k-1} + \mathbf{E}_{k-1} \cdot \mathbf{\Omega} - \mathbf{\Omega} \cdot \mathbf{E}_{k-1}, \tag{1.11}$$

for $k = 1, 2, \ldots$, with

$$\mathbf{E} \equiv \mathbf{D}(t) \equiv \tfrac{1}{2}\mathbf{A}_1(t).$$

Here, $\mathscr{D}/\mathscr{D}t$ denotes the corotational derivative. A set of stress derivatives $\mathbf{T}_k$ for (1.3) can now be written in an obvious way in terms of this derivative. It is well known that the tensors $\mathbf{E}_1, \mathbf{E}_2, \ldots$ can be directly related to the tensors $\mathbf{A}_1, \mathbf{A}_2, \ldots$. The corotational derivative, in contrast to the various

convected derivatives, enjoys all the algebraic properties of ordinary time
derivatives (Goddard and Miller, 1966; Goddard, 1967), and, in view of
the fact that it embodies the *minimal* description of material objectivity,
there is little mathematical need for the introduction of other special time
derivatives, except as a notational convenience.

2. *Integral Models*

Integral representations of "hystersis" or history dependence of the type
implied by (1.1), especially in the context of linear viscoelasticity, date back
more than a century to Boltzmann's superposition principle, an idea which
has recently been commemorated by Markovitz (1977). Most modern
theory, or rheological "modeling," borrows its foundations from the early
works of Green and Rivlin, Coleman and Noll, and associates, on Fréchet-
series expansions about the rest state ($\mathbf{G} \equiv \mathbf{0}$), which, for later reference, we
recall to have the general form

$$\mathscr{F}\{\mathbf{G}(\cdot)\} = \delta\mathscr{F}\{\mathbf{G}(\cdot)\} + \delta^2\mathscr{F}\{\mathbf{G}(\cdot),\mathbf{G}(\cdot)\} + O(\|\mathbf{G}(\cdot)\|^3), \quad (1.12)$$

where notation for the support of $\mathbf{G}(s)$, $s \geq 0$, has been suppressed. Here,
$\delta\mathscr{F}$ is a linear form in $\mathbf{G}(s)$, $\delta^2\mathscr{F}$ is a bilinear form, etc., which represent
functional derivatives, and $\|\mathbf{G}(\cdot)\|$ denotes a suitable functional norm for
$\mathbf{G}(\cdot)$. Based on assumptions of smoothness, $\delta\mathscr{F}, \delta^2\mathscr{F}, \dots$ can be represented, respectively, as a single integral (in the time variable s), a double
integral (in two time variables), and so forth, all of which involve kernels or
"memory functions" that depend only on the time variable in the arguments
of $\mathbf{G}(\cdot)$.

There has been a sustained and vigorous activity over the past decade to
develop explicit integral models which are, in some sense, "mathematically
simple" and simultaneously adequate to describe the observed rheological
behavior of specific polymer or classes of polymers, especially in rheological
tests involving steady and transient or oscillatory flow behavior. To attempt
a complete review of the somewhat amorphous body of literature on
"rheological model building" would be well beyond the capacities and
proclivities of this reviewer, and, in any case, there are a number of comprehensive textbooks and review articles which address themselves to such
questions (Bird *et al.*, 1974, 1977a). The intent here is, rather, to characterize,
and in some instances to generalize, certain of the more clear-cut hypotheses
and methods which emerge from the literature.

a. Integral expansions and generalized strain measures. Because of their
ostensible complexity, and the potential number of memory functions involved, multiple-integral expansions based on the form (1.12) have not
enjoyed a wide-spread acceptance. Moreover, as pointed out some time ago

(Goddard, 1967) and reemphasized in the more recent literature (Bird *et al.*, 1974, 1977a), certain of the formal expansions based on the usual strain measure, of the type (1.2), may provide slowly convergent, or even non-convergent representations of real fluid behavior, because of intrinsically strong nonlinearities or long time scales in fluid response. Thus, in their simplest forms, most of existing models of finite linear viscoelasticity, corresponding to the first term in (1.12), fail to provide a good approximation to viscometric behavior (e.g., "shear thinning" viscosity) over any extended range of shear rates. On logical grounds, one cannot demand more of these representations, in view of their underlying or mathematical derivations. As one possible empirical stratagem to remedy these defects, which is based on the physical idea of substructural or network "slip," Johnson and Segalman (1977), Thien and Tanner (1977), PhanThien (1978) and Tanner (1979) have proposed a class of fluid models, termed "nonaffine" by Johnson and Segalman, which involve the use of alternative strain measures. In line with the remarks surrounding (1.7) above, the author proposes to designate these as *nonholonomic* or *generalized** strains. The basic idea underlying the models is that the microstructural strains which serve to determine stress do not correspond simply to the imposed macroscopic strain, an idea which to some extent is already embodied in certain continuum "mixture theories," and also in various microstructural theories. Thus, one defines a nonholonomic strain tensor $\hat{\mathbf{G}}(s)$ according to the usual recipe

$$\hat{\mathbf{G}}(s) = \hat{\mathbf{C}}_t(t') - \mathbf{1} \qquad \text{with} \qquad \hat{\mathbf{C}}_t(t') = \hat{\mathbf{F}}_t^T(t') \cdot \hat{\mathbf{F}}_t(t'). \tag{1.13}$$

The "effective" deformation gradient $\hat{\mathbf{F}}$ is given by the usual relations

$$\frac{d\hat{\mathbf{F}}(t)}{dt} = \hat{\mathbf{L}}(t) \cdot \hat{\mathbf{F}}(t) \qquad \text{or} \qquad \frac{d\hat{\mathbf{F}}_t(t')}{dt} = \hat{\mathbf{L}}(t') \cdot \hat{\mathbf{F}}_t(t') \tag{1.14}$$

but, now, $\hat{\mathbf{L}}(t)$ is an *effective* velocity gradient which differs from the actual velocity gradient $\mathbf{L}$ and which is determined entirely by the instantaneous kinematics of the material at "present" time t. More precisely, we assume that $\mathbf{L}$ can be expressed as an isotropic tensor function of a set of kinematical tensors, $\mathbf{E}_1, \mathbf{E}_2, \ldots$, such as (1.5) or (1.10), together with the vorticity $\mathbf{\Omega}$. In order to satisfy the assumption of material objectivity, it is convenient to express this function in terms of separate functions for the (effective) deformation rate $\hat{\mathbf{D}}$ and vorticity $\hat{\mathbf{\Omega}}$, which are defined by affixing carets

---

* The time-worn term, "generalized," has already been used on numerous occasions, most recently by Phillips (1977), to designate a simple isotropic function of the usual strain measure of the type discussed generally by Truesdell and Toupin (1960). In line with the present discussion, these may be called "holonomic" strains.

to (1.8). Thus, we have the separate, "slip" (or "*dislocation rate*") equations:

$$\hat{D} - D = d(E_1, E_2, \ldots) \quad \text{and} \quad \hat{\Omega} - \Omega = w(E_1, E_2, \ldots) \quad (1.15)$$

where d, symmetric, and w, antisymmetric, are isotropic tensor functions of their arguments $E_1, E_2, \ldots$, the total number and degree of which determines the *complexity* of the model. The relations (1.15) represent a generalization of those cited above, for which $w \equiv 0$.

The formulas (1.15) can be rendered explicit through expansions of the Rivlin–Ericksen variety, thereby, providing various "orders" of approximation for slip. Thus, to terms of second-order in the velocity gradient,

$$d = d_1 E_1 + d_2 E_1^2 + d_2' E_2 + d_2''(tr\ E_1^2)1 + 0(|L|^3)$$

and

$$w = w_3[E_1 \cdot E_2 - E_2 \cdot E_1] + 0(|L|^4), \quad (1.16)$$

where $d_1, d_2, \ldots; w_3, \ldots$, are material constants, with d_2'' accounting for a compressible microstructure. The models employed by the workers cited above appear to correspond to "first-order" theories, with Johnson and Segalman employing a parameter $a = 1 + d_1$ in their analysis. As indicated by the latter authors, the usual strain measures are recovered by choosing $d_1 = 0$, while, in the limit $d_1 \to -1$, one obtains the corotational strain as,

$$\hat{G}(s) \to (1 + d_1)E_t(t'), \quad t' = t - s.$$

Hence, in order to obtain a nonvanishing strain effect in the subsequent rheological model, one must either incorporate a constant factor $(d_1 + 1)^{-1}$ in the definition of the strain measure (1.13), or else, in the relevant integrals of $\hat{G}(s)$, which is the procedure employed by Johnson and Segalman.

As suggested in the original development of the corotational-strain expansion (Goddard, 1967), which was actually developed in terms of the *strain rate* $\dot{E}_t(t')$, the introduction of a nonholonomic strain measure amounts to a change of functional variables in (1.1). Accordingly, the substitution of $\hat{G}(s)$ for $G(s)$ in the expansion (1.12) provides an alternative integral expansion, and, while all such expansions should be equivalent for $\|G(\cdot)\| \to 0$, up to terms of a given order in any reasonably smooth norm $\|G(\cdot)\|$, certain expansions may exhibit much more desirable convergence properties than others.

For rotationally steady motions, i.e. "motions with constant stretch history" it is an easy matter to obtain explicit representations for the generalized strain (1.13) even when terms beyond the first order are retained in (1.16). Also, for the practically important case of steady or unsteady (orthogonal) extensional flow, the rotational slip w must vanish identically, to terms of any order, on grounds of symmetry alone, so that the present treatment is equivalent to that of the previous work, to first-order terms in the velocity gradient. In view of this, one can immediately make certain important distinctions between the particular, limiting case of the corotational strain and the more general form of (1.13)–(1.16), as applied to steady extensional flow. In particular, one encounters exponentially large strains $\hat{\mathbf{G}}(s)$ for $s \to \infty$, in the general form (1.13), unless higher order, nonlinear terms are retained in the expression (1.16) for $\mathbf{d}$. Hence, it appears that the first-order models of the previous works *do not* lead to integral representations which are radically different from the usual theory in their ability to discriminate between "weak" and "strong" flows, a subject which has taken on a considerable importance in the recent literature (Tanner, 1975a; Tanner and Huilgol, 1975; Pipkin and Tanner, 1977). With the microstructural interpretation, one can say that a higher order model, or the corotational limit of the first-order model, is necessary to allow for limiting states wherein the microstructural strain remains bounded (algebraically in s) for strong flows.

b. Single-integral models. It appears that most empirically motivated, explicit models proposed to date can be cast into the form

$$\mathbf{T}(t) + p\mathbf{1} = \int_0^\infty \mathscr{L}\{I(\cdot); \mathbf{G}(s), s\} \, ds, \tag{1.17}$$

where the tensor $\mathscr{L}$ is simultaneously (1) *an isotropic tensor function of* $\mathbf{G}(s)$, (2) *a function of the time variable* s, and (3) *a functional defined on the scalar invariants* $I_1(s), I_2(s), I_3(s)$ of $\mathbf{G}(s)$ or else of, $\mathbf{D}(t-s)$, which are denoted here collectively by $I(s)$. For the incompressible fluids at hand there are, of course, only two independent invariants.

As applied to the form (1.17), the terminology "single integral" is a loose one, since the integrand involves a functional which, more often than not, would itself be represented by integrals. In the framework of rational continuum mechanics, (1.17) can at best be regarded as a truncation of the general representation established by Wineman and Pipkin (1964: cf., Rivlin and Sawyer, 1971). As such, it must be regarded as mathematically restrictive in its description of spatial stress patterns for arbitrary deformation histories and in its description of response to perturbations on a given deformation history, as represented by Fréchet series of the type (1.12).

Within the limitations of the model (1.17), the functional dependence on the scalar invariants I can be visualized as representing the cumulative effect of prior $(s' > s)$ and subsequent $(s' < s)$ deformations on the "relaxation" of the fluid, i.e., on its response to strains $\mathbf{G}(s)$ imposed at past time $t' = t - s$. This is one of a number of devices employed by various workers to account for strain-induced modifications of linear viscoelastic response. The models of Bogue (1966) and Carreau (1972; Bird and Carreau, 1968), with relaxation time depending on a time-average strain-rate invariant, fall into this category.

At the same time, however, certain of the above models, such as that of Carreau, involve a singular form of the functional dependence on $I(\cdot)$ exhibiting an "atom at time lapse s," such that an invariant of the deformation rate $\mathbf{D}(t - s)$ appears in the argument of $\mathscr{L}$. Astarita and co-workers (Marrucci and Astarita, 1974; Astarita and Marrucci, 1974; Astarita and Jongschaap, 1978) have criticized such models on the grounds that they do not satisfy the Coleman–Noll (1961) principle of fading memory, especially as concerns the fluid response to strain discontinuities and its frequency-dependent departures from linear viscoelasticity in small-amplitude oscillatory tests.

With no intent here to defend any particular empirical model, several comments regarding the above criticism are nevertheless in order.

First of all, there is the well-known practical issue of whether one wishes to reject the Newtonian fluid model on the above grounds. Quite apart from arguments as to the overall consequences for classical fluid mechanics, the answer must be resoundingly negative, because the issue rests basically on the question of time scales, as indeed does much of the motivation for studying nonclassical fluids.

In the present context, the question of classical fluid response is relevant, since many plausible microstructal or molecular theories involve the (good) approximation of a Newtonian solvent, or "viscous" environment, characterized by relaxation times much shorter than those of the microstructural time scales of interest, the latter of which are comparable to a *suitably restricted* set of "laboratory" time scales. Thus, a Newtonian ("solvent") contribution to stress emerges directly from many such theories, and, offhand, the author knows of no general reason why related effects should not be manifest to some extent in the associated non-Newtonian contributions. While such microstructurial theories may be amenable to modification, allowing for ("high-frequency") viscoelastic response of a nominally Newtonian solvent, these would constitute for most applications a needless embellishment and a hopeless complexity, which would hardly be defensible in the present *milieu* of "single-integral models" and like approximations.

As a theoretical issue, there is also the question as to precisely what class of models that incorporate strain rate in a single integral representation fail to exhibit fading memory, and, second, as to the strength of the resultant singularities. One potentially useful class of models which appear worthy of investigation are those endowed with a *Lagrangian density* in (1.17) of the form

$$\mathscr{L} = \mathscr{L}\{\mathbf{G}(s), \dot{\mathbf{G}}(s), s\}. \tag{1.18}$$

By way of a final comment on the issue at hand, it will be noted that the Coleman–Noll (1961) definition of fading memory, based on the norm $\|\mathbf{G}(\cdot)\|$, is only one of a number possible, some of which may well incorporate *deformation rates*, thereby leading to modified theories of fading memory. Without a full understanding of the implications for various thermodynamic theories, and with no particular intent of launching an investigation, this reviewer will nevertheless remark that the notion of *rate* as a measure of irreversibility has considerable appeal.

Returning, now, to the form (1.17), we recall that the tensor $\mathscr{L}$ can be expressed as

$$\mathscr{L} = \mathscr{M}_1\{I(\cdot); I_G(s), s\}\mathbf{G}(s) + \mathscr{M}_2\{I(\cdot); I_G(s), s\}\mathbf{G}^2(s), \tag{1.19}$$

where $\mathscr{M}_1$ and $\mathscr{M}_2$ are scalars, which are endowed with the properties (2) and (3) listed immediately after (1.17), and which, as indicated, are also functions of the scalar invariants of $\mathbf{G}(s)$, denoted here collectively by $I_G(s)$.

As pointed out in numerous works (e.g., Rivlin and Sawyer, 1971; Astarita and Marrucci, 1974), if one drops the functional dependence on $I(\cdot)$ from (1.17) and (1.19), one obtains a model resembling the well-known "BKZ" fluid, of Bernstein *et al.* (1964):

$$\mathscr{L} = M_1(I_G, s)\mathbf{G}(s) + M_2(I_G, s)\mathbf{G}^2(s), \tag{1.20}$$

where M_1 and M_2 are (memory) functions of I_G and s alone. The BKZ fluid emerges, in a slightly different form from the customary one, upon the assumption that $\mathscr{L}$ be derivable from a scalar "potential."

With or without the last-mentioned assumption, the fluid model defined by (1.20) exhibits a strain effect on relaxation, through the dependence of the memory functions on $I_G(s)$. This leads one to question whether there is any compelling necessity for incorporation of the history dependences on $I(\cdot)$ in (1.19), unless such dependence can be inferred from plausible microstructural or molecular theories.

Whatever the empirical status of (1.20), the resulting fluid model (1.17) can be shown to follow from Noll's functional (1.1), under the assumption of *additivity of functionals* or what I would call "*temporal separability in* $\mathbf{G}(s)$," together with smoothness necessary for integral representations (Martin and Mizel, 1964; Chacon and Friedman, 1965; cf. Rivlin and Sawyer, 1971; Bird *et al.*, (1974). In a certain sense, this means that "strains in the past do not interact with one another" (Rivlin and Sawyer, 1971; Bird *et al.*, 1974). However, in strict analogy to separability of variables in multivariable calculus, the above condition of separability depends strongly on the variables adopted, that is to say, on the strain measure employed. This may suggest some justification for the substitution in (1.20) of a generalized strain $\hat{\mathbf{G}}(s)$ of the type (1.13), in place of $\mathbf{G}(s)$. Indeed, as pointed out by Bird *et al.*

(1974), the corotational strain (rate) model of Leroy and Pierrard (1973) represents one such possibility.

For the sake of completeness, another type of "separability" assumption should be mentioned here. In particular, the author refers to the "separation of relaxation and strain effects" in (1.20) which is achieved by taking the *ratio* of M_1 and M_2 to be independent of the time lapse s. The resulting model thus dictates that the linear viscoelastic response (or memory function) for the fluid, together with a "generalized" (*holonomic*) strain measure, is sufficient to characterize its general response. Wagner (1977, 1978) indicates that such a model is capable of representing data on both steady and transient simple shear and simple extension of polyethylene melts. However, in view of the underlying restrictive assumptions, which contradict those embodied in various fluid models discussed above, the latter type of model should be greeted with appropriate caution.

As a final perspective on the whole subject of empirical "single-integral" models, it is worthwhile to note that essentially a decade of effort has been expended on the term $I(\cdot)$ in (1.17). This should surely lead one to question the prospects for the coming decade, if more systematic and fundamental approaches are still lacking. It appears that many of the current experimental methods, much as with simple viscosity measurements in the early literature, do not provide severe or critical tests of rheological models. This suggests a fertile ground for activity, involving "strong" flows and rapid deformation processes of the kind already envisioned early on by several workers (Metzner *et al.*, 1966; cf. Walters, 1975).

C. Microstructural and Molecular Theories

There has been much attention given in the past decade to various molecular and microstructural models for viscoelasticity in fluids. The recent textbook of Bird *et al.* (1977b) provides the most comprehensive and up-to-date summary of various kinetic-theory calculations for macromolecular systems, including polymer solutions and melts, while numerous review articles, such as those of Brenner (1974), Batchelor (1974), and Jeffrey and Acrivos (1976), as well as the recent textbook of Schowalter (1978), treat the rheology of particle suspensions in Newtonian liquids. [The treatise of Bird *et al.* has itself been reviewed in the recent literature by Middleman (1978) and Goddard (1978a).]

Although the present review does not address itself directly to suspension rheology in the usual sense, it should nevertheless be recognized that much of the work of the last decade shows a number of remarkable similarities between results which emerge from particle-suspension theories and those of macromolecular theories. This should perhaps not be surprising, given certain underlying physical similarities between the systems together with the general continuum-mechanical principles which dictate the form that their response must take. Thus, in many of the suspension theories one has viscous forces of the suspending fluid coupled with the "entropic" elasticity arising from Brownian rotations of (small) rigid particles, or elasticity arising from the interface of deformable liquid droplets (cf. Schowalter, 1978). Furthermore, most of the tractable macromolecular theories, such as the "hydrodynamic" theory of polymer solutions, are effectively quasicontinuum models, in which a macromolecule is represented as an assemblage of "entropically" elastic elements or flexibly connected rigid elements which are subject to Brownian motions and viscous or frictional drag arising from their environment. Thus, many of the assumptions on which molecular theories of classical continua rest, particularly the notions of widely differing time scales, are embodied in the above macromolecular theories. Also, in both the particulate and macromolecule theories one may anticipate a near-equilibrium or "slow-flow" regime, wherein elastic restoring forces or Brownian motions dominate over frictional forces, and which is characterized by a progression through various "second-order" and higher fluids, based on the magnitude of a characteristic (relaxation) time parameter, reflecting the importance of viscous forces relative to elastic forces. In elementary elastic-particle theories, this parameter appears as the ratio of a suspending fluid (Newtonian) viscosity to the effective elastic modulus of the particle. For numerous macromolecules theories it is generally recognized that this same parameter provides a rationalization of the temperative dependence in viscoelasticity and of various, more or less empirical "time–temperature" superposition principles," which could more properly be called "scaling" rules (cf. Morland and Lee, 1960; Ferry, 1970; Sarti, 1977; Crochet and Naghdi, 1978). Given this similarity in the particulate and macromolecular theories, we may anticipate that any significant differences in their continuum rheology can be expected to arise only in the highly nonlinear or "strong-flow" regime. It will therefore be interesting to observe in future years the continued interplay of macromolecular theory with experimental and phenomenological rheology.

In attempting to summarize certain highlights of recent developments in the macromolecular theory of polymer fluids, it is perhaps not overly trite, or unfair, to observe that there has been little change in our understanding and description of the basic molecular processes, which date back 30 to 40 years to the pioneer labors of a number of physical chemists and fluid mechanicists whose names need no citation here.

As in several areas of science, much of the progress has consisted of a systematic reformulation of the basic macromolecular models, followed by recent attempts to extend the kinetic theory calculations beyond linear viscoelasticity into the highly nonlinear regime, and, to cast the results into the form of continuum-mechanical models for rheology. This overall effort is epitomized by the treatise of Bird and co-workers (1977a,b).

For the present summary, we consider only a few variations on the most elementary macromolecular models. One of the most elementary and well understood is the (Kuhn-and-Kuhn) "bead-spring" or "flexible dumbbell" model illustrated in Fig. 1c. Here, the deformed macromolecule in dilute solution is represented by an elastic spring connecting two beads which constitute centers of a systematic, localized hydrodynamic resistance and a random "large-scale" or "low-frequency" Brownian motion. The spring itself represents an intramolecular restoring force which is largely thermal, or "entropic," in origin and arises from "small-scale" or "high-frequency" Brownian motion. In its usual form, this model is taken to represent an isolated macromolecule in highly dilute solution in a Newtonian liquid, where the localized drag exerted by the solvent on the beads is of the linear (Stokesian) variety. However, the same type of model has been employed extensively to represent concentrated polymer solutions, polymer melts and even polymer solids, with frictional drag arising from "entanglement" points or temporary "network junctions." In essence, the model and its variants provide a somewhat more concrete realization of the phenomenological "spring-dashpot" representations of viscoelasticity in the early rheological literature.

While there have been recent attempts to improve the hydrodynamic picture involved, it appears that the underlying statistical mechanical approximations have gone largely unchallenged although they have been catalogued by Bird *et al.* (1977b, Vol. II Chap. 14).

In the simple dumbbell model, the end-to-end vector l provides a "molecular" analog of the directors intrinsic to certain micromechanical continuum theories, notably the anisotropic fluid of Ericksen (1960, 1961) and later variants thereof (cf. Leslie, 1968; Hand, 1962). However, in contrast to such phenomonological director theories which involve *deterministic* equations for director orientation, the presence of large-scale Brownian motion in the molecular model necessitates a *stochastic* or statistical–mechanical treatment, at least if one intends to pursue the molecular theory from its foundations. A statistical–mechanical or kinetic-theory treatment turns on the solution of the relevant Fokker–Planck or "configurational convection–diffusion" equation governing the statistical distribution function for l. The general form of such equations is now well-documented (Bird *et al.*, 1977b) and need not be repeated here. Suffice it to say that the relevant Fokker–Planck equation has a structure similar to the convective-diffusion equation for passive scalars in classical fluid mechanics and, similarly, reflects the competing effects of (Brownian) diffusion and flow on the statistical distribution of l. As shown in several recent works and summarized by Bird *et al.* (1977b, Vol. II), the stress tensor in a dilute solution is determined by a relation of the form

$$\mathbf{T} = \mathbf{T}' + 2\eta_s\mathbf{D} - p\mathbf{1}, \tag{1.21}$$

where η_s is the (Newtonian) solvent viscosity and

$$\mathbf{T}' = n[\langle \mathbf{lf} \rangle - \langle \mathbf{lf} \rangle_0] \tag{1.22}$$

is the stress contribution of the particles or macromolecules, with n denoting their number per unit volume of solution and $\mathbf{f} = \mathbf{f}(l)$ the spring force. The brackets $\langle \ \rangle$ and $\langle \ \rangle_0$ denote statistical averages based, respectively, on the actual distribution function, i.e., for the dynamic state, and on that for the equilibrium state. The absence of external couples, together with the balance of angular momentum, dictates the symmetry of (1.22). For the equilibrium state

$$\langle \mathbf{lf} \rangle_0 = H\langle \mathbf{ll} \rangle_0 = \frac{Hl_0^2}{3}\mathbf{1},$$

where

$$l_0^2 = \mathrm{tr}\langle \mathbf{ll} \rangle_0. \tag{1.23}$$

H is the effective Hookean spring constant, and l_0 the root-mean-square radius of gyration for l. (For a "Gaussian" macromolecule having large number $N \gg 1$ of freely orienting rigid segments, of length b, we recall that $l_0^2 = Nb^2$ and $Hl_0^2 = 3\kappa T$, where κT is the Boltzmann factor.) The governing equation for the gyration tensor $\langle \mathbf{ll} \rangle$ is found to be

$$\frac{\mathscr{D}}{\mathscr{D}t}\langle \mathbf{ll} \rangle = \mathbf{D} \cdot \langle \mathbf{ll} \rangle + \langle \mathbf{ll} \rangle \cdot \mathbf{D} + \frac{4}{\zeta}[\langle \mathbf{fl} \rangle_0 - \langle \mathbf{fl} \rangle], \tag{1.24}$$

where ζ is the Stokesian drag coefficient for the individual "beads." For perfectly Hookean dumbbells, with $\mathbf{f} = H\mathbf{l}$, in nonequilibrium extended states, (1.24) provides a directly soluble differential equation for $\langle \mathbf{ll} \rangle$. We recall that (1.21) and (1.22) then lead exactly to a well-known "single-integral" model

$$\mathbf{T} + p\mathbf{1} = 2\eta_s\mathbf{D} + \frac{\mu}{\theta}\int_0^\infty e^{-s/\theta}[\mathbf{C}_t^{-1}(t-s) - \mathbf{1}]\,ds, \tag{1.25}$$

where

$$\mu = \frac{1}{3}nHl_0^2 (= n\kappa T)$$

and

$$\theta = \zeta/4H\left(= \frac{\zeta l_0^2}{12\kappa T} \right)$$

denote, respectively, an elastic modulus and a relaxation time for the fluid. As shown by several workers (see Bird *et al.*, 1977b), the above model can be generalized to bead-spring assemblages, resulting in a discrete spectrum of relaxation times. The greatest virtue of the above model and its general-izations lies perhaps in the direct estimates they provide for constitutive parameters like μ and θ, in terms of a few fundamental constants and molec-ular dimensions.

We recall that (1.25) also corresponds exactly to the most elementary form of the (Lodge) *network theory*, with a somewhat different molecular inter-pretation of the parameters involved (cf. Graessley, 1974; Lodge, 1964, 1974).

The exact correspondence, in the form of (1.25) between the two micro-structural models, which appear superficially to represent widely diver-gent physical pictures, suggests an evident danger in various rheological "confirmations" of such models, unless the microstructural parameters are calculable from first principles or else derivable from independent measurements.

It is instructive for later purposes to consider the basic reason for the tractability of the above dumbbell model, which is best illustrated by the *Langevin* formation of the dumbbell model. This involves an equation of motion of the form

$$m\ddot{l} + \zeta(\dot{l} - \mathbf{L} \cdot l) + 2\mathbf{f} = \mathbf{b}(t) \tag{1.26}$$

representing a balance between acceleration, viscous drag, the spring force **f**, and a stochastic Brownian force $\mathbf{b}(t)$. It can be cast into a form involving dyadic products:

$$m(\ddot{l}l + l\ddot{l}) + \left[\frac{\mathscr{D}}{\mathscr{D}t} ll - \mathbf{D} \cdot ll - ll \cdot \mathbf{D} \right] = -2(\mathbf{f}l + l\mathbf{f}) + (l\mathbf{b} + \mathbf{b}l). \tag{1.27}$$

Then, by the usual time averaging of (1.27), together with the "erogodic" hypothesis and the assumption of zero statistical correlation between **b** and *l*, one arrives at an equation of the form (1.24), wherein the "equipartion" value of an average kinetic-energy term, of the form $m\langle \ddot{l}l \rangle$, is properly related to (isotropic) terms of the form $\langle \mathbf{f}l \rangle_0$.

It becomes immediately apparent then that in the case of a Hookean dumbbell one can effectively discard the Brownian terms in (1.27) to arrive at a deterministic equation for the dyad *ll* which bears a strong similarity to "director" equations arising in anisotropic fluid models. Upon integration, together with a suitable statistical averaging, one arrives again at the constitutive equation (1.25). Thus, Brownian motion effectively drops out of the above *linear* model, and a similar kind of simplification arises in the corresponding "network" model (Lodge, 1964; Bird *et al.*, 1977b).

A well-known defect of the Hookean dumbbell model, and generalizations thereof, is their infinite extensibility in certain types of strong flows, such as steady simple extension, and the consequent catastropic growth of dumbbell length and stress at finite strain rates $O(\theta^{-1})$. This catastrophic response in stress is of course dictated by any model of the form (1.25), whatever the underlying microstructural model. Also, such models fail to predict many important nonlinear effects such as shear-dependent viscosity in viscometric flows. Hence, there are numerous adequate grounds for concluding that models of the form (1.25) are restricted basically to linear viscoelasticity, which provides much of the motivation for theoretical investigation of nonlinear dumbbell models, or else of various rigid bead-rod assemblages and flexible-thread models (Hinch, 1976; Christiansen and Bird, 1977–78; Bird *et al.*, 1977b). Unfortunately, the kinetic theory of such models is much more difficult, and in most cases it has not been possible to carry out detailed computations for general flows beyond second-order terms in strain rates, which correspond to terms $(\|G(\cdot)\|^2)$ and, hence, "second-order viscoelasticity" in general continuum models. However, for the important special case of steady (orthogonal) extensional flows, the viscous forces often degenerate to a symmetric form derivable from a (pseudo) potential, and one may therefore employ the methods of equilibrium statistical mechanics to write down the relevant distribution function and the stress in explicit closed forms (Brenner, 1974; Bird *et al.*, 1977b). One can anticipate, also, that for these flows the stress will be given by a Reiner–Rivlin equation, with coefficients which are calculable from the molecular theory. These facts alone suffice to elevate steady extensional flow to a paramount position in rheological tests of specific microstructural models.

In the recent literature, Phan-Thien *et al.* (1978; cf. Hinch, 1975a,b) have resorted to a Langevin approach in order to discuss various nonlinear effects, including a dependence of the effective drag coefficient ζ on the macromolecular extension. A rigorous statistical–mechanical treatment of such models would involve further *l*-dependent terms, in either the Langevin or Fokker–Planck equations, and ζ must be "brought inside" certain averages $\langle \rangle$ in equations of the type (1.24). Because of the resultant complexity, and with a view to application in the theory of turbulent drag reduction in polymer solutions, Hinch (1975a, 1977) has resorted to various approximate, quasi-deterministic models, based essentially on (1.27), with Brownian motion dropped out. While this may be defensible for applications of the type contemplated, it is not clear at this time whether the results are much more reliable for "order-of-magnitude" estimates than various less complex models, and even less clear that they can provide an adequate prediction of macromolecular structure effects. In any event, one might hope that unwarranted pursuit of such models, beyond the applications presently envisioned, will not lead to a general revival of rheological "spring-dashpottery" cloaked in nonlinear forms.

To complete the discussion of definite microstructural models, it is worthwhile to mention certain important differences between the type of flexible-dumbbell models discussed above and various rigid-dumbbell or rigid-particle models which have also been considered in considerable detail (Brenner, 1974; Hinch and Leal, 1976; Bird *et al.*, 1977b). There is now a body of opinion to the effect that such models, which exhibit a limiting orientation and a concomitant "saturation" in the stress level for certain strong flows, may provide a more realistic portrayal of certain macromolecular systems, such as polymer solutions. Hence, they may also point the way to an asymptotic kinetic theory for flexible-particle models in certain flows, as suggested by the analysis of Hinch (1976) for flexible threads.

There is already a considerable literature dealing with dilute solutions of rigid particles subject to rotary Brownian motion (cf. Brenner, 1974; Rallison, 1978), and Hinch and Leal (1976) have recently suggested certain approximate constitutive equations for rigid axisymmetric particles in general flows. Because these systems involve particle rotations and Brownian torques, a Langevin equation in the form of an angular momentum balance replaces the linear momentum balance of (1.26). Accordingly, the relevant Fokker–Planck equations involves the distribution of angular orientation. This leads naturally to the appearance of averages involving not only the second-rank "structure" tensors $\langle ll \rangle$, but also fourth-rank tensors $\langle llll \rangle$, which are essentially angular averages since l now has constant magnitude (cf. Bird *et al.*, 1977b, Vol. 2, pp. 530ff.) Barthès-Biesel and Acrivos (1974) have already signaled the role of such "structure" tensors in various rigid-particle and deformable-particle models of suspensions, as well as their relation to the continuum theory of "anisotropic" fluids of Hand (1962).

Hinch and Leal (1976) suggest various "closure" approximations, like those of turbulence theory, which allow one to express the higher order moment $\langle llll \rangle$ in terms of $\langle ll \rangle$ without precise knowledge of the angular distribution function, thereby leading to an explicit differential equation for $\langle ll \rangle$ which is reminiscent of anisotropic-fluid or related phenomenological "structure" theories. Their work suggests the appropriateness of such theories which, involving differential equations as they do, may not be easily converted to the customary explicit integral forms for stress in terms of strain history. As such, these theories bear a strong resemblance to phenomenological continuum models based on "generalized" strain measures, of the type discussed above in Section I,B. Thus, in view of the possible relevance

of such continuum models to network theories for concentrated macromolecular systems (Phan-Thien, 1978), they may warrant further investigation.

In closing here, it should be noted that the "structure theory," proposed by
Acierno *et al.* (1977) has not been discussed at length here. While motivated
by considerations of entanglement dynamics in polymer networks, these
theories rest so heavily on *ad hoc* microstructural kinetic equations that it is
not clear to this author whether they amount to much more than phenomenological continuum models.

II. Kinematics and Stress for Commutative Motions of Simple Fluids

The use of symmetry arguments to reduce the form of general constitutive
equations for restricted classes of material deformations is a well-established
technique in continuum mechanics, particularly for (Noll) simple materials
(Coleman, 1962b; Coleman and Noll, 1962; Coleman *et al.*, 1966; Huilgol,
1969). Some time ago, in an unpublished talk (1971), the author pointed
out that essentially all those deformations which are readily amenable to
such reduction and which, in fact, serve as the basis of most common
rheological tests appear to fall into a general category of deformations which
may be called *commutative motions*. These include steady and unsteady
simple shear or viscometric flows, steady and unsteady extensions, and the
so-called "motions of constant stretch history" together with certain related
unsteady flows. Subsequently and independently, Zahorski (1972a) has
given a discussion of such motions, which he calls "superposable motions
of proportional stretch history." However, while he suggests a general
representation for their kinematics, he establishes neither a complete classification nor the associated minimal parametric representation of the kinematics, which is the main subject of the present work. Also, as a second
objective, the intent is to employ these parametric representations, together
with symmetry arguments, to obtain the maximal reduction of the constitutive equation for stress in an (isothermal) simple fluid; that is, to derive the
minimal set of independent scalar functionals defined on the history of the
kinematic parameters. In this and the following section, II,A, we consider
the kinematic representations and, in Section II,B the reduction of constitutive equations.

In what is essentially standard notation, we let

$$\mathbf{F}_*(t) = \mathbf{F}_*(\mathbf{X}, t) = \partial \boldsymbol{\chi} / \partial \mathbf{X} \tag{2.1}$$

denote the deformation gradient of a material particle, which is at position
$\mathbf{x} = \boldsymbol{\chi}(\mathbf{X}, t)$ in the present configuration and at position $\mathbf{x} = \mathbf{X}$ in a given

reference configuration (*) of a material. In the following, we shall suppress as usual the notation for dependence on "the particle" **X**. We say, then, that the deformation at **X** is a "motion with commutative history" or, more briefly, "*a commutative motion*," provided there exists a reference configuration (∗) such that

$$\mathbf{F}_*(t') = \mathbf{Q}(t') \cdot \mathbf{F}(t') \tag{2.2}$$

for all past times $t' = t$, where $\mathbf{Q}(t')$ is an arbitrary, time-dependent orthogonal tensor, and $\mathbf{F}(t')$ is a real nonsingular tensor, such that

$$[\mathbf{F}(t_1), \mathbf{F}(t_2)] = 0 \tag{2.3}$$

for $t_1, t_2 \le t$, with

$$[\mathbf{A}, \mathbf{B}] = \mathbf{A} \cdot \mathbf{B} - \mathbf{B} \cdot \mathbf{A} \tag{2.4}$$

denoting the *commutator* or *Lie product* of two arbitrary tensor **A**, **B**. Otherwise stated, there exists a reference configuration and an observer frame such that the corresponding deformation gradients $\mathbf{F}(t')$ evaluated at any two arbitrary times represent commutative linear transformations. We shall call any frame of reference for which (2.3) holds a "canonical" frame.

Thus, with

$$\mathbf{F}_{t_1}(t_2) = \mathbf{F}(t_2) \cdot \mathbf{F}^{-1}(t_1) \tag{2.5}$$

denoting the *relative* deformation gradient in a canonical frame, one has, by (2.3), that

$$\mathbf{F}_{t_1}(t_2) = \mathbf{F}_\tau(t_2) \cdot \mathbf{F}_{t_1}(\tau) = \mathbf{F}_{t_1}(\tau) \cdot \mathbf{F}_\tau(t_2). \tag{2.6}$$

That is, the deformation from the state at time t_1 to that at t_2 can be achieved by reversing the order of the deformations associated with an arbitrary intermediate state, where $t_1 < \tau < t_2$. This is an evident property of a number of the elementary homogeneous deformations mentioned above. In the general case, the one parameter family of commutative deformation-gradient tensors $\mathbf{F}(t')$, $t' = t$, represents a subset of a one-parameter (continuous) *Abelian group*, which can be expressed in the form

$$\mathbf{F}(t') = e^{\mathbf{M}(t')}, \tag{2.7}$$

where the second-rank tensor $\mathbf{M}(t')$ is also endowed with the commutative property

$$[\mathbf{M}(t_1), \mathbf{M}(t_2)] = \mathbf{0} \tag{2.8}$$

for all $t_1, t_2 \le t$. The one-parameter set $\mathbf{M}(t')$, whose members are *not* restricted to nonsingular tensors, represents, then, a subset of an algebra of commutative second-rank tensors (i.e., linear transformations on a three-dimensional linear vector space) which we shall call the *generator algebra*, $\mathscr{M}$ for $\mathbf{F}(t')$ and which is, of course, isomorphic to an algebra of commutative three-by-three matrices.

When $\mathbf{F}(t')$ and $\mathbf{M}(t')$ are differentiable, with material time derivatives $\dot{\mathbf{F}}(t')$ and $\dot{\mathbf{M}}(t')$, respectively, the corresponding material *velocity gradient* $\mathbf{L}(t')$ in the canonical frame is given by

$$\mathbf{L}(t') = \dot{\mathbf{F}}(t') \cdot \mathbf{F}^{-1}(t') = \dot{\mathbf{M}}(t') \tag{2.9}$$

and, hence, belongs to the same commutative algebra as $\mathscr{M}$.

A. Classification and Representation of Generator Subalgebras

Here, we rely heavily on known results for the algebra of commutative linear operators on finite (n) dimensional linear vector spaces, or equivalently the algebra of $n \times n$ square matrices, which are summarized in the treatise of Suprunenko and Tyshkevich (1968). In particular (cf. Theorem 2, p. 42), we recall that all the elements $[\mathbf{M}]$ of an algebra $\mathscr{M}$ of pairwise commutative square matrices of order n can be simultaneously reduced to a quasi-diagonal form:

$$[\mathbf{M}] = \text{diag}\{[\mathbf{M}]_1, [\mathbf{M}]_2, \ldots, [\mathbf{M}]_s\}$$

where each diagonal block or submatrix $[\mathbf{M}]_i$ is a square matrix of order n_i, with $n_1 + n_2 + \cdots + n_s = n$. The matrix $[\mathbf{M}]_i$ represents the restriction of $[\mathbf{M}]$ to an invariant subspace of dimension n_i and also represents a commutative matrix algebra $\mathscr{M}_i$ of $n_i \times n_i$ matrices. Furthermore, for $i \neq j$, $[\mathbf{M}]_i$ is *not equivalent* to $[\mathbf{M}]_j$ (that is, not related by an equivalence or similarity relation). Each $[\mathbf{M}]_i$ is a quasi-triangular matrix with diagonal blocks or submatrices which are all equivalent matrices representing the restriction of $[\mathbf{M}]_i$ to the irreducible invariant subspaces of $[\mathbf{M}]_i$ (and, hence, of $[\mathbf{M}]$). Furthermore, the algebras $\mathscr{M}_i$ can be constructed from the maximal nilpotent subalgebra of degree $k = n_i$; that is, the subalgebra $\mathscr{N}$ such that $\mathbf{N}^k = \mathbf{0}$ for every $\mathbf{N} \in \mathscr{N}$. Thus, the problem of describing the maximal commutative (matrix) algebra $\mathscr{M}$ of order n reduces to that of describing the maximal commutative (matrix) subalgebras of order $m \leq n$ with equivalent irreducible parts (i.e., equivalent representations on the irreducible invariant subspaces), and the latter problem is equivalent to that of constructing the maximal commutative *nilpotent* subalgebras of degree $m \leq n$ (Suprunenko and Tyshkevich, 1968, pp. 43–45). The nilpotent subalgebras for $n = 3$ are given in Suprunenko and Tyshkerich (1968, p. 134), and, based on their listing, we may state that *there are exactly six distinct subalgebras of commutative second-rank tensors, and each subalgebra has one of six distinct, irreducible* 3×3 *matrix representations, relative to an appropriate fixed basis* $\{\mathbf{g}_i\}$, *which we shall call a "canonical basis."*

To summarize the resulting subalgebras concisely in terms of matrix representations and to be consistent with the usual rules of matrix multiplication, we take matrix element of the ith row and jth column of the matrix $[\mathbf{M}]$ for a tensor $\mathbf{M}$ to correspond to the (mixed) tensor component $M^i_{\cdot j}$, $i, j = 1, 2, 3$, relative to an arbitrary, basis $\{\mathbf{g}_i\}$ with metric tensor g_{ij}, reciprocal basis $\{\mathbf{g}^j\}$ and inverse metric g^{ij}, such that, with the usual summation convention,

$$\mathbf{M} = M^i_{\cdot j}\mathbf{g}_i\mathbf{g}^j. \tag{2.10}$$

Here, we list the matrix representations relative to a canonical basis (the eigenvectors of $\mathbf{L}$, $\mathbf{M}$, or $\mathbf{F}$) in terms of parameters associated with the velocity gradient $\mathbf{L} = \mathbf{M}$; and, without loss of generality we can restrict attention to the isochoric case, where

$$\operatorname{tr}\mathbf{L}(t) \equiv \operatorname{tr}\dot{\mathbf{M}}(t) = 0 \qquad \text{and} \qquad \det\mathbf{F}(t) \equiv 1 \tag{2.11}$$

since commutativety is affected neither by addition of an isotropic term, of the form $c(t)\mathbf{1}$, to $\mathbf{L}(t)$ or $\mathbf{M}(t)$, nor by multiplication of $\mathbf{F}(t)$ by an associated scalar multiple. We now consider the properties of the distinct matrix subalgebras $\mathcal{N}$, which will be listed according to their degree of nilpotency. We note that this same type of scheme, employed first by Noll (1962) and later by Huilgol (1971) to classify motions of constant stretch history, was subsequently adopted by Zahorski (1972b) in his discussion of certain unsteady analogs, "motions of superposable stretch history." However, as will be seen below, the present classification exhibits certain important differences.

1. *Nilpotent Algebras of Degree Two* ($\mathbf{L}^2 \equiv \mathbf{0}$): *Simple Shear*

Here, all three eigenvalues of $\mathbf{L}(t)$ are identically zero, and we have the distinct cases

$$[\mathbf{L}] = \begin{pmatrix} 0 & \kappa_1 & \kappa_2 \\ 0 & 0 & 0 \\ 0 & 0 & 0 \end{pmatrix} \tag{2.12}$$

and

$$[\mathbf{L}] = \begin{pmatrix} 0 & 0 & \kappa_1 \\ 0 & 0 & \kappa_2 \\ 0 & 0 & 0 \end{pmatrix}, \tag{2.13}$$

where $\kappa_i = \kappa_i(t)$, $i = 1, 2$, are "shear rates." Here as in the following we make no distinction between past time t' and present time t.

In general, we denote by (unsteady) *simple shear* a motion having velocity gradient of the form

$$\mathbf{L}(t) = \kappa(t)\mathbf{P}(t) \cdot \mathbf{N} \cdot \mathbf{P}^T(t), \tag{2.14}$$

where $\mathbf{N}$ is the *dyad*

$$\mathbf{N} = \mathbf{ij}, \qquad \mathbf{N}^2 = \mathbf{0},$$

with $\mathbf{i}$, $\mathbf{j}$ being an orthonormal pair of vectors, and $\mathbf{P}(t)$ is an orthogonal tensor. We call $\kappa(t)$ the *shear rate*, $\mathbf{P}(t) \cdot \mathbf{i}$ the *flow direction*, and $\boldsymbol{P}(t) \cdot \mathbf{j}$ the *gradient direction*; and, we call the plane containing both the *plane of shear* and the normal to this plane the *neutral axis of shear*. If both $\mathbf{P}(t)$ and $\kappa(t)$ are constants, we call the simple shear *steady*. If $\mathbf{P}(t)$ is constant, we call the simple shear *uniaxial*. Finally, we say that it is *steady uniaxial simple shear* if $\kappa(t)$ is constant and, otherwise, we say *unsteady uniaxial shear*. As we can now show, the motions (2.12) and (2.13) are both *unsteady simple shears*.

We note that both (2.12) and (2.13) can be expressed in terms of a pair of constant, nilpotent basis tensors or "shears", $\mathbf{N}_1, \mathbf{N}_2$, as

$$\mathbf{L}(t) = \kappa_1(t)\mathbf{N}_1 + \kappa_2(t)\mathbf{N}_2$$

with

$$\mathbf{N}_i \cdot \mathbf{N}_j = \mathbf{0} \qquad \text{for } i, j = 1, 2. \tag{2.15}$$

A specific realization of this representation is implied by (2.12) and (2.13), in which $\mathbf{N}_1$ and $\mathbf{N}_2$ are the basic dyads:

$$\mathbf{N}_1 = \mathbf{g}_1\mathbf{g}^2 \qquad \text{and} \qquad \mathbf{N}_2 = \mathbf{g}_1\mathbf{g}^3 \quad \text{in (2.12)}$$

or

$$\mathbf{N}_1 = \mathbf{g}_1\mathbf{g}^3 \qquad \text{and} \qquad \mathbf{N}_2 = \mathbf{g}_2\mathbf{g}^3 \quad \text{in (2.13)}. \tag{2.16}$$

The motion represented by (2.12) can aptly be termed *cylindrical shear*, since any cylinder with generator parallel to $\mathbf{g}_1$ is invariant under the motion. It represents a simple shear having constant flow direction ($\mathbf{g}_1$), and a generally variable gradient direction and shear rate. This follows from the representation (2.12), which can be written

$$\mathbf{L}(t) = \mathbf{g}_1\mathbf{g}(t),$$

where

$$\dot{\mathbf{g}}(t) = \kappa_1(t)\mathbf{g}^2 + \kappa_2(t)\mathbf{g}^3 \qquad \text{and} \qquad \mathbf{g}_1 \cdot \mathbf{g}(t) = 0. \tag{2.17}$$

Here, $\mathbf{g}(t)$ represents a gradient vector, which has generally a variable direction but always lies in the plane perpendicular to the constant basic vector

$\mathbf{g}_1$. It follows, then, that $\mathbf{L}(t)$ has a representation of the form (2.14), with

$$\left.\begin{aligned}
\mathbf{i} &= \mathbf{g}_1/g_{11}^{1/2}, \ \mathbf{P}(t) \cdot \mathbf{i} \equiv \mathbf{i}, \\
\mathbf{g}(t) &= \kappa(t)\mathbf{P}(t) \cdot \mathbf{j}, \\
\kappa(t) &= |\mathbf{g}(t)|,
\end{aligned}\right\} \qquad (2.18)$$

and with $\mathbf{P}(t)$ being the proper orthogonal tensor which represents that rotation about $\mathbf{i}$ necessary to bring the constant vector $\mathbf{j}$ into alignment with $\mathbf{g}(t)$. Figure 3 provides a schematic illustration of this motion together with those described below.

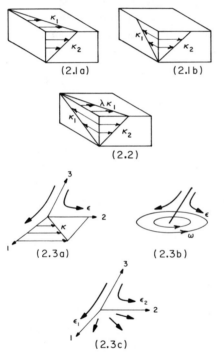

FIG. 3. Schematic representation of commutative motions: (2.1) simple shears: (a) cylindrical shear, (b) laminar shear; (2.2) double shear; (2.3) extensional flows, with (a), (b), and (c) corresponding to (2.27), (2.28), and (2.29), respectively.

The motion represented by (2.13) may be termed (unsteady) *laminar shear*, since planes or "laminae" perpendicular to $\mathbf{g}^3$ move rigidly and remain parallel to one another (Fig. 3). That is, the flow direction is generally variable, while the gradient direction ($\mathbf{g}^3$) is time-independent. This follows from (2.16), which can be written

$$\mathbf{L}(t) = \mathbf{g}(t)\mathbf{g}^3,$$

with

$$\mathbf{g}(t) = \kappa_1(t)\mathbf{g}_1 + \kappa_2(t)\mathbf{g}_2. \qquad (2.19)$$

Once again, a representation of the form (2.14) follows, with

$$\mathbf{j} = \mathbf{g}^3/(g^{33})^{1/2}, \qquad \mathbf{P}(t) \cdot \mathbf{j} \equiv \mathbf{j},$$
$$\mathbf{g}(t) = \kappa(t)\mathbf{P}(t) \cdot \mathbf{i}, \tag{2.20}$$

where now $\mathbf{P}(t)$ represents the rotation about $\mathbf{j}$ necessary to align $\mathbf{i}$ with $\mathbf{g}(t)$.

The motions represented by (2.12) and (2.13) provide the unsteady analogs of Huilgol's (1971) "superposable viscometric flows," and they reduce to unsteady uniaxial shears whenever $\kappa_1(t)$ and $\kappa_2(t)$ stand in constant ratio and, further, to steady uniaxial shears or "viscometric flows" whenever κ_1 and κ_2 are constants, as in Huilgol's work (1971).

2. Nilpotent Algebra of Degree Three ($\mathbf{L}^3 \equiv \mathbf{0}$, $\mathbf{L}^2 \not\equiv \mathbf{0}$): Double Shear

Here as above, the eigenvalues of $\mathbf{L}$ are identically zero but we have only one type of subalgebra, with

$$[\mathbf{L}] = \begin{pmatrix} 0 & \kappa_1 & \kappa_2 \\ 0 & 0 & \kappa_1 \\ 0 & 0 & 0 \end{pmatrix}, \tag{2.21}$$

where $\kappa_i = \kappa_i(t)$, and Fig. 3 depicts the associated motion. All the elements of the present algebra can be expressed in the form

$$\mathbf{L}(t) = \kappa_1(t)\mathbf{N} + \kappa_2(t)\mathbf{N}^2 \qquad (\text{with } \mathbf{N}^2 = \mathbf{N} \cdot \mathbf{N}), \tag{2.22}$$

where $\mathbf{N}$ is a constant, nilpotent tensor of degree 3, i.e.,

$$\mathbf{N}^3 = \mathbf{0}, \qquad \mathbf{N}^2 \neq \mathbf{0}. \tag{2.23}$$

The matrix representation (2.21) provides one possible realization, namely,

$$\mathbf{N} = \mathbf{g}_1\mathbf{g}^2 + \mathbf{g}_2\mathbf{g}^3. \tag{2.24}$$

It is evident that this motion represents the superposition of *three* simple shears, which includes the so-called "doubly superposed viscometric flows" discussed by Huilgol (1971) and Zahorski (1972b) and, hence, the "helical torsional" motion of Oldroyd (1965). On an orthogonal cartesian system, with basis vectors $\mathbf{i}$, $\mathbf{j}$, $\mathbf{k}$, the motion can be represented in the alternative form

$$[\mathbf{L}] = \begin{pmatrix} 0 & \lambda\kappa_1 & \kappa_2 \\ 0 & 0 & \kappa_1 \\ 0 & 0 & 0 \end{pmatrix}, \tag{2.25}$$

where λ is a geometric constant, independent of time, and $\kappa_1(t)$, $\kappa_2(t)$ are linearly related to, but generally distinct from the parameters denoted by the same symbols in (2.21). Moreover, the representation (2.22) carries over with

$$\mathbf{N} = \lambda\mathbf{ij} + \mathbf{jk} \tag{2.26}$$

which once again satisfies the nilpotency condition (2.23).

3. Nonnilpotent Algebras $(\mathbf{L}^k \not\equiv \mathbf{0}, k = 1, 2, \ldots)$: Extensional Flows

Here, we have generally nonzero eigenvalues of $\mathbf{L}$ and three distinct subalgebras corresponding to the multiplicity and complexity of these eigenvalues, say λ_i, $i = 1, 2, 3$. The deviatoric parts of the elements of these algebras are given, respectively, by

$$[\mathbf{L}] = \begin{pmatrix} \varepsilon & \kappa & 0 \\ 0 & \varepsilon & 0 \\ 0 & 0 & -2\varepsilon \end{pmatrix} \tag{2.27}$$

with $\lambda_1 = \lambda_2 = -\lambda_3/2 = \varepsilon$;

$$[\mathbf{L}] = \begin{pmatrix} \varepsilon & -\omega & 0 \\ \omega & \varepsilon & 0 \\ 0 & 0 & -2\varepsilon \end{pmatrix} \tag{2.28}$$

with $\lambda_1 = \varepsilon + i\omega$, $\lambda_2 = \bar{\lambda}_1 = \varepsilon - i\omega$, and $\lambda_3 = -2\varepsilon$; and

$$[\mathbf{L}] = \begin{pmatrix} \varepsilon_1 & 0 & 0 \\ 0 & \varepsilon_2 & 0 \\ 0 & 0 & -(\varepsilon_1 + \varepsilon_2) \end{pmatrix} \tag{2.29}$$

with $\lambda_1 = \varepsilon_1$, $\lambda_2 = \varepsilon_2$ and $\lambda_3 = -(\varepsilon_1 + \varepsilon_2)$. The parameters $\varepsilon, \kappa, \omega, \ldots$, are once again understood to be prescribed functions of time.

It can be seen that case (2.27) is a quasi-diagonal form involving a quasi-diagonal matrix having one 2×2 block. This corresponds to a two-dimensional irreducible subspace, with repeated eigenvalue $\lambda = \varepsilon$, while (2.29) correspond to "complete reducibility," distinct eigenvalues, and "semisimple" algebras (Suprunenko and Tyshkevich, 1968, pp. 10–14). The 2×2 block associated with ω in (2.28) defines an algebraic field which is isomorphic with the field of complex numbers. The matrix could be diagonalized over the field of complex numbers which, however, would introduce complex basis vectors.

Our terminology "extensional flow" is suggested by the special case where $\{\mathbf{g}_i\}$ reduces to an orthogonal basis. Indeed, all the representations (2.12)–(2.13), (2.21), and (2.27)–(2.29) are equivalent, in the sense of linear transformations, to commutative motions involving an orthogonal basis, through an *equivalence relation* or *similarity transformation* of the form

$$\mathbf{F} = \mathbf{H}^{-1} \cdot \mathbf{F}^* \cdot \mathbf{H}, \qquad \mathbf{L} = \mathbf{H}^{-1} \cdot \mathbf{L}^* \cdot \mathbf{H}, \dots, \tag{2.30}$$

where $\mathbf{H}$ is a real, constant nonsingular tensor which represents a deformation transforming the basis $\{\mathbf{g}_i\}$ to an orthogonal basis $\{\mathbf{h}_i\}$, according to

$$\mathbf{h}_i = \mathbf{H} \cdot \mathbf{g}_i, \qquad i = 1, 2, 3. \tag{2.31}$$

Here $\mathbf{F}^*, \mathbf{L}^*, \dots$, have components, $F^i_{\cdot j}, L^i_{\cdot j}, \dots$, referred to $\{\mathbf{h}_i\}$, which are identical with those of $\mathbf{F}, \mathbf{L}, \dots$, referred to $\{\mathbf{g}_i\}$. However, the "similar" motions represented respectively by $\mathbf{F}$ and $\mathbf{F}^*$ are, generally speaking, quite different in their effects on material elements, since the transformation (2.30), and the associated obliquity of the coordinate system $\{\mathbf{g}_i\}$, involves additional material strains beyond those associated with the description in terms of tensor components on an orthogonal coordinate system.

This is easily seen, for example, in the case of (2.28) where, for an underlying orthogonal system, the motion consists of a simple extension with superimposed rigid body rotation about the symmetry axis of extension at rate $\omega(t)$. Otherwise, the terms involving $\omega(t)$, as well as those involving $\varepsilon(t)$, are easily seen to give rise to shears superimposed on a simple extension.

Despite such differences, there remains a degree of kinematical similarity, in that the equivalence relation (2.30) preserves the eigenvalues of the kinematic tensors involved and, hence, our classification scheme is invariant under changes of the underlying basis. As a consequence, certain asymptotic properties of motions remain invariant. In particular, with

$$|\mathbf{A}| = [\mathrm{tr}(\mathbf{A}^T \cdot \mathbf{A})]^{1/2} \tag{2.32}$$

defining the *modulus* of a real second rank $\mathbf{A}$, then the modulus of the *Cauchy–Green tensor*,

$$\mathbf{C}_t(t') = \mathbf{F}_t^T(t') \cdot \mathbf{F}_t(t'),$$

which is related to the stretching of material elements, is given by

$$|\mathbf{C}_t(t')| = |\mathbf{F}_t(t')|^2. \tag{2.33}$$

For commutative motions having finite degree of nilpotency, i.e., the simple or double shears of Sections II,A,1 and 2, this quantity will be bounded algebraically for $s = t - t' \to \infty$, i.e., bounded by a finite polynomial in s, whenever $|\mathbf{L}(t')|$ is also bounded algebraically. This follows directly from the representations (2.7) and (2.9) and the fact that the tensor $e^{\mathbf{M}}$ is algebraic in the kinematic parameters $\kappa_1(t'), \kappa_2(t')$ defining $\mathbf{L}(t')$. On the other hand, no such algebraic bound is generally possible for the nonnilpotent case, i.e., the extensional flows of Section 3.

If then, the asymptotic stretching of material elements in the past is employed as a basis for classifying motions (or flows) as either "weak" or "strong," in a scheme similar to that proposed by Tanner (1975) and by Tanner and Huilgol (1975), it therefore follows that the *extensional motions represented by* (2.27)–(2.29) *must include all those strong flows that are also commutative motions.* (It should be noted, however, that Tanner, in the article referred to, incorporates a material relaxation time in his scheme.)

As with the nilpotent algebras of degree *three,* all the algebras of (2.27)–(2.29) are generated by a basic tensor **N** through an expression of the form

$$\mathbf{L}(t) = \alpha(t)\mathbf{1} + \beta(t)\mathbf{N} + \gamma(t)\mathbf{N}^2, \tag{2.34}$$

where **N** is a constant tensor such that now

$$\mathbf{N}^k \neq 0 \qquad \text{for} \quad k = 1, 2, \ldots, \tag{2.35}$$

and we recall that only the deviatoric parts of **L** are displayed in (2.27)–(2.29). Specifically realizations, providing one possible (but nonunique) correspondance with (2.27)–(2.29) are given in Table I.

TABLE I

BASIC TENSORS FOR EXTENSIONAL FLOWS

Representative matrix [**L**]	Possible form of **N**	Corresponding matrix [**N**]
Eq. (2.27)	$\mathbf{g}_3\mathbf{g}^3 + \mathbf{g}_1\mathbf{g}^2$	$\begin{pmatrix} 0 & 1 & 0 \\ 0 & 0 & 0 \\ 0 & 0 & 1 \end{pmatrix}$
Eq. (2.28)	$\mathbf{g}_1\mathbf{g}^2 - \mathbf{g}_2\mathbf{g}^1$	$\begin{pmatrix} 0 & 1 & 0 \\ -1 & 0 & 0 \\ 0 & 0 & 0 \end{pmatrix}$
Eq. (2.29)	$\mathbf{g}_1\mathbf{g}^1 - \mathbf{g}_2\mathbf{g}^2$	$\begin{pmatrix} 1 & 0 & 0 \\ 0 & -1 & 0 \\ 0 & 0 & 0 \end{pmatrix}$

This completes the list of generating subalgebras for commutative motions.

B. REDUCTION OF THE STRESS FUNCTIONAL FOR INCOMPRESSIBLE SIMPLE FLUIDS

In the case of an (isothermal) simple fluid we can represent the present stress **T**(t) formally as a tensor-valued functional on the history of the velocity gradient,

$$\mathbf{T}(t) = \mathscr{F}\{\mathbf{L}(\cdot)\} \equiv \mathop{\mathscr{F}}_{-\infty}^{t}\{\mathbf{L}(t')\}, \tag{2.36}$$

where notation for dependence on present mass density is suppressed. Also, as done in the first equality here, we shall suppress in the following the explicit notation for the dependence of functionals on past history of functions, writing $(\cdot)$ for the argument (t'), with the support $-\infty < t' \le t$ understood. By the usual assumption of material frame indifference, the functional $\mathscr{F}$ must satisfy

$$\mathscr{F}\{\mathbf{Q}(\cdot) \cdot \mathbf{L}(\cdot) \cdot \mathbf{Q}^T(\cdot)\} = \mathbf{Q}(t) \cdot \mathscr{F}\{\mathbf{L}(\cdot) - \mathbf{W}(\cdot)\} \cdot \mathbf{Q}^T(t) \ldots, \quad (2.37)$$

where

$$\mathbf{W}(t) = -\mathbf{W}^T(t) = \mathbf{Q}^T(t) \cdot \dot{\mathbf{Q}}(t) \ldots \quad (2.38)$$

with $\mathbf{Q}(t)$ denoting an arbitrary time-dependent orthogonal tensor.

Given the various matrix representations of the velocity gradient relative to the canonical basis $\{\mathbf{g}_i\}$, for the various subalgebras of commutative motions, it is evident that the six independent stress components $T_{ij} = T_{ji}$ relative to this same basis represent a set of material functionals which are functionals defined on the history of the relevant pair of kinematical parameters discussed above, (κ_1, κ_2), (κ, ε), *etc.* These functionals are also functions of the (constant) metric tensor g_{ij} for the canonical basis. The following discussion will be restricted to the case of incompressible fluids, or else isochoric motions, for which there can be at most *five* independent functionals representing the stresses T_{ij}. If one so wishes, the more general case of compressible fluids undergoing nonisochoric commutative motions can be treated by the obvious incorporation of an additional scalar parameter.

Based on the various parametric representations for the kinematics of commutative motion discussed above, it is now possible to develop various representations for the stress in commutative motions and to consider, in the usual way, their symmetries under constant orthogonal transformations as represented by the time-independent form of (2.37)–(2.38) with $\dot{\mathbf{Q}} \equiv 0$ and, hence, $\mathbf{W} \equiv \mathbf{0}$.

As with any other special class of isochoric motions, we demand to know whether degeneracies or symmetries of the motions at hand allow for a reduction to a number of material functionals less than the maximal number, five. It will now be shown that for commutative motions which are nilpotent of degree 2, that is, simple shear, the answer is in the affirmative and for this case there are only three independent functionals. However, for the remaining types of commutative motions, it appears that reduction to less than five functionals is possible only for those restricted subclasses of motions in which there exists underlying symmetries in the velocity gradient, as reflected by corresponding symmetries of the canonical basis.

We consider now the various subclasses of commutative motions defined by the various subalgebras discussed above in Section II,A.

1. *Simple Shear*

We consider first the cylindrical shear of (2.12). Since the basis vectors $\mathbf{g}^2$, $\mathbf{g}^3$ are both orthogonal to $\mathbf{g}_1$ they can be replaced by another pair of mutually orthogonal vectors lying in their plane, and Eqs. (2.12) and (2.16) carry over. Hence, no loss in generality is incurred by supposing the canonical basis to be orthonormal, and this will greatly simplify certain of the arguments to follow.

On the basis of the representation (2.15), it is evident that (2.37) can be reduced to a form already postulated by Zahorski (1973):

$$\mathbf{T}(t) = \mathbf{f}(\kappa_1(\cdot), \kappa_2(\cdot); \mathbf{N}_1, \mathbf{N}_2) - p\mathbf{1}, \tag{2.39}$$

where $\mathbf{f}$ is a functional on the history of the scalar parameters $\kappa_1(t')$, $\kappa_2(t')$ and, also, a function of the constant tensors $\mathbf{N}_1$, $\mathbf{N}_2$. The representation (2.39) can be further reduced by expressing it, in a well-known way, as a linear combination of the elements of the integrity basis for symmetric tensors (Spencer and Rivlin, 1959) generated by $\mathbf{N}_1$ and $\mathbf{N}_2$, with scalar coefficients which are functionals on κ_1 and κ_2. For the case of an orthonormal basis, where the $\{\mathbf{g}_i\}$ are identical with the $\{\mathbf{g}^i\}$, it becomes evident from (2.16) that the integrity basis consists of the five linearly independent elements

$$\mathbf{N}_1 + \mathbf{N}_1^T, \quad \mathbf{N}_2 + \mathbf{N}_2^T, \quad \mathbf{N}_1^T \cdot \mathbf{N}_1, \quad \mathbf{N}_2^T \cdot \mathbf{N}_2, \quad \mathbf{N}_1^T \cdot \mathbf{N}_2 + \mathbf{N}_2^T \cdot \mathbf{N}_1 \tag{2.40}$$

together with the unit tensor $\mathbf{1}$. The expression of $\mathbf{T}$ as a linear combination of these involves five scalar functionals on κ_1, κ_2. However, there is an immediately evident symmetry under the orthogonal transformation defined by $\mathbf{g}_3 \to \mathbf{g}_3$ and $\mathbf{g}_1 \rightleftarrows \mathbf{g}_2$, for which $\mathbf{N}_1 \rightleftarrows \mathbf{N}_2$, so that we may reduce the representation further to

$$\begin{aligned}
\mathbf{T} + p\mathbf{1} = {}& \mathscr{F}_1(\mathbf{N}_1 + \mathbf{N}_1^T) + \mathscr{F}_2(\mathbf{N}_2 + \mathbf{N}_2^T) \\
& + \mathscr{S}_1\mathbf{N}_1^T \cdot \mathbf{N}_1 + \mathscr{S}_2\mathbf{N}_2^T \cdot \mathbf{N}_2 \\
& + \mathscr{C}(\mathbf{N}_1^T \cdot \mathbf{N}_2 + \mathbf{N}_2^T \cdot \mathbf{N}_1),
\end{aligned} \tag{2.41}$$

where p is an arbitrary pressure and where, in terms of components on the orthonormal basis $\{\mathbf{g}_i\}$,

$$\mathscr{F}_1 \equiv T_{12} = \mathscr{F}\{\kappa_1(\cdot), \kappa_2(\cdot)\}, \qquad \mathscr{F}_2 = T_{13} = \mathscr{F}\{\kappa_2(\cdot), \kappa_1(\cdot)\},$$

$$\mathscr{S}_1 \equiv T_{22} - T_{11} = \mathscr{S}\{\kappa_1(\cdot), \kappa_2(\cdot)\}, \qquad \mathscr{S}_2 \equiv T_{33} - T_{11} = \mathscr{S}\{\kappa_2(\cdot), \kappa_1(\cdot)\}, \tag{2.42}$$

and

$$\mathscr{C} \equiv T_{23} = \mathscr{C}\{\kappa_1(\cdot), \kappa_2(\cdot)\} \equiv \mathscr{C}\{\kappa_2(\cdot), \kappa_1(\cdot)\}.$$

These relations involve only *three* distinct scalar functionals, a "primary" shear stress functional $\mathscr{F}$, a "normal" stress $\mathscr{S}$, and a "secondary" shear (or "coupling") stress $\mathscr{C}$. By considering various reflections in the coordinate planes one can further deduce the parity of $\mathscr{F}$, $\mathscr{S}$, $\mathscr{C}$, in their arguments, which can be summarized as: $\mathscr{F}\{-,+\}$, $\mathscr{S}\{+,+\}$, and $\mathscr{C}\{-,-\}$, where the negative sign denotes odd parity and the positive, even parity.

There is yet another important symmetry deriving from the inherent symmetry about the $\mathbf{g}_1$ axis or flow direction. In particular, we consider rotations through an arbitrary angle θ about $\mathbf{g}_1$, represented by constant orthogonal tensor $\mathbf{Q}$ having matrix

$$[\mathbf{Q}] = \begin{pmatrix} 1 & 0 & 0 \\ 0 & C & -S \\ 0 & S & C \end{pmatrix}, \tag{2.43}$$

where $C = \cos\theta$, $S = \sin\theta$. Then (2.37)–(2.38), and the ordinary rules of tensor transformation applied to (2.41), yield relations of the form

$$\mathscr{F}\{\kappa_1(\cdot), \kappa_2(\cdot)\} = \ell\{\kappa^2(\cdot)|\kappa_1(\cdot)\},$$
$$\mathscr{S}\{\kappa_1(\cdot), \kappa_2(\cdot)\} = \mathscr{s}\{\kappa^2(\cdot)|\kappa_1^2(\cdot)\} + \mathscr{i}\{\kappa^2(\cdot)\}, \tag{2.44}$$
$$\mathscr{C}\{\kappa_1(\cdot), \kappa_2(\cdot)\} = \mathscr{s}\{\kappa^2(\cdot)|\kappa_1(\cdot)\kappa_2(\cdot)\},$$

where

$$\kappa^2(t) \equiv [\kappa(t)]^2 = \kappa_1^2(t) + \kappa_2^2(t), \tag{2.45}$$

and where ℓ, $\mathscr{s}$, $\mathscr{i}$ are functionals, with arguments indicated and with vertical bars signifying that ℓ and $\mathscr{s}$ are *linear* in their arguments; i.e., if c_1 and c_2 are constants and $a(t')$, $b(t')$ are functions in the domain of ℓ, then

$$\ell\{a(\cdot)|c_1 b_1(\cdot) + c_1 b_2(\cdot)\} = c_1 \ell\{a(\cdot)|b_1(\cdot)\} + c_2 \ell\{a(\cdot)|b_2(\cdot)\} \tag{2.46}$$

with a similar equation for $\mathscr{s}$. In accordance with (2.15), we now let

$$\mathbf{K}(t) = \kappa_1(t)\mathbf{N}_1 + \kappa_2(t)\mathbf{N}_2 \tag{2.47}$$

denote the velocity gradient $\mathbf{L}(t)$ in a canonical frame (to distinguish it from a general frame). Then,

$$\text{tr}\{\mathbf{K}^T(t) \cdot \mathbf{K}(t)\} = \kappa^2(t), \tag{2.48}$$

where $\kappa(t)$ is defined in (2.45), and we see that (2.41), (2.42), and (2.44) are equivalent to a tensor equation of the form

$$\mathbf{T} + p\mathbf{1} = \ell\{\kappa^2(\cdot)|\mathbf{K}(\cdot) + \mathbf{K}^T(\cdot)\} + \mathscr{n}_1\{\kappa^2(\cdot)|\mathbf{K}(\cdot) \cdot \mathbf{K}^T(\cdot)\}$$
$$+ \mathscr{n}_2\{\kappa^2(\cdot)|\mathbf{K}^T(\cdot) \cdot \mathbf{K}(\cdot)\}, \tag{2.49}$$

where ℓ and $\mathscr{n}_2$ are identical, respectively, with the functionals ℓ and $\mathscr{s}$ of (2.44) extended to tensor arguments via relations of the form (2.46), while the functional $\mathscr{i}$ of (2.44) is given in terms of $\mathscr{n}_1$ by

$$\mathscr{i}\{\kappa^2(\cdot)\} = -\mathscr{n}_1\{\kappa^2(\cdot)|\kappa^2(\cdot)\}. \tag{2.50}$$

In (2.49) all the functionals are linear in their second arguments. However, by inspection of (2.50) it becomes evident that the functional n_1 introduced formally in (2.49) is not completely determined by the stress functionals (2.44) and, hence, not determined by a knowledge of material response in cylindrical shear. The representation (2.49) is nevertheless useful since it is a properly invariant form which remains valid whatever the choice of the basis tensors $\mathbf{N}_1$, $\mathbf{N}_2$ and whatever the underlying set of basis vectors $\{\mathbf{g}_i\}$. Furthermore, (2.49) immediately yields up a set of scalar stress functionals for the other possible type of commutative simple shear, namely, the laminar shear of (2.13). Indeed, by an argument similar to that just employed for cylindrical shear, one arrives at a set of relations like (2.41), obtained by replacing $\mathbf{N}_i$ everywhere by $\mathbf{N}_i^T$ for $i = 1, 2$; then, a set like (2.42), obtained by the cyclic permutation of indices $3 \to 2, 2 \to 1, 1 \to 3$ on the stress components T_{ij}; and, finally, a relation of the form (2.49). Hence, we have the following:

THEOREM.   *For a simple fluid, the stress in a commutative simple shear can be reduced to the form (2.49).*

We cannot, however, conclude that the functionals for cylindrical shear in (2.49) are generally the same as their counterparts for laminar shear, since the present method of derivation involves functionals (2.41) which depend on the symmetric integrity basis for $\mathbf{N}_1$ and $\mathbf{N}_2$. Therefore, in attempting to treat both cylindrical shear and laminar shear simultaneously, one would be forced to adjoin to the set (2.40) the three *new* elements which are generated by substituting $\mathbf{N}_i^T$ for $\mathbf{N}_i$ in (2.40), for $i = 1, 2$.

Of course, in certain instances cylindrical and laminar shear may reduce to the same motion, an important example being steady simple shear or viscometric flow, where $\mathbf{K}$ and κ are both constant. Indeed, the notation adopted in (2.49) is intended to suggest here the material functions appropriate to this very case, namely, the viscometric shear stress, primary normal stress and secondary normal stress functions:

$$\tau(\kappa) = \ell\{\kappa^2 | \kappa\}, \qquad \sigma_1(\kappa) = \jmath_1\{\kappa^2 | \kappa^2\}, \qquad \sigma_2 = \jmath_2\{\kappa^2 | \kappa^2\} \qquad (2.51)$$

with $N_1 = \sigma_1 - \sigma_2$ and $N_2 = \sigma_2$ being the usual functions.

The material functions for yet another important motion can be derived from functionals of the type (2.49) for laminar shear. In particular, the steady circular shear, which is usually associated with the Maxwell–Chartoff or "orthogonal" rheometer (Goddard, 1978b), corresponds to the periodic laminar shear defined by taking κ, the antisymmetric tensor

$$\mathbf{\Omega} = \dot{\mathbf{P}}(t) \cdot \mathbf{P}^{-1}(t)$$

and, hence, its *vector*

$$\text{Vec}\,\mathbf{\Omega} = \omega\mathbf{j}, \qquad (2.52)$$

all to be independent of time in (2.20). By a change of frame, corresponding to the change of orthonormal basis vectors, **i**, **j**, **k**,

$$\mathbf{i}^* = \mathbf{P}(t) \cdot \mathbf{i} = e^{\Omega t} \cdot \mathbf{i} = \cos \omega t \, \mathbf{i} + \sin \omega t \, \mathbf{k},$$

$$\mathbf{k}^* = \mathbf{P}(t) \cdot \mathbf{k} = e^{\Omega t} \cdot \mathbf{k} = -\sin \omega t \, \mathbf{i} + \cos \omega t \, \mathbf{k},$$

and

$$\mathbf{j}^* = \mathbf{j}, \tag{2.53}$$

the above motion can be rendered spatially steady. Then, with

$$\mathbf{g}_1^* = \mathbf{i}^*, \qquad \mathbf{g}_2^* = \mathbf{k}^*, \qquad \mathbf{g}_3^* = \mathbf{j}^*, \tag{2.54}$$

which corresponds to the usual orthogonal basis, the stress components are found to be given as *material functions*:

$$
\begin{aligned}
T_{13}^* &= \kappa \, \ell\{\kappa^2 | \cos \omega s\}, \\
T_{23}^* &= -\kappa \, \ell\{\kappa^2 | \sin \omega s\}, \\
T_{12}^* &= -\kappa^2 n_1\{\kappa^2 | \cos \omega s \sin \omega s\}, \\
T_{11}^* - T_{33}^* &= \kappa^2 [n_1\{\kappa^2 | \cos^2 \omega s\} - n_2\{\kappa^2 | 1\}], \\
T_{22}^* - T_{33}^* &= \kappa^2 [n_1\{\kappa^2 | \sin^2 \omega s\} - n_2\{\kappa^2 | 1\}],
\end{aligned}
\tag{2.55}
$$

all depending on ω, κ. Here, the functionals ℓ, n_1, n_2 are those of (2.49) for laminar shear, and we employ notation of the form

$$\ell\{\kappa^2 | \cos \omega s\} = \ell\{\kappa^2 | \cos \omega [t - (\cdot)]\} = \mathop{\ell}_{t' = -\infty}^{t} \{\kappa^2 | \cos \omega(t - t')\} \quad (2.56)$$

for ℓ and n_1, with $t' = t - s$ and $s \geq 0$. The relations (2.55), which can be employed to deduce certain of the properties of the material functions for circular shear [derived by a different method by Goddard (1978b)], also follow from Caswell's (1967; also private communication, 1971) treatment of the unsteady laminar shear near a rigid, no-slip surface.

2. Double Shear

In view of (2.22), we may write for this motion

$$\mathbf{T}(t) = \mathbf{f}\{\kappa_1(\cdot), \kappa_2(\cdot); \mathbf{N}\} - p\mathbf{1}, \tag{2.57}$$

where $\mathbf{f}$ is a functional on $\kappa_1(t')$, $\kappa_2(t')$ and a function of $\mathbf{N}$, the constant tensor of (2.23) and (2.24).

As with (2.39), we may reduce (2.57) to a linear combination of the elements of a symmetric integrity basis for the tensor $\mathbf{N}$. By means of the matrix representation (2.25), one has the further representation

$$[\mathbf{N}^2] = \begin{pmatrix} 0 & 0 & \lambda \\ 0 & 0 & 0 \\ 0 & 0 & 0 \end{pmatrix}, \qquad [\mathbf{N}^T \cdot \mathbf{N}] = \begin{pmatrix} 0 & 0 & 0 \\ 0 & 1 & 0 \\ 0 & 0 & \lambda^2 \end{pmatrix},$$

$$[\mathbf{N} \cdot \mathbf{N}^T] = \begin{pmatrix} 1 & 0 & 0 \\ 0 & \lambda^2 & 0 \\ 0 & 0 & 0 \end{pmatrix}, \qquad [\mathbf{N} \cdot \mathbf{N}^T \cdot \mathbf{N}] = \begin{pmatrix} 0 & 1 & 0 \\ 0 & 0 & \lambda^3 \\ 0 & 0 & 0 \end{pmatrix},$$

$$(2.58)$$

relative to an orthonormal basis. Therefore, by inspection, it is easy to see that for $\lambda \neq 1$ the set of five symmetric tensors

$$\mathbf{N} + \mathbf{N}^T, \quad \mathbf{N}^2 + \mathbf{N}^{2T}, \quad \mathbf{N}^T \cdot \mathbf{N}, \quad \mathbf{N} \cdot \mathbf{N}^T, \quad \mathbf{N} \cdot \mathbf{N}^T \cdot \mathbf{N} + \mathbf{N}^T \cdot \mathbf{N} \cdot \mathbf{N}^T, \quad (2.59)$$

together with the unit tensor **1**, form a complete basis for the (six-dimensional linear) space of symmetric tensors. Hence, for the case of nonorthogonal eigenvectors where, $\lambda \neq 1$, it is evident that the expression of (2.57) as a linear combination of the set (2.59) will involve, apart from an arbitrary isotropic term, exactly five functionals with arguments $\kappa_1(\cdot)$, $\kappa_2(\cdot)$.

Although no rigorous proof will be attempted here, it appears that there are no inherent symmetries for $\lambda \neq 1$ which would allow for reduction of the above five functionals to a smaller set. On the other hand, for $\lambda = 1$, corresponding to orthogonal eigenvectors, it is evident from (2.58) that the set (2.59) degenerates to only *four* linearly independent members which can be chosen as the first four in (2.59). Furthermore, by a consideration of the matrix representations of higher products, $\mathbf{N}^2 \cdot \mathbf{N}^T \cdot \mathbf{N}$, $\mathbf{N}^2 \cdot \mathbf{N}^T \cdot \mathbf{N} \cdot \mathbf{N}^T, \ldots$, it can be shown, inductively, that the first four members of (2.59) provide a complete symmetric integrity basis for $\mathbf{N}$, and that the associated material functionals can be made to correspond to the distinct stresses $T_{12} = T_{23}$, T_{13}, $T_{11} - T_{22}$, $T_{11} - T_{33}$. Once again, no further reductions on the basis of symmetry are readily evident.

3. *Extensional Flows*

Because of (2.22), the representation (2.57) is once again applicable. However, except for special cases in which two, or all, members of the canonical basis $\{\mathbf{g}_i\}$ are mutually orthogonal, there are no readily apparent symmetries which would allow for reduction to fewer than five material functionals. On the other hand, in the case of a completely orthogonal basis, the motions (2.27) and (2.29) can be reduced either to "orthogonal" extensions of the type already considered in general by Coleman and Noll (1962) or else to special cases of the "sheared extensions" treated by Coleman (1968). In that regard, it appears that, of all the classes of restricted kinematics considered to date, only the latter do not qualify as commutative motions, except in special cases.

As a final comment here, it will be noted that there is an interesting but open question as to the "automorphisms" within the class of commutative motions represented by changes of frame. For example, the circular shear (or Maxwell-orthogonal flow) of (2.52)–(2.55) is a commutative motion that is unsteady in one frame but represents a flow of constant stretch history, hence, a commutative motion in a second frame.

III. Kinematics and Dynamics for Generals Flows

Of all the areas of research on polymer fluids, the area of fluid dynamics remains perhaps the most underdeveloped from a theoretical point of view. This state of affairs is largely due to the lack of generally accepted constitutive equations which are simultaneously reliable for application to general flows and tractable by existing analytical or numerical methods, when coupled with the equations of momentum and continuity. As a consequence, most existing treatments of boundary-value problems are restricted to (1) *elementary flows*, such as simple shear or simple extension, where the kinematics and rheology can be delimited or specified approximately, in advance of an actual solution to the dynamical equations, or (2) *slightly perturbed flows*, arising from weak disturbances on a more-or-less elementary flow.

An excellent survey of elementary flows and associated rheological models, many associated with rheometrical tests, have been given in several review articles and textbooks (Pipkin and Tanner, 1972; Lodge, 1974; Tanner, 1975a; Walters, 1975; Bird, 1976; Bird *et al.*, 1977a; Schowalter, 1978). There are many well-documented applications in polymer processing for such analyses, including, flow in dies, surface and wire coating devices, liquid filament or "fiber" spinning, etc., many of which involve important heat transfer effects (Bird *et al.*, 1977a). In many cases, an elementary flow structure arises as an *asymptotic* geometric limit, in problems with disparate length scales, the most important examples being "lubrication flows," which were long ago recognized and analyzed kinematically by Pearson (1967), as well as the flow in thin filaments and films having free surfaces. As such, the mechanics can often be treated systematically be a formal perturbation analysis based on a geometric parameter (e.g., an aspect ratio) of the type suggested by Pearson (1967) and by Pearson and Matovich (1969). Such a method has also been used recently, in the form of a slender-body theory, to analyze the near-field parallel flow around elongated rigid inclusions in viscoelastic fluids (Goddard, 1976).

In this brief review, attention is focused mainly on the mechanics of perturbed flows, wherein an assumed elementary, *base* motion of the material is altered by weak boundary disturbances, inertially driven instabilities, or else by small rheological incompatibilities. This type of flow often allows for rather general rheological approximations.

The question of material response to small perturbations on a restricted class of kinematics, for which the rheology and mechanics are more-or-less completely defined, is not new. Indeed, the progression in *amplitude* of past deformations from a state of rest through linear viscoelasticity and, thence, to nonlinear effects is already embodied in the expansion (1.2). There is a similar well-known progression through "frequency" or rapidity of deformation via well-known "slow-flow" or retarded motion expansions (Truesdell, 1966; Pipkin, 1972 (Chap. 8), from the Newtonian fluid model to the second-order and higher order Rivlin–Ericksen fluids.

Owing to the explicit nature of the latter expansions, which involve well defined sets of material constants, they have historically received the most attention by those fluid mechanicists whose main concern is with concrete boundary-value calculations. Apart from the well-understood singularities in their "high-frequency" response, such expansions, and the resultant calculations, are severely limited in the description of many flows of interest for applied rheology, basically because of their inadequate representation of nonlinear effects in real polymer fluids.

In the absence of more comprehensive and tractable rheological models, which as Pipkin suggests (1972, Chap. 8) may be contradictory qualities, and without the associated and mathematical methods of application to flow problems, there is considerable motivation for studying the rheology and mechanics of flows or motions which consist of "small" perturbations on more-or-less well-defined and elementary base flows.

Section III,B, below, provides, as a generalization of Pipkin and Owen's (1967) analysis of viscometric flows, a treatment of perturbations on "rotationally steady flows" or, as usually called, "flows with constant stretch history." In Section III,A, immediately following, a review is given of the status of a number of outstanding problems, which basically involve perturbation analyses and which have come to represent benchmarks of progress in the mechanics of viscoelastic fluids.

A. Mechanics of Weakly Perturbed Flows

Viscometric flows represent an important class of flows for which the study of mechanical perturbations may yield useful rheological information. We consider here a few salient examples.

1. *The Hole Pressure*

A now classic problem is that involving the hole-pressure "effect," wherein the static pressure at the bottom of a deep pressure tap, in the form of a finite hole or cross-stream slot in a rigid wall bounding a flow, differs from the normal wall stress that would exist in the absence of the hole. The original analysis of Tanner and Pipkin (1969), which has been generalized in a more approximate form by others (cf. Pipkin and Tanner, 1972, 1977; Higashitani and Lodge, 1975; Bird *et al.*, 1977a) indicates that the concave deflection of the streamlines over the hole, combined with the presence of an effective streamline "tension" in the form of the primary normal stress N_1, gives rise to a pressure defect p_H at the bottom of a hole proportional to N_1. What was once regarded as a pressure-tap "error" has more recently been viewed as an "effect" allowing for the measurement of N_1.

The only exact theory for p_H is that based on a second-order (or "second-grade") fluid model for which $p_H = -N_1{}^{(0)}/4$ in the case of a slot, with $N_1^{(0)}$ denoting the primary normal stress in a hypothetical unperturbed wall shear flow. A recent photographic tracer study of Hou *et al.* (1977) tends to confirm the basic mechanical picture. In striving for a more exact mathematical treatment, it might be interesting and useful to carry the analysis beyond second-order fluid theory, by means of the theory of perturbations on viscometric flows having finite shear rates. This would no doubt introduce new material parameters into the interpretation of the measurement.

2. *Secondary Flow in Straight Ducts*

It is well known that rectilinear motion of simple fluids in finite ducts of arbitrary cross section is not generally possible. This can be attributed directly to the presence of a general secondary normal stress N_2 which in a hypothetical rectilinear and, hence, viscometric flow, would generate unbalanced stresses, except for the special case of flow in circular ducts or between parallel planes. Thus, a "secondary" or "circulatory" motion is necessary.

Within the framework of retarded motion expansions, this problem has received several treatments (see Truesdell, 1966; Rivlin and Sawyers, 1971; Rivlin, 1976). Based on the analysis given by Truesdell (1966, Lectures 10 and 25, in work attributed to Langlois and Rivlin, and Noll) the secondary flow appears at the "fourth-order", i.e., for a Rivlin–Ericksen fluid of "grade" four, and the amplitude is proportional to a parameter δ, in Truesdell's notation (Eq. 25.28), which a bit of algebra shows to be given by

$$\delta = \left[\eta \frac{d}{d\kappa^2} \left(\frac{N_2}{\eta\kappa^2} \right) \right]_{\kappa=0},$$

where $\eta(\kappa)$ and $N_2(\kappa)$ are, respectively, the viscosity and normal stress functions for the fourth-order fluid. This result is in accordance with the classical Ericksen–Stone result that rectilinear flow is generally possible only if $N_2/\eta\kappa^2$ is a constant for the fluid.

There is much experimental evidence to indicate that the ratio $N_2/\kappa\eta$ should be small for many polymer fluids, suggesting a weak secondary flow. References to past works and the recent work of Walters and co-workers are to be found in Townsend *et al.* (1976). In this article, the authors report on experiment observations and numerical computations, based on an approximate rheological and dynamical model, for the secondary flow in a duct of square cross section. Their computations and experiments do show a weak secondary flow, consisting of a pair of stationary counterrotating eddies in opposite corners with one diagonal of the cross section serving as axis of reflectional symmetry. Once again, this type of problem, involving a "weak" rheological incompatibility of an assumed flow should be amenable to an analysis of perturbations on a viscometric flow.

Also, in closing, it should be noted that such incompatibilities may have important implications in other contexts, such as the flow in slender slots discussed by Kearsley (1970) and "exterior" flows along elongated slender bodies (Goddard, 1976).

3. *Rod Climbing*

Joseph and co-workers have in recent times performed a comprehensive analysis and experiments on the classical "Weissenberg climbing effect," which concerns the shape of the free-surface of a fluid confined in the annular gap between vertical coaxial circular cylinders, with the inner cylinder rotating (see Beavers and Joseph, 1975; Joseph and Beavers, 1976, 1977a,b). Their analyses, based on a perturbation about the state of rest, have in some instances been carried through to fourth-order terms in the Rivlin–Ericksen expansion, with account taken of inertia and static surface tension, as well as oscillatory rotation of the inner cylinder. Within the framework of their perturbation technique, which basically involves rotational speed as the parameter in a simultaneous set of perturbations on rheology, surface shape and inertia, their work provides a classic and definitive study. However, viewed as a potential viscometric measurement, this type of system obviously involves many complicating effects. Furthermore, because of the potentially short range of applicability of the retarded motion expansions for many polymer fluids, it can be anticipated that the analysis will be limited practically to small shear rates and, hence, slow rotations. Indeed, there is experimental evidence to suggest that the nonlinear variations of viscosity $\eta(\kappa)$ in wide-gap instruments may have a strong effect on the resultant radial distribution of shear rates, normal stresses and, hence, free-surface shapes (Chan, 1972). Also, there is experimental evidence showing that the free surface develops peculiar unsteady instabilities at modest shear rates, which may already be signaled by the apparent secondary flows that emerge from the analyses of Joseph *et al.* (Joseph and Beavers, 1977b). In any case, there is reason to believe that one should be cautious in treating this free-surface flow as a weakly perturbed viscometric flow, whenever the shear rates become large i.e., outside the "retarded-motion" regime. In the absence of a free surface, there still remains the possibility of inertially driven perturbations on a viscometric flow, which we now consider.

4. *Taylor–Couette Stability*

Over a decade ago, a number of investigators took up the classical problem of G. I. Taylor (1923), involving the centrifugal instability of a fluid confined to the annular gap between coaxial circular cylinders, with the intent of extending this analysis to viscoelastic fluids. Here one is dealing with a well-defined case of a perturbed viscometric flow, from which one might hope to glean certain rheological information. A number of theoretical analyses of infinitesimal stability were given, the most complete being those of Rivlin and co-workers (Lockett and Rivlin, 1968; Smith and Rivlin, 1972) and a heretofore unpublished work of Miller (1967), both of which are valid for quite general simple fluids. Denn and co-workers (Ginn and Denn, 1969; Denn and Roisman, 1969; cf. Petrie and Denn, 1976) have subsequently reviewed the work on this problem, including a simplified theory and some experiments which suggest the importance of the secondary normal stress N_2 as a principal determinant of infinitesimal stability. However, as recognized by Denn *et al.*, the general stability analyses of Lockett and Rivlin and of Miller indicate a potentially large number of "new" or auxiliary rheological parameters, not strictly derivable from viscometric flow data, which may be relevant to the stability problem.

Here, attention will be confined to the *neutral, stationary stability* as analyzed by Lockett and Rivlin and Miller for the *narrow-gap limit*. In the infinitesimal or linear stability analysis it suffices to consider the linearized rheological response to small perturbations on the base flow. The analysis of Lockett and Rivlin (1968) proceeds from the assumption of a general Rivlin–Ericksen model, whereas the original analysis of Miller (1967) is based on the general rheology of a simple fluid subject to small perturbations on a viscometric flow and, as such, can in principle be related to the perturbation theory of Pipkin and Owen (1967). In view of the comprehensive nature of the above theoretical analyses and a number of important differences which emerge in the interpretation and the apparent number of parameters

involved, Miller has subsequently reworked his analysis,* starting from the
same basic rheological model as that of Lockett and Rivlin. Given the
potential interest and the lack of previous publication, the resulting analysis,
which includes a commentary on many other works involving specialized
fluid models, is reproduced as a matter of record in the Appendix of this
review. The interested reader is referred to the Appendix for details, as the
present summary will only cover the highlights citing, where appropriate,
equations listed in the Appendix.

   We recall, on the assumption of a spatially periodic disturbance of the
Taylor-vortex form, having dimensionless axial wave number k and defined
by (A.63) and (A.66), the relevant "disturbance" equations (A.57)–(A.58) can
be reduced in the classical way to two ordinary differential equations (A.68).
These involve several dynamic, kinematic and geometric parameters defined
in (A.61)–(A.63), including two *Reynolds numbers* R and R' which combine
to yield a Taylor number T in (A.63), as well as a set of twelve rheological
parameters n_i $(i = 1, 2, 3)$ and β_i $(i = 1, 2, \ldots, 9)$. As explained in the Appendix, *five* of these parameters n_i $(i = 1, 2, 3)$, β_i $(i = 8, 9)$ are derivable from the
viscometric functions for the fluid, whereas *seven* are entirely *new* and related
to the response of a general (simple) fluid to perturbations on a viscometric
flow. However, the linearized perturbation characteristic of the assumed
Taylor vortex structure appear to be sufficiently special in nature, so that a
knowledge of the fluid response in (*some possibly restricted class of*) "*steady
flows*" ("*motion with constant stretch history*") *suffices completely to determine all the pertinent rheological parameters and hence Taylor–Couette
stability.* As indicated in the Appendix, this may have interesting rheological
consequences. In any event, the above findings imply that certain of the
nine rheological parameters derived by Lockett and Rivlin (account taken of
one omitted term equivalent to β_7) as coefficients in a Rivlin–Ericksen type
expansion, can be related to viscometric behavior. Moreover, a previous
sensitivity analysis of Miller (1967) suggests that the *seven* "nonviscometric"
parameters may possibly be replaced to a good approximation by *five*.

   A commentary on several previous analyses is also given in the Appendix
and material functions are summarized in Tables IIA and B. As pointed out
in the Appendix, many of the previous works can be cast into a physically
more meaningful and a mathematically more economical form by basing
shear-dependent constitutive parameters and derived constants on the unperturbed gap shear rate $\kappa^{(0)}$.

   Also, the limiting case of *plane Couette flow* and *inertia-free flow* are considered and exact equations emerge for certain novel types of instability

---

* Although unpublished, his analysis has received a rather widespread circulation and
literature citation as a thesis (Miller, 1967) and as a technical report (Miller and Goddard,
1967).

TABLE IIA

MATERIAL FUNCTIONS OF PREVIOUS ANALYSES[a]

Investigator	Model	η	σ_1	σ_2
1. Graebel (1961)	Reiner–Rivlin	α_1	$\alpha_2\kappa^2$	$\alpha_2\kappa^2$
2. Graebel (1964)	Bingham plastic	$\mu\left(1+\dfrac{v}{\mu|\kappa|}\right)$	0	0
3. Thomas and Walters (1964a,b)	Walters liquid B'	η_0	$2K_0\kappa^2$	0
4. Chan-Man Fong (1965)	Walters liquid A'	η_0	0	$-2K_0\kappa^2$
5. Datta (1964)	Second-order fluid	α_1	$\alpha_2\kappa^2$	$(\alpha_2+2\alpha_3)\kappa^2$
6. Davies (1965)	Modified Walters B'	$\eta_0+\mu_0 K_0\kappa^2$	$2K_0\kappa^2$	0
7. Rubin et al. (1968)[b]	Liquid C'	η_0	$K_0\kappa^2$	$-K_0\kappa^2$
8. Leslie (1964)[b]	Ericksen Ainsotropic	$\mu+\mu_3+\mu_2 n_1^2 n_2^2$	$n_1 n_2(\mu_2 n_2^2+2\mu_3)|\kappa|$	$n_1 n_2(\mu_2 n_1^2+2\mu_3)|\kappa|$
9. Lockett and Rivlin (1968)	Rivlin–Ericksen	$C_{53}=C_{45}$	$C_{43}\kappa$	$C_{55}\kappa$

[a] Presented in the original nomenclature.
[b] Expressions for material functions cited in the original work have been corrected here.

TABLE IIB

MATERIAL FUNCTIONS OF PREVIOUS ANALYSES

*	β_1	β_2	β_4	β_5	β_7
1.	$\alpha_2\kappa/\alpha_1$	$\alpha_2\kappa/\alpha_1$	1	1	0
2.	0	0	1	1	0
3.	$K_0\kappa/\eta_0$	$K_0\kappa/\eta_0$	1	1	Negligible
4.	$-K_0\kappa/\eta_0$	$-K_0\kappa/\eta_0$	$1+\dfrac{3S_0\kappa^2}{2\eta_0}$	$1+\dfrac{3S_0\kappa^2}{2\eta_0}$	Negligible
5.	$(\alpha_2+\alpha_3)\kappa/\alpha_1$	$(\alpha_2+\alpha_3)\kappa/\alpha_1$	1	1	0
6.	$K_0\kappa\left[1-\dfrac{3\mu_0 S_0\kappa^4}{\eta_0 K_0}\right]$	$K_0\kappa^2$	1	1	Negligible
7.	0	0	1	1	Negligible
8.	0	$\dfrac{n_1}{n_2}\left(1-\dfrac{\mu}{\eta}\right)$	$4\mu+n_1^2(1-2n_1^2)\mu_2/4\eta$	$(\mu+n_1^4\mu_2+2n_1^2\mu_3)/\eta$	Negligible
9.	$\dfrac{g}{(1+f)}$	$-\dfrac{h}{(1+f)}$	$1+a/2$	$1+b$	Neglected

[a] See Table IIA for listing of investigator.

(Giesekus, 1966b) some of which are *purely-rheological* in origin and, hence, reminiscent of rheological "incompatibilities" of the type discussed in Section III,A,2 above.

Finally, a numerical solution of the relevant differential equations has been carried out (Miller, 1967) for the oft-used "generalized Newtonian fluid," which is characterized here solely by a "shear-dependent" power-law viscosity function. The results for the critical Taylor number T_c and the critical wave number k_c versus effective power-law or flow behavior index are presented in Figs. 4 and 5.

5. *Other Flows*

All of the preceding examples have dealt with perturbed shear flows, that is to say "weak flows," according to recent classifications of kinematics (cf., Section II above.) However, there are a number of theoretically and practically important problems involving "strong flows" with large stretching of material elements, which have received much attention in the recent literature and which will be mentioned briefly here.

Steady extensional flows, and certain unsteady versions thereof, are relevant to a number of applications, such as fiber spinning and convergent entry flow into dies and channels. There has been a considerable body of

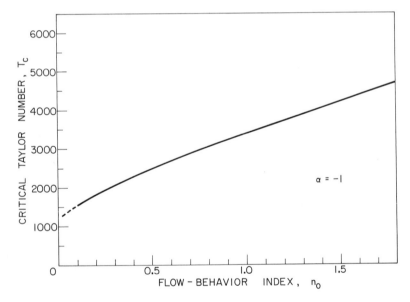

FIG. 4. Critical Taylor number vs. flow behavior ("power-law") index for a generalized Newtonian fluid.

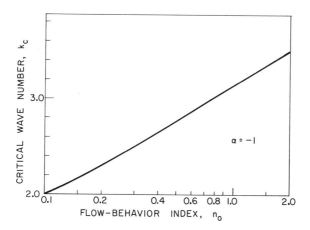

FIG. 5. Critical wave number vs. flow behavior ("power-law") index for a generalized Newtonian fluid.

analysis and experiment over the past decade on the mechanical stability of the extensional flow of liquid filaments with free surface. Much of this work was prompted by the early analysis of Pearson and Matovich (1969), who showed that one particularly important form of filament instability, the "draw resonance," which is characterized by violent diameter fluctuations, could be predicted even for purely Newtonian fluids. Fisher and Denn (1976), among others, have pursued this analysis to more complex fluid rheology, including elastic response and "strain-thickening" viscosity, and Petrie and Denn (1976) have given a comprehensive review of this area of work. Although the stability analysis of Fisher and Denn indicates that filament stability may be drastically altered by non-Newtonian effects, the more recent experimental work of Weinberger *et al.* (1976) suggests that the issue is far from settled.

Flow in converging channels, or in ducts with abrupt change of section, represents an interesting and important class of problems involving widely variable modes of deformation. Thus, near the channel walls one may anticipate a simple shearing motion or "weak flow," whereas near the centerline one expects an extensional motion, or "strong flow." Dating from the early Langlois–Rivlin retarded-motion analysis of "sink flow" (cf. Rivlin and Sawyer, 1971) which already indicates a curious recirculating eddy structure on the centerline, this type of problem has continued to command the attention of rheologists fluid mechanicists [Black *et al.*, 1974; Cable and Boger, 1978; cf. Schowalter (1978) and Bird *et al.* (1977d) who show several interesting photographs, due to Giesekus and others]. It appears that at elevated flow rates, the intense centerline stretching may give rise to an "elastic core" structure, with associated, large entry pressure effects and, possibly, flow instabilities that are manifest in certain polymer processing operations (Petrie and Denn, 1976; Cable and Boger, 1978). There remain several outstanding questions about the convergence of iterative solution methods for viscoelastic flow in contractions (Black *et al.*, 1974), and recent attempts at numerical solutions reportedly show interesting computational (convergence) problems at certain more-or-less fixed values of the "Deborah" or "Weissenberg" number (Caswell and Viriyayuthakorn, 1978; Bernstein and Malkus, 1978). At this time, it is not clear whether such effects are purely computational, and whether their origin lies in the core structure or in the flow around sharp corners in abrupt contractions. This kind of problem represents a potentially interesting area for application of kinematic perturbation techniques. To some extent, the convergent axisymmetric channel represents an "inside-out" version of far-field extensional flow around aligned slender bodies, where such techniques have already been applied (Goddard, 1976).

As a final example, in which macromolecular deformation associated with strong flow may be important, some mention should be made of the "Toms" phenomenon, or turbulent-drag reduction (Virk, 1975). It is now accepted by numerous workers that the potentially powerful mechanical effects exerted in suspensions of filamentary particles or solutions of macromolecules subject to extensional flows may account for the supression of turbulent dissipation in dilute polymer solutions. While this is suggested by the work of Batchelor (1971) on rigid-particle systems, the success of this mechanism, as an explanation of macromolecular drag reduction, depends on whether fluctuating turbulent extension rates are large enough and of a sufficiently long duration to cause significant macromolecular alignment. Recent attempts have been made to model extensible macromolecules in turbulent flow fields (Hinch, 1977, in Frenkiel *et al.*, 1977) by accounting for various microstructural features mentioned in Section I,C of this review.

As opposed to the situation of only a few years ago, it may become possible to measure routinely the rheological properties of drag-reducing polymers in independent rheological tests (Tsai and Darby, 1978). However, if the above type of mechanism is correct, then the usual studies of viscometric and "weak-flow" response may have at best an indirect relation to drag reduction.

B. Perturbed Kinematics—Nearly Steady Flow

The idea of small strains superimposed on finite deformations is a well-established one in the solid-mechanics literature (Green and Zerna, 1975). However, for fluids with memory, the paper of Pipkin and Owen (1967) on "nearly viscometric flows" provides the most complete implementation of this method. Here, the intent is to outline a somewhat more general version of their method which suggests a possible approach to more complex base flows. In particular, a treatment is given here of perturbations on the more general class of motions, *rotationally steady flows* or "*motions with constant stretch history,*" for incompressible fluids. For the present purpose, the abbreviated terminology "*steady flow*" is adopted, which given the assumption of frame indifference, should not be the subject of misinterpretations in the present context.

Thus, following Pipkin and Owen (1967) and, as a generalization of (1.2), we consider a given, *base* motion having specified strain history $\mathbf{G}^{(0)}(\cdot)$ in (1.2), and stress $\mathbf{T}^{(0)}(t)$, given by (1.1) up to an arbitrary dynamic pressure p, together with a *perturbed* history

$$\mathbf{G}(\cdot) = \mathbf{G}^{(0)}(\cdot) + \mathbf{G}^{(1)}(\cdot) \tag{3.1}$$

with $\mathbf{G}^{(1)}(s)$, $s \geq 0$, representing the strain perturbation and its history $\mathbf{G}^{(1)}(\cdot)$. Based on an assumption of fading memory at the history $\mathbf{G}^{(0)}(\cdot)$, the *perturbed stress* is written as

$$\mathbf{T}(t) = \mathbf{T}^{(0)}(t) + \mathbf{T}^{(1)}(t), \tag{3.2}$$

where the *stress perturbation* $\mathbf{T}^{(1)}(t)$ is determined by

$$\mathbf{T}^{(1)}(t) = \delta \mathscr{F} \{ \mathbf{G}^{(0)}(\cdot) | \mathbf{G}^{(1)}(\cdot) \} + 0(\|\mathbf{G}^{(1)}(\cdot)\|^2). \tag{3.3}$$

Here, $\delta \mathscr{F}$ is a functional on $\mathbf{G}^{(0)}(\cdot)$ and $\mathbf{G}^{(1)}(\cdot)$, linear $\mathbf{G}^{(1)}(\cdot)$, which represents the Frechet derivative of (1.1) at the base history $\mathbf{G}^{(0)}(\cdot)$. Also, $\| \ \|$ denotes a suitable norm of the perturbation $\mathbf{G}^{(1)}(\cdot)$ which will in general depend on the base history in question.

As discussed by Pipkin and Owen, the usual assumptions of smoothness leads to an integral representation for $\delta \mathscr{F}$, with kernel or memory function expressible as a fourth-rank tensor. As evident from their analysis for viscometric flow, it is far from easy to give a corresponding, explicit representation of $\delta \mathscr{F}$ which eliminates certain degeneracies associated with the case where the perturbed history $\mathbf{G}(\cdot)$ belongs to the same class of flows as the base history $\mathbf{G}^{(0)}(\cdot)$. It appears that the main difficulties in this type of analysis lie in finding the appropriate formulation of the difference history of perturbation $\mathbf{G}^{(1)}(\cdot)$ as an appropriate *distance* (*in the space of tensor functions*) *from the manifold of motions of a given class of histories*, e.g., viscometric flows, rather than simply as the distance from one particular element of the manifold, as done by Pipkin and Owen. Therefore, it is proposed here to employ the former idea to treat the more general class of base flow, *steady flow*, which includes viscometric flow as a special case.

We recall that for a steady flow the stress is given, by Wang's Theorem (Truesdell, 1966), as

$$\mathbf{T}(t) = \mathbf{f}[\mathbf{E}_1(t), \mathbf{E}_2(t), \mathbf{E}_3(t)], \tag{3.4}$$

where **f** denotes an isotropic tensor function of a set of three kinematic tensor $\mathbf{E}_k(t)$, of the type defined in Section I,B,1. However, rather than the customary choice of the Cauchy strain $\mathbf{G}(s)$ of (1.2) and the Rivlin–Ericksen tensors of (1.6), we shall employ a description based on the corotational strain $\mathbf{E}_t(t')$ of (1.7) and its associated kinematic tensors, that is, the co-rotational derivatives of (1.10), since this description appears to provide for a simpler analytical treatment.

From the usual definitions of steady flows, or motions of constant stretch history (Truesdell, 1966), it is an easy matter to derive an alternative relation, in the form of the characteristic equation for the corotational strain rate for all past times $t' \leq t$,

$$\dot{\mathbf{E}}_t^3(t') - \mathrm{I}\,\dot{\mathbf{E}}_t^2(t') + \mathrm{II}\,\dot{\mathbf{E}}_t(t') - \mathrm{III}\,\mathbf{1} = \mathbf{0}. \tag{3.5}$$

This is equivalent to the characteristic equation for the deformation rate $\mathbf{D}(t')$ for $t' \leq t$,

$$\mathbf{D}^3(t') - \mathrm{I}\,\mathbf{D}^2(t') + \mathrm{II}\,\mathbf{D}(t') - \mathrm{III}\,\mathbf{1} = \mathbf{0}, \tag{3.6}$$

where I, II, III are the *constant*, principal scalar invariants of the present rate, $\mathbf{D}(t) \equiv \dot{\mathbf{E}}_t(t)$. For incompressible fluids one can of course immediately set $\mathrm{I} = 0$.

The relations (3.5) and (3.6) simply state that, for steady flows, the three principal invariants of the past strain or deformation rates at time $t' \leq t$ are identical with those evaluated at the present time t, which are constants, independent of t. Furthermore, by power series expansion of (3.5) in terms of the time lapse $s = t - t'$, or else by successive differentiations with respect to t' followed by evaluation of the resulting higher order derivatives at $t' = t$, one obtains a set of algebraic restrictions connecting the kinematic tensors $\mathbf{E}_k$, $k = 1, 2, \dots$. Thus, at the zeroth order, (3.5) or (3.6) provides, identically, the characteristic equation for $\mathbf{E}_1 \equiv \mathbf{D}(t)$, while at the first order one obtains the relation between $\mathbf{E}_1$ and $\mathbf{E}_2$:

$$\mathbf{E}_1^2 \cdot \mathbf{E}_2 + \mathbf{E}_1 \cdot \mathbf{E}_2 \cdot \mathbf{E}_1 + \mathbf{E}_2 \cdot \mathbf{E}_1^2 - \mathrm{I}(\mathbf{E}_1 \cdot \mathbf{E}_2 + \mathbf{E}_2 \cdot \mathbf{E}_1) - \mathrm{II}\,\mathbf{E}_2 = \mathbf{0}. \tag{3.7}$$

Similarly, at second order, one obtains a relation involving $\mathbf{E}_1, \mathbf{E}_2, \mathbf{E}_3$, and so forth for higher order terms, none of which will be recorded here. It appears that such relations might provide an alternative to the usual set of defining relations for steady flow and, possibly, an alternative pathway to Wang's theorem (Truesdell, 1966), relating higher order kinematic tensors to the first three.

Whatever the status of (3.5) as a definition of steady flow, we see that for general motions the kinematic tensor $\mathbf{H}(s)$, defined for $s = t - t' \geq 0$ by

$$\mathbf{H}(s) = \dot{\mathbf{E}}_t^3(t') - \mathrm{I}(t)\dot{\mathbf{E}}_t^2(t') + \mathrm{II}(t)\dot{\mathbf{E}}_t(t') - \mathrm{III}(t)\mathbf{1} \tag{3.8}$$

with I, II, III denoting now the generally *time-dependent* invariants of $\mathbf{D}(t)$, represents an objective tensor whose history $\mathbf{H}(\cdot)$ *is null on the manifold of steady flows.* As such it provides one possible measure of distance from this manifold.

If one so wishes, it is also possibly to define an alternative form to (3.8) that is homogeneous of degree *one* in $\dot{\mathbf{E}}_t(t')$, and an associated "strainlike" measure, through integrals of the form

$$\int_0^s \mathbf{H}(s')\,ds' \tag{3.9}$$

which, like those discussed in Section I,B,2, represents yet another "generalized" strain.

Without developing all the formal apparatus here, it is apparent that, for perturbation on a steady flow defined by the kinematic tensors $\mathbf{E}_1^{(0)}$, $\mathbf{E}_2^{(0)}$, $\mathbf{E}_3^{(0)}$, the perturbation formulas of (3.1)–(3.3), with $\mathbf{T}^{(0)}$ given in terms of $\mathbf{E}_1^{(0)}$, $\mathbf{E}_2^{(0)}$, $\mathbf{E}_3^{(0)}$ by (3.4), can be replaced by

$$\mathbf{T}^{(1)} = \sum_{k=1}^{3} [\partial f/\partial \mathbf{E}_k]^{(0)}{:}\mathbf{E}_k^{(1)}(t) + \mathscr{G}\{\mathbf{E}_1^{(0)}, \mathbf{E}_2^{(0)}, \mathbf{E}_3^{(0)}\,|\,\mathbf{H}(\cdot)\}, \tag{3.10}$$

essentially, by means of a functional change of variables. Here, $\mathscr{G}$ denotes an isotropic tensor function of $\mathbf{E}_1^{(0)}$, $\mathbf{E}_2^{(0)}$, $\mathbf{E}_3^{(0)}$ and *linear* functional of $\mathbf{H}(\cdot)$, which can be made to correspond to the appropriate Fréchet derivative of a particular stress functional. Of course, the error term in (3.3) is to be replaced by one involving an appropriate norm in $\mathbf{H}(\cdot)$, depending generally on $\mathbf{E}_1^{(0)}$, $\mathbf{E}_2^{(0)}$, $\mathbf{E}_3^{(0)}$. Once again, we observe that a formulation is possible in which a strain-like measure, such as (3.9), replaces $\mathbf{H}(\cdot)$ in (3.10). Also, we note that for actual application to a "small" perturbation $\dot{\mathbf{E}}_t^{(1)}(t')$ on a steady flow, having corotational strain rate $\dot{\mathbf{E}}_t^{(0)}(t')$, that the tensor function defined in (3.8), in terms of

$$\dot{\mathbf{E}}_t(t') = \dot{\mathbf{E}}_t^{(0)}(t') + \dot{\mathbf{E}}_t^{(1)}(t')$$

should in principle be replaced by its linearized form in $\dot{\mathbf{E}}_t^{(1)}(t')$.

However, as the key point of the present analysis we note that, as it stands, the representation (3.10) is *free of any "consistency" restrictions*, of the type necessitated by Pipkin and Owen's (1967) perturbation analysis, since the term $\mathscr{G}$ (3.10) automatically produces a *null* stress contribution to $\mathbf{T}^{(1)}$ for any perturbed flow which is also a materially steady flow. This means that (3.10) can be cast into explicit forms, most notably, a linear integral in $\mathbf{H}(\cdot)$, by the usual representations for isotropic tensor functions, and without regard for consistency requirements.

No attempt will be made here to set down such representations in full, generality, since it can be anticipated that the most likely applications would involve elementary types of steady flow, for which it is vastly simpler to reduce the form of (3.10) directly.

For example, in the case of a simple extension, all higher corotational strain rates beyond $E_1^{(0)}$ vanish. The latter has the diagonal matrix representation (2.29), with $2\varepsilon_1 = -\varepsilon_2 = -\varepsilon_3$, and therefore the functional (3.10) can be expressed in terms of a complete set of independent elements of the symmetric, isotropic integrity basis for $E_1^{(0)}$ and H (Spencer and Rivlin, 1959, 1960; Smith, 1965) which are also linear in H, *viz.*,

$$H, \quad (\mathrm{tr}\,H)E, \quad E \cdot H + H \cdot E, \quad \mathrm{tr}(E \cdot H)E. \tag{3.11}$$

Thus, an integral representation of (3.11) would involve *four* scalar memory functions, dependent on the time lapse s and the scalar invariants of $E_1^{(0)}$ or, equivalently, on s and ε_1.

On the other hand, for the (isochoric) perturbations on viscometric flows already considered by Pipkin and Owen (1967), there are only two independent kinematic tensors for the base flow, say $E_1^{(0)}$ and $E_2^{(0)}$. Their properties have been discussed elsewhere (Goddard, 1967), and from these it appears that a corresponding integrity basis, linear in H, is contained within the set of *fifteen* tensors, consisting of the symmetric parts of the first *three* members of the set:

$$H, \quad H \cdot E_1, \quad H \cdot E_2, \quad H \cdot E_1^2, \tag{3.12}$$

together with the set of *twelve* tensors obtained by multiplying the respective traces of all four members of (3.12) by each of the three tensors $E_1 \equiv E_1^{(0)}$, $E_2 \equiv E_2^{(0)}$ and E_1^2. Although it should be possible, the author has not been able at the time of this writing, to reduce this set to a number smaller than *fifteen* and, hence, to bring the resultant number of memory functions for an integral representation into correspondence with the number, *thirteen*, indicated by Pipkin and Owen (1967) without imposition of their later consistency requirements (cf. Schowalter, 1978).

The analysis of Taylor–Couette stability discussed in Section III,A represents one important example of a perturbation on a viscometric flow. The treatment given in the Appendix, which predates considerably the main body of the review, is based on a somewhat different and more direct approach than the general one discussed above, and it provides a good example of the direct use of "consistency" relations. Also, it illustrates the extent to which certain kinematic assumptions about the specific form of perturbations may lead to immediate simplifications in the expressions for perturbed stresses. At the time of this writing, no attempt has been made to compare the resultant perturbation formulas with those generated by the general analysis discussed above in this section.

For the sake of wider application to various flow problems there may be some merit in considering more general *base* kinematics than the example of steady flow considered above. Some of the "commutative motions," discussed in Section II, provide a possibly important subclass of motions for such investigation, which would involve interesting theoretical questions as to the optimal kinematic representation of perturbations. From the point of view of application, however, the kind of analyses discussed above suggest that one may expect to encounter a profusely large number of material functions or functionals.

Appendix: Analysis of the Linear, Neutral
Taylor–Couette Stability in Simple Fluids*

The motion existing prior to the onset of instability in shear between concentric rotating cylinders is known as *circular Couette flow*, which is a member of the viscometric class of motions.

In order to consider the stability of circular Couette flow analytically, one must carry out a perturbation analysis on the system, which requires determination of the effects of small perturbations on viscometric flows of viscoelastic fluids. General techniques for handling such motions have been discussed in detail by Pipkin and Owen (1967). For the present purposes, some special assumptions will first be made as to the basic features of the perturbed flow, which allows one to obtain somewhat more manageable expressions for direct application to the Taylor stability problem.

We assume, following Lockett and Rivlin (1968), that the past history of the deformation may be expressed as a sufficiently smooth, analytic function of time (following a particle) so that the Noll fluid is described by the Rivlin and Ericksen equation:

$$\mathbf{T} + p\mathbf{1} = \mathbf{f}[\mathbf{A}_1, \mathbf{A}_2, \dots, \mathbf{A}_n, \dots], \qquad (A.1)$$

where $\mathbf{T}$ is the stress tensor, $\mathbf{1}$ is the identity tensor, p is an arbitrary scalar with units of stress, and $\mathbf{A}_n$ are the Rivlin–Ericksen tensors, defined by

$$\mathbf{A}_1 = \nabla\mathbf{v} + (\nabla\mathbf{v})^T, \qquad (A.2)$$

$$\mathbf{A}_{n+1} = \frac{\partial \mathbf{A}_n}{\partial t} + \mathbf{v} \cdot \nabla\mathbf{A}_n + (\nabla\mathbf{v})^T \cdot \mathbf{A}_n + \mathbf{A}_n \cdot (\nabla\mathbf{v}) \qquad (A.3)$$

in which $\mathbf{v}$ denotes the velocity at position $\mathbf{x}$ in the flow field, and ∇ represents the gradient operator. The function $\mathbf{f}$ in (A.1) is an isotropic tensor function of the arguments $\mathbf{A}_n$.

In terms of the flows to be considered in the present stability analysis, it will be shown presently that the kinematics of the motion permits the use of (A.1) to represent the constitutive equation for a Noll simple fluid.

---

* Chester Miller is principal author of this work.

As indicated above, the base flow, circular Couette flow, belongs to the well-known class of viscometric flows, for which

$$\left.\begin{array}{l} \mathbf{A}_n = \mathbf{0} \qquad \text{for } n \geq 3, \\[6pt] \mathbf{A}_1^3 = \kappa^2 \mathbf{A}_1, \\[6pt] \mathbf{A}_2^3 = 2\kappa^2 \mathbf{A}_2^2, \\[6pt] \mathbf{A}_1 \cdot \mathbf{A}_2 + \mathbf{A}_2 \cdot \mathbf{A}_1 = 2\kappa^2 \mathbf{A}_1, \end{array}\right\} \qquad (A.4)$$

and

where the shear rate κ is given by

$$\kappa^2 = \operatorname{tr}(\mathbf{A}_1^2/2),$$

where tr denotes the trace of a tensor.

We recall that the stress for a simple fluid in a viscometric flow may be written in terms of the Rivlin–Ericksen tensors as

$$\mathbf{T} + p\mathbf{I} = \eta\mathbf{A}_1 + \frac{\sigma_1}{\kappa^2}\mathbf{A}_1^2 + \frac{(\sigma_2 - \sigma_1)}{2\kappa^2}\mathbf{A}_2, \qquad (A.5)$$

where η, σ_1, and σ_2 are viscometric functions. Specifically, η is the viscosity function, while $\sigma_1 - \sigma_2 = N_1$ and $\sigma_2 = N_2$ are the usual "primary and secondary normal stress differences." All three are even functions of the shear rate κ.

Let $\mathbf{v}^{(0)}$ and $\mathbf{T}^{(0)}$ represent the velocity and associated stress fields for a viscoelastic fluid in an arbitrary base flow, and let these fields be disturbed by a small perturbation on the motion. Our objective is to derive the linearized equations which must be satisfied in order for the resulting motion to also be compatible with the equations of motion and continuity and the rheological constitutive equation for the fluid (A.1). For this purpose, we write,

$$\mathbf{v} = \mathbf{v}^{(0)} + \varepsilon\mathbf{v}^{(1)}, \qquad (A.6)$$

where $\varepsilon \ll 1$ and where $\mathbf{v}^{(1)}$ is a spatially smooth velocity field having the same order of magnitude as $\mathbf{v}^{(0)}$. The quantity ε is, in the usual way, regarded as an expansion parameter in the analysis. Substituting Eq. (A.6) into Eq. (A.1), one obtains

$$\mathbf{T} = \mathbf{T}^{(0)} + \varepsilon\mathbf{T}^{(1)} + \cdots +, \qquad (A.7)$$

where

$$\mathbf{T}^{(0)} = -p^{(0)}\mathbf{1} + \mathbf{f}(\mathbf{A}_1^{(0)}, \mathbf{A}_2^{(0)}, \mathbf{A}_3^{(0)}, \ldots) \qquad (A.8)$$

and where the *stress perturbation* $\mathbf{T}^{(1)}$ is given by

$$\mathbf{T}^{(1)} = -p^{(1)}\mathbf{1} + \sum_{n=1}^{\infty} \mathbf{g}_n(\mathbf{A}_1^{(0)}, \mathbf{A}_2^{(0)}, \mathbf{A}_3^{(0)}, \ldots, |\mathbf{A}_n^{(1)}) \qquad (A.9)$$

with, $\mathbf{g}_n$, a function linear in its last argument, defined formally by

$$\mathbf{g}_n = (\partial \mathbf{f}/\partial \mathbf{A}_n)^{(0)} : \mathbf{A}_n^{(1)}, \qquad n = 1, 2, \ldots . \tag{A.10}$$

Also,

$$
\left.
\begin{aligned}
\mathbf{A}_1^{(0)} &= \nabla \mathbf{v}^{(0)} + \left[\nabla \mathbf{v}^{(0)} \right]^T, \\
\mathbf{A}_{n+1}^{(0)} &= \mathscr{D}^{(0)} \mathbf{A}_n^{(0)}, \\
\mathbf{A}_1^{(1)} &= \nabla \mathbf{v}^{(1)} + \left[\nabla \mathbf{v}^{(1)} \right]^T, \\
\mathbf{A}_{n+1}^{(1)} &= \mathscr{D}^{(0)} \mathbf{A}_n^{(1)} + \mathscr{D}^{(1)} \mathbf{A}_n^{(0)},
\end{aligned}
\right\} \tag{A.11}
$$

with

$$
\left.
\begin{aligned}
\mathscr{D}^{(0)} \mathbf{B} &= \frac{D^{(0)}}{Dt} \mathbf{B} + \left[\nabla \mathbf{v}^{(0)} \right]^T \cdot \mathbf{B} + \mathbf{B} \cdot \left[\nabla \mathbf{v}^{(0)} \right], \\[2mm]
\mathscr{D}^{(1)} \mathbf{B} &= \frac{D^{(1)}}{Dt} \mathbf{B} + \left[\nabla \mathbf{v}^{(1)} \right]^T \cdot \mathbf{B} + \mathbf{B} \cdot \left[\nabla \mathbf{v}^{(1)} \right], \\[2mm]
\frac{D^{(0)}}{Dt} \mathbf{B} &= \frac{\partial \mathbf{B}}{\partial t} + \mathbf{v}^{(0)} \cdot \nabla \mathbf{B},
\end{aligned}
\right\} \tag{A.12}
$$

and

$$
\frac{D^{(1)}}{Dt} \mathbf{B} = \mathbf{v}^{(1)} \cdot \nabla \mathbf{B}
$$

for arbitrary $\mathbf{B}$.

We demonstrate now that, as a consequence of the particular form of the kinematic relationships that characterize viscometric flows, (A.4), the expressions obtained above for the stress fields $\mathbf{T}^{(0)}$ and $\mathbf{T}^{(1)}$ can be greatly simplified.

Since the undisturbed velocity field $\mathbf{v}^{(0)}$ represents a viscometric flow, one has from (A.5)

$$\mathbf{T}^{(0)} + p^{(0)} \mathbf{1} = \eta \mathbf{A}_1^{(0)} + \frac{\sigma_1}{\kappa^{(0)2}} \mathbf{A}_1^{(0)2} + \frac{(\sigma_2 - \sigma_1)}{2\kappa^{(0)2}} \mathbf{A}_2^{(0)}, \tag{A.13}$$

where the functions η, σ_1, and σ_2 are evaluated at the undisturbed shear rate $\kappa^{(0)}$, with

$$\kappa^{(0)} = \left[\operatorname{tr} \mathbf{A}_1^{(0)2}/2 \right]^{1/2}. \tag{A.14}$$

Considering next the stress perturbation $\mathbf{T}^{(1)}$, we recall that the kinematic tensors $\mathbf{A}_n^{(0)}$ ($n = 1, 2, \ldots$) must satisfy (A.4) identically, in viscometric motions. Thus, (A.9) reduces to the form

$$\mathbf{T}^{(1)} + p^{(1)} \mathbf{1} = \sum_{n=1}^{\infty} \hat{g}_{\mathbf{n}}(\mathbf{A}_1^{(0)}, \mathbf{A}_2^{(0)} | \mathbf{A}_n^{(1)}), \tag{A.15}$$

where

$$\hat{g}_n(\mathbf{A}_1^{(0)}, \mathbf{A}_2^{(0)} | \mathbf{A}_n^{(1)}) = \mathbf{g}_n(\mathbf{A}_1^{(0)}, \mathbf{A}_2^{(0)}, 0, 0, \ldots, | \mathbf{A}_n^{(1)}) \tag{A.16}$$

is a function of only three tensors $\mathbf{A}_1^{(0)}$, $\mathbf{A}_2^{(0)}$, and $\mathbf{A}_n^{(1)}$ and, of course, is linear with respect to $\mathbf{A}_n^{(1)}$.

It is possible also to simplify the expression presented earlier for the tensors $\mathbf{A}_n^{(1)}$. In particular, substitution of the first member of (A.4) into (A.11) leads to the result that

$$\mathbf{A}_{n+1}^{(1)} = \mathscr{D}^{(0)} \mathbf{A}_n^{(1)} \qquad (n \geq 3). \tag{A.17}$$

If, as is sometimes assumed, (A.5) were taken to represent the general rheological expression of a viscoelastic fluid, it is readily demonstrated that the functions $\hat{g}_n$ would reduce simply to

$$\hat{\mathbf{g}}_1 = [A_1^{(1)} + n_0 \gamma A_1^{(0)}] \eta + (n_1 - 2) \sigma_1 \gamma (A_1/\kappa^{(0)})^2$$

$$+ \frac{\sigma_1}{\kappa^{(0)2}} [\mathbf{A}_1^{(0)} \cdot \mathbf{A}_1^{(1)} + \mathbf{A}_1^{(1)} \cdot \mathbf{A}_1^{(0)}]$$

$$+ [(n_2 - 2)\sigma_2 - (n_1 - 2)\sigma_1] \frac{\gamma}{2} \frac{A_2^{(0)}}{\kappa^{(0)2}} \tag{A.18}$$

$$\hat{\mathbf{g}}_2 = \frac{(\sigma_2 - \sigma_1)}{2} \frac{A_2^{(1)}}{\kappa^{(0)2}}, \tag{A.19}$$

$$\hat{\mathbf{g}}_n = \mathbf{0}, \qquad \text{for } n \geq 3, \tag{A.20}$$

where the viscometric functions η, σ_1, and σ_2 are evaluated at the undisturbed shear rate $\kappa^{(0)}$ and where

$$n_0 = 1 + \frac{d \ln \eta}{d \ln \kappa}, \qquad \text{at } \kappa = \kappa^{(0)},$$

$$n_1 = \frac{d \ln \sigma_1}{d \ln \kappa}, \qquad \text{at } \kappa = \kappa^{(0)},$$

$$n_2 = \frac{d \ln \sigma_2}{d \ln \kappa}, \qquad \text{at } \kappa = \kappa^{(0)}, \tag{A.21}$$

$$\gamma = \kappa^{(1)}/\kappa^{(0)},$$

with

$$\kappa^{(1)} = \frac{1}{2\kappa^{(0)}} \text{tr}[A_1^{(0)} \cdot A_1^{(1)}].$$

Obviously, (A.18)–(A.21) cannot generally describe the stress perturbation in a simple fluid unless the kinematics of the flow perturbation are of a very specific nature. In particular, the disturbance to the viscometric velocity field $\mathbf{v}^{(0)}$ must be such that the perturbed velocity field $\mathbf{v}$ is also viscometric. In this case, it is obvious $\mathbf{v}$ must satisfy the kinematic relationships (A.4). Substituting (A.6) into (A.4), we can readily establish necessary conditions for the velocity perturbation to produce the required flow. Thus we find, after discarding the nonlinear terms in ε, that

$$
\left.
\begin{aligned}
&\mathbf{A}_n^{(1)} = \mathbf{0}, \qquad \text{for } n \geq 3 \\[6pt]
&\mathbf{A}_1^{(0)^2} \cdot \mathbf{A}_1^{(1)} + \mathbf{A}_1^{(0)} \cdot \mathbf{A}_1^{(1)} \cdot \mathbf{A}_1^{(0)} + \mathbf{A}_1^{(1)} \cdot \mathbf{A}_1^{(0)^2} \\
&\qquad = \kappa^{(0)^2} \mathbf{A}_1^{(1)} + 2\kappa^{(0)} \kappa^{(1)} \mathbf{A}_1^{(0)}, \\[6pt]
&\mathbf{A}_2^{(0)^2} \cdot \mathbf{A}_2^{(1)} + \mathbf{A}_2^{(0)} \cdot \mathbf{A}_2^{(1)} \cdot \mathbf{A}_2^{(0)} + \mathbf{A}_2^{(1)} \cdot \mathbf{A}_2^{(0)^2} \\
&\qquad = 2\kappa^{(0)^2} \big[\mathbf{A}_2^{(0)} \cdot \mathbf{A}_2^{(1)} + \mathbf{A}_2^{(1)} \cdot \mathbf{A}_2^{(0)} \big] + 4\kappa^{(0)} \kappa^{(1)} \mathbf{A}_2^{(0)^2},
\end{aligned}
\right\} \quad \text{(A.22)}
$$

and

$$
\begin{aligned}
&\big[\mathbf{A}_1^{(1)} \cdot \mathbf{A}_2^{(0)} + \mathbf{A}_2^{(0)} \cdot \mathbf{A}_1^{(1)} \big] + \big[\mathbf{A}_1^{(0)} \cdot \mathbf{A}_2^{(1)} + \mathbf{A}_2^{(1)} \cdot \mathbf{A}_1^{(0)} \big] \\
&\qquad = 2\kappa^{(0)^2} \mathbf{A}_1^{(1)} + 4\kappa^{(0)} \kappa^{(1)} \mathbf{A}_1^{(0)},
\end{aligned}
$$

where $\kappa^{(1)}$ is given in (A.21).

The stress pattern present in a viscometric flow is in general determined by (A.5), and this relationship, of course, must also provide a description of the stress to $O(\varepsilon)$ in situations where (A.22) are satisfied, since in that case, the overall velocity field $\mathbf{v}$ is viscometric. If we put (A.6) into (A.5), we obtain an expression for the stress $\mathbf{T}$, in which the stress perturbation $\mathbf{T}^{(1)}$ is given by (A.15)–(A.20). In their analysis of "nearly viscometric flows" Pipkin and Owen (1967) derive a set of equations, referred to as "consistency relations," which are equivalent to our (A.22) taken together with (A.15)–(A.20). In comparison to Pipkin and Owen's results, the present relations are somewhat more easily applicable to flows where (A.1) is assumed to hold.

Before departing from the subject of perturbations on viscometric flows, let us consider one final point. Recall that the stress functions $\hat{\mathbf{g}}_n$ in (A.15) for the stress perturbation are linear in the perturbed Rivlin–Ericksen tensors $\mathbf{A}_n^{(1)}$. Let us assume that in a specific physical situation, the tensors $\mathbf{A}_n^{(1)}$ can be represented in the form

$$
\left.
\begin{aligned}
\mathbf{A}_n^{(1)} &= \bar{\mathbf{A}}_n^{(1)} + \tilde{\mathbf{A}}_n^{(1)}, \qquad \text{for } n = 1, 2, \\
\mathbf{A}_n^{(1)} &= \tilde{\mathbf{A}}_n^{(1)}, \qquad \text{for } n \geq 3,
\end{aligned}
\right\} \quad \text{(A.23)}
$$

where the $\bar{\mathbf{A}}_n^{(1)}$ satisfies (A.22) identically, while $\tilde{\mathbf{A}}_n^{(1)}$ *does not*. If we substitute (A.23) into the Eq. (A.15), we find that

$$
\mathbf{T}_1^{(1)} + p^{(1)} \mathbf{I} = \sum_{n=1}^{2} \hat{\mathbf{g}}_n \big[\mathbf{A}_1^{(0)}, \mathbf{A}_2^{(0)} \big| \bar{\mathbf{A}}_n^{(1)} \big] + \sum_{n=1}^{\infty} \hat{\mathbf{g}}_n \big[\mathbf{A}_1^{(0)}, \mathbf{A}_2^{(0)} \big| \tilde{\mathbf{A}}_n^{(1)} \big], \quad \text{(A.24)}
$$

where the first summation in this expression corresponds to the "viscometric" portion of the stress perturbation and the second represents the "nonviscometric" portion. Since the $\bar{\mathbf{A}}_n^{(1)}$ ($n = 1, 2$) satisfy (A.22) identically, the contribution to the stress perturbation arising from these tensors must now be given by (A.18)–(A.19), with the $\mathbf{A}_n^{(1)}$ replaced by $\bar{\mathbf{A}}_n^{(1)}$. This follows from the fact that the stress functions $\hat{\mathbf{g}}_n$ are linear in the $\mathbf{A}_n^{(1)}$. Thus, even in situations where the overall flow $\mathbf{v}$ is nonviscometric, a certain portion of the deformation and stress may in the above sense be associated with viscometric response.

We turn now to a consideration of the base Couette flow, wherein the fluid is contained in the annulus between two long coaxial cylinders of radii r_1 and r_2 which rotate at steady angular velocities Ω_1 and Ω_2, respectively. We employ cylindrical polar coordinates to describe the geometry of the system with the axis of the cylinders being the z-axis. The undisturbed motion consists of a velocity field $\mathbf{v}^{(0)}$ given by

$$\mathbf{v}^{(0)} = V\mathbf{e}_\theta, \tag{A.25}$$

where $\mathbf{e}_\theta$ is the unit vector in the θ-direction and V is a function of r only.

The motion described by (A.25) is a viscometric flow and, as such, exhibits the kinematics and stresses discussed above. Rivlin–Ericksen tensors for this flow may be expressed in the form

$$\mathbf{A}_1^{(0)} = \kappa^{(0)}(\mathbf{e}_r\mathbf{e}_\theta + \mathbf{e}_\theta\mathbf{e}_r),$$

$$\mathbf{A}_2^{(0)} = 2\kappa^{(0)2}\mathbf{e}_r\mathbf{e}_r, \tag{A.26a}$$

$$\mathbf{A}_n^{(0)} = \mathbf{0} \qquad \text{for } n \geq 3,$$

where the indicated products of unit vectors represent dyadics, and where

$$\kappa^{(0)} = r\frac{d(V/r)}{dr}. \tag{A.26b}$$

The stress pattern existing in Couette flow, obtained by substituting the above expressions into (A.13) is

$$\sigma_{r\theta}^{(0)} = \eta\kappa^{(0)},$$

$$\sigma_{\theta\theta}^{(0)} - \sigma_{zz}^{(0)} = \sigma_1, \tag{A.27}$$

$$\sigma_{rr}^{(0)} - \sigma_{zz}^{(0)} = \sigma_2$$

in which η, σ_1, and σ_2 are, we recall, evaluated at $\kappa = \kappa^{(0)}$.

For this stress pattern to be compatible with the equations of motion we recall that the following relationship must hold for the shear rate $\kappa^{(0)}$:

$$\frac{d\ln\kappa^{(0)}}{d\ln r} = -\frac{2}{n_0}, \qquad \text{where } n_0 = \frac{d\ln\eta\kappa}{d\ln\kappa}\bigg]_{\kappa(0)}. \tag{A.28}$$

One will note that the parameter n_0 is merely the slope of the shear stress-shear rate curve on a logarithmic basis, i.e., the familiar "flow behavior index."

The no-slip boundary conditions are assumed to apply at the walls of the inner and outer cylinders:

$$V = \Omega_1 r_1, \qquad r = r_1,$$
$$V = \Omega_2 r_2, \qquad r = r_2. \tag{A.29}$$

The exact solution to (A.28)–(A.29) for the velocity profile in the annular gap will not be further considered in detail here. Instead we turn attention to an approximate, narrow-gap treatment of the problem, which provides simple analytic expressions for the local velocity and shear rate. We begin by noting that, over the typical experimental range of shear rates in the gap, the flow behavior index n_0 can for many real fluids be treated as nearly constant. This enables one to integrate the equations directly, to yield

$$V = r\Omega_1 \left\{ 1 + \alpha \, \frac{\left[1 - (r_1/r)^{2/n_0} \right]}{\left[1 - (r_1/r_2)^{2/n_0} \right]} \right\} \tag{A.30}$$

for the velocity profile and

$$\kappa^{(0)} = \frac{2\alpha\Omega_1 (r_1/r)^{2/n_0}}{n_0 \left[1 - (r_1/r_2)^{2/n_0} \right]} \tag{A.31}$$

for the shear rate, where

$$\alpha = (\Omega_2 - \Omega_1)/\Omega_1. \tag{A.32}$$

For many fluids, especially polymer fluids, these relations can be expected to provide a reasonable estimate of the flow prior to the onset of Taylor instability. Furthermore, (A.30) and (A.31) serve to provide convenient analytical expressions for both V and $\kappa^{(0)}$ in the narrow-gap limit, where the gap width becomes small relative to the radii.

Returning to the stability analyses, we assume a rotationally symmetric disturbance having the form

$$\mathbf{v}^{(1)} = u\mathbf{e}_r + v\mathbf{e}_\theta + w\mathbf{e}_z, \tag{A.33}$$

where u, v, w are analytic functions of r and z, independent of time t. Thus we are confining attention to the *neutral, nonoscillatory*, i.e., *stationary*, mode of stability, where the disturbance neither grows, decays, nor oscillates with time (cf. Beard *et al.*, 1966).

In view of the assumptions embodied in (A.6), (A.25), and (A.33), it is plausible that the deformation occurring in the neighborhood of each material particle can be represented consistently as an analytic function of time, following the particle. Therefore, our use of the Rivlin–Ericksen fluid model to describe general material constitutive behavior is to this extent justified.

To obtain an expression for the stress perturbation $\mathbf{T}^{(1)}$ for the disturbed flow, we must first determine the perturbations of the Rivlin–Ericksen tensors, which are provided by (A.11) and (A.17) as:

$$\mathbf{A}_1^{(1)} = 2 \frac{\partial u}{\partial r} \mathbf{e}_r \mathbf{e}_r + 2 \frac{u}{r} \mathbf{e}_\theta \mathbf{e}_\theta - 2 \left[\frac{\partial u}{\partial r} + \frac{u}{r} \right] \mathbf{e}_z \mathbf{e}_z + r \frac{\partial (v/r)}{\partial r} (\mathbf{e}_r \mathbf{e}_\theta + \mathbf{e}_\theta \mathbf{e}_r)$$

$$+ \left[\frac{\partial w}{\partial r} + \frac{\partial u}{\partial z} \right] (\mathbf{e}_r \mathbf{e}_z + \mathbf{e}_z \mathbf{e}_r) + \frac{\partial v}{\partial z} (\mathbf{e}_\theta \mathbf{e}_z + \mathbf{e}_z \mathbf{e}_\theta), \tag{A.34}$$

$$\mathbf{A}_2^{(1)} = \kappa^{(0)} \left\{ 4r \frac{\partial (v/r)}{\partial r} \mathbf{e}_r \mathbf{e}_r + \left[\frac{\partial u}{\partial r} + \frac{u}{r} (2 + n_0) \right] (\mathbf{e}_r \mathbf{e}_\theta + \mathbf{e}_\theta \mathbf{e}_r) \right.$$

$$\left. + \frac{\partial u}{\partial z} (\mathbf{e}_\theta \mathbf{e}_z + \mathbf{e}_z \mathbf{e}_\theta) + 2 \frac{\partial v}{\partial z} (\mathbf{e}_r \mathbf{e}_z + \mathbf{e}_z \mathbf{e}_r) \right\}, \tag{A.35}$$

$$\mathbf{A}_3^{(1)} = 3\kappa^{(0)2} \left\{ 2 \left[\frac{\partial u}{\partial r} + n_0 \frac{u}{r} \right] \mathbf{e}_r \mathbf{e}_r + \frac{\partial u}{\partial z} (\mathbf{e}_r \mathbf{e}_z + \mathbf{e}_z \mathbf{e}_r) \right\}, \tag{A.36}$$

$$\mathbf{A}_n^{(1)} = \mathbf{0} \qquad \text{for } n > 3, \tag{A.37}$$

where use has been made here of (A.28) and the condition of incompressibility to simplify the form of certain terms.

As can be seen from (A.37), *only the first three Rivlin–Ericksen tensors suffice to describe the perturbed kinematics.* We recall that, by Wang's theorem (Truesdell, 1966), the first three Rivlin–Ericksen tensors are sufficient to characterize the response of a simple fluid in *motions of constant stretch history.* Thus, since the description of kinematics for the present problem involves, to $O(\varepsilon)$, only the first three Rivlin–Ericksen tensors, we may conclude that *a knowledge of the fluid response in a general motion of constant stretch history is sufficient* (but not strictly necessary) *to determine its stationary Taylor–Couette stability.* This important fact suggests that it is possible to devise an experimental program for measuring the pertinent material parameters independently of the stability experiment itself.

For the present analysis, (A.34)–(A.37) be expressed in the form (A.23), involving a viscometric part and a nonviscometric part. The specific representation for the viscometric part is

$$\bar{\mathbf{A}}_1^{(1)} = r \frac{\partial (v/r)}{\partial r} (\mathbf{e}_r \mathbf{e}_\theta + \mathbf{e}_\theta \mathbf{e}_r) + \frac{\partial w}{\partial r} (\mathbf{e}_r \mathbf{e}_z + \mathbf{e}_z \mathbf{e}_r) + \frac{\partial v}{\partial z} (\mathbf{e}_\theta \mathbf{e}_z + \mathbf{e}_z \mathbf{e}_\theta),$$

$$\bar{\mathbf{A}}_2^{(1)} = \kappa^{(0)} \left[4r \frac{\partial (v/r)}{\partial r} \mathbf{e}_r \mathbf{e}_r + 2 \frac{\partial v}{\partial z} (\mathbf{e}_r \mathbf{e}_z + \mathbf{e}_z \mathbf{e}_r) \right] \tag{A.38}$$

while the nonviscometric portion may be written in the form

$$\tilde{\mathbf{A}}_n^{(1)} = \sum_{j=1}^{3} \tilde{\mathbf{A}}_{n,j}^{(1)}, \tag{A.39a}$$

where

$$\tilde{A}_{1,1}^{(1)} = 2\frac{\partial u}{\partial r}[e_r e_r - e_z e_z],$$

$$\tilde{A}_{1,2}^{(1)} = 2\frac{u}{r}[e_\theta e_\theta - e_z e_z],$$

$$\tilde{A}_{1,3}^{(1)} = \frac{\partial u}{\partial z}(e_r e_z + e_z e_r),$$

$$\tilde{A}_{2,1}^{(1)} = \kappa^{(0)}\frac{\partial u}{\partial r}(e_r e_\theta + e_\theta e_r),$$

$$\tilde{A}_{2,2}^{(1)} = \kappa^{(0)}(2 + n_0)\frac{u}{r}[e_r e_\theta + e_\theta e_r],$$

$$\tilde{A}_{2,3}^{(1)} = \kappa^{(0)}\frac{\partial u}{\partial z}(e_\theta e_z + e_z e_\theta),$$

$$\tilde{A}_{3,1}^{(1)} = 6\kappa^{(0)2}\frac{\partial u}{\partial r}e_r e_r,$$

$$\tilde{A}_{3,2}^{(1)} = 6\kappa^{(0)2}n_0\frac{u}{r}e_r e_r,$$

and

$$\tilde{A}_{3,3}^{(1)} = 3\kappa^{(0)2}(e_r e_z + e_z e_r)\frac{\partial u}{\partial z}.$$

(A.39b)

As can be seen from (A.39), only the radial component of the disturbance u contributes to the nonviscometric part of the flow and, conversely, the angular and axial components v and w are manifest only in the viscometric perturbation. For the latter, there is a physical interpretation of these results, in that the angular component of the disturbance v merely alters the local shear rate in the gap between the cylinders, but does not affect the basic parallel streamline character of the motion; likewise, the axial velocity w provides for a helical component to the deformation, and the resulting helical flow again represents another type of viscometric motion.

At any rate, an expression for the stress perturbation $\mathbf{T}^{(1)}$ can now be obtained by substituting (A.37) and (A.39) into (A.24):

$$\mathbf{T}^{(1)} + p^{(1)}\mathbf{1} = \sum_{n=1}^{2}\hat{\mathbf{g}}_n[A_1^{(0)}, A_2^{(0)}|\tilde{A}_n^{(1)}] + \sum_{j=1}^{3}\sum_{n=1}^{3}\hat{\mathbf{g}}_n[A_1^{(0)}, A_2^{(0)}|\tilde{A}_{n,j}^{(1)}], \quad (A.40)$$

where the order of summation has been interchanged in the double summation. It will be convenient for purposes of the forthcoming development to *re*express one of the summations in (A.40) as

$$\sum_{n=1}^{3} \mathbf{g}_n[\mathbf{A}_1^{(0)}, \mathbf{A}_2^{(0)} | \tilde{\mathbf{A}}_{n,j}^{(1)}] = \mathbf{h}_j[\mathbf{A}_1^{(0)}, \mathbf{A}_2^{(0)} | \tilde{\mathbf{A}}_{1,j}^{(1)}, \tilde{\mathbf{A}}_{2,j}^{(1)}, \tilde{\mathbf{A}}_{3,j}^{(1)}], \qquad \text{(A.41)}$$

where the $\mathbf{h}_j$ are isotropic tensor functions which are linear in the perturbation components shown. Substituting (A.41) into (A.40), we obtain

$$\mathbf{T}^{(1)} + p^{(1)}\mathbf{1} = \sum_{n=1}^{2} \hat{\mathbf{g}}_n[\mathbf{A}_1^{(0)}, \mathbf{A}_2^{(0)} | \bar{\mathbf{A}}_n^{(1)}]$$

$$+ \sum_{j=1}^{3} \mathbf{h}_j[\mathbf{A}_1^{(0)}, \mathbf{A}_2^{(0)} | \tilde{\mathbf{A}}_{1,j}^{(1)}, \tilde{\mathbf{A}}_{2,j}^{(1)}, \mathbf{A}_{3,j}^{(1)}]. \qquad \text{(A.42)}$$

The viscometric portion of the stress perturbation tensor is obtained directly from (A.18) and (A.19) by substituting $\bar{\mathbf{A}}_1^{(1)}$ and $\bar{\mathbf{A}}_2^{(1)}$ for $\mathbf{A}^{(1)}$ and $\mathbf{A}_2^{(1)}$. One thus finds that

$$\sum_{n=1}^{2} \hat{\mathbf{g}}_n[A_1^{(0)}, A_2^{(0)} | \bar{\mathbf{A}}_n^{(1)}] = r\frac{\partial(v/r)}{\partial r}\left[\frac{n_2\sigma_2}{\kappa^{(0)}}\,\mathbf{e}_r\mathbf{e}_r + \frac{n_1\sigma_1}{\kappa^{(0)}}\,\mathbf{e}_\theta\mathbf{e}_\theta\right.$$

$$\left. + n_0\eta(\mathbf{e}_r\mathbf{e}_\theta + \mathbf{e}_\theta\mathbf{e}_r)\right] + \left[\eta\frac{\partial w}{\partial r} + \frac{\sigma_2}{\kappa^{(0)}}\frac{\partial v}{\partial z}\right](\mathbf{e}_r\mathbf{e}_z + \mathbf{e}_z\mathbf{e}_r)$$

$$+ \left[\frac{\sigma_1}{\kappa^{(0)}}\frac{\partial w}{\partial r} + \eta\frac{\partial v}{\partial z}\right](\mathbf{e}_\theta\mathbf{e}_z + \mathbf{e}_z\mathbf{e}_\theta). \qquad \text{(A.43)}$$

The nonviscometric part of the stress perturbation is more difficult to determine, and moreover is not necessarily related (except in certain limiting situations) to the viscometric functions η, σ_1, and σ_2. Let us undertake consideration of the nonviscometric stress by confining attention initially to the determination of the stress function $\mathbf{h}_1$ appearing in (A.42). Recall that this function is linear in the perturbation components $\tilde{\mathbf{A}}_{1,1}^{(1)}$, $\tilde{\mathbf{A}}_{2,1}^{(1)}$, and $\tilde{\mathbf{A}}_{3,1}^{(1)}$. Furthermore one will note by referring to (A.39) that each of these perturbation components contains the velocity gradient $\partial u/\partial r$ as a factor. Therefore, it is apparent that $\mathbf{h}_1$ must, for the present flow, be directly proportional to $\partial u/\partial r$.

In order to satisfy our basic material isotropy requirements, the function $\mathbf{h}_1$ must also be an *isotropic* tensor function of its arguments, such that the following identity applies for all arbitrary orthogonal tensors $\mathbf{Q}$:

$$\mathbf{h}_1[\mathbf{A}^*,\mathbf{B}^* | \mathbf{C}^*, \mathbf{D}^*, \mathbf{E}^*] = \mathbf{h}_1^*[\mathbf{A}, \mathbf{B} | \mathbf{C}, \mathbf{D}, \mathbf{E}], \qquad \text{(A.44)}$$

where

$$\mathbf{G}^* = \mathbf{Q} \cdot \mathbf{G} \cdot \mathbf{Q}^T$$

for an arbitrary second-rank tensor. If we now choose to consider the particular situation where the tensor $\mathbf{Q}$ is taken to be

$$\mathbf{Q} = \mathbf{e}_r\mathbf{e}_r + \mathbf{e}_\theta\mathbf{e}_\theta - \mathbf{e}_z\mathbf{e}_z \tag{A.45}$$

we find, for the kinematic tensors given in (A.26) and (A.39), that

$$\mathbf{A}_n^{(0)*} = \mathbf{A}_n^{(0)} \qquad \text{for } n = 1, 2$$

and

$$\tilde{\mathbf{A}}_{n,1}^{(1)*} = \tilde{\mathbf{A}}_{n,1}^{(1)} \qquad \text{for } n = 1, 2, 3. \tag{A.46}$$

Substitution of these results into (A.43) then leads to the condition that

$$\mathbf{h}_1[\mathbf{A}, \mathbf{B} | \mathbf{C}, \mathbf{D}, \mathbf{E}] = \mathbf{Q} \cdot \mathbf{h}_1[\mathbf{A}, \mathbf{B} | \mathbf{C}, \mathbf{D}, \mathbf{E}] \cdot \mathbf{Q}^T. \tag{A.47}$$

If one resolves (A.47) into component form, one finds that certain components of $\mathbf{h}_1$ must vanish. In particular,

$$(\mathbf{h}_1)_{rz} = (\mathbf{h}_1)_{\theta z} = 0. \tag{A.48}$$

Furthermore, since the stress perturbation $\mathbf{T}^{(1)}$ is determined uniquely only to within an arbitrary isotropic stress $-p^{(1)}\mathbf{1}$, we can choose

$$(\mathbf{h}_1)_{zz} = 0 \tag{A.49}$$

with no loss of generality. Our expression for the stress function $\mathbf{h}_1$ may then be written in the form

$$\mathbf{h}_1 = \frac{\partial u}{\partial r}\left[4\eta_1\mathbf{e}_r\mathbf{e}_r + 2\eta_2\mathbf{e}_\theta\mathbf{e}_\theta + \frac{\sigma_3}{2\kappa^{(0)}}(\mathbf{e}_r\mathbf{e}_\theta + \mathbf{e}_\theta\mathbf{e}_r)\right], \tag{A.50}$$

where η_1, η_2, and σ_3 are *new* material functions that depend on the undisturbed shear rate $\kappa^{(0)}$. (The dependence on $\kappa^{(0)}$ can be attributed to the dependence on $\mathbf{A}_1^{(0)}$ and $\mathbf{A}_2^{(0)}$ in the general expression for $\mathbf{h}_1$.)

Our purpose in representing the new material functions η_1, η_2, σ_3 in the above manner will become apparent shortly. Suffice it to say at this point that by again making use of the isotropic property (A.44) for the specific instance where the tensor $\mathbf{Q}$ is given by

$$\mathbf{Q} = -\mathbf{e}_r\mathbf{e}_r + \mathbf{e}_\theta\mathbf{e}_\theta + \mathbf{e}_z\mathbf{e}_z \tag{A.51}$$

we can readily demonstrate that η_1, η_2, and σ_3 are even functions of $\kappa^{(0)}$.

A treatment similar to that presented above may also be employed to determine the form of the functions $\mathbf{h}_2$ and $\mathbf{h}_3$ appearing in (A.42). The details of the derivations are omitted from the present development, but the final result for the complete stress perturbation $\mathbf{T}^{(1)}$ is

$$T_{rr}^{(1)} - T_{zz}^{(1)} = 4\eta_1 \frac{\partial u}{\partial r} + 2\eta_3 \frac{u}{r} + r \frac{\partial(v/r)}{\partial r} \frac{n_2 \sigma_2}{\kappa^{(0)}},$$

$$T_{\theta\theta}^{(1)} - T_{zz}^{(1)} = 2\eta_2 \frac{\partial u}{\partial r} + 4\eta_4 \frac{u}{r} + r \frac{\partial(v/r)}{\partial r} \frac{n_1 \sigma_1}{\kappa^{(0)}},$$

$$T_{r\theta}^{(1)} = \frac{\sigma_3}{\kappa^{(0)}} \frac{\partial u}{\partial r} + \frac{\sigma_4}{\kappa^{(0)}} \frac{u}{r} + n_0 \eta r \frac{\partial(v/r)}{\partial r}, \qquad \text{(A.52)}$$

$$T_{rz}^{(1)} = \frac{\partial w}{\partial r} \eta + \frac{\partial v}{\partial z} \frac{\sigma_2}{\kappa^{(0)}} + \frac{\partial u}{\partial z} \eta_5,$$

$$T_{\theta z}^{(1)} = \frac{\partial w}{\partial r} \frac{\sigma_1}{\kappa^{(0)}} + \frac{\partial v}{\partial z} \eta + \frac{\partial u}{\partial z} \frac{\sigma_5}{\kappa^{(0)}},$$

where $\eta_1, \eta_2, \eta_3, \eta_4, \eta_5, \sigma_3, \sigma_4,$ and σ_5 are all new material functions, which are distinct from the viscometric functions and are even functions of the undisturbed shear rate $\kappa^{(0)}$.

We recall that the response of simple fluids with fading memory in the limit of very slow deformations approaches that of a "second-grade fluid," for which

$$\mathbf{T} + p\mathbf{1} = \alpha_1 \mathbf{A}_1 + \alpha_2 \mathbf{A}_1^2 + \alpha_3 \mathbf{A}_2, \qquad \text{(A.53)}$$

where the constants $\alpha_1, \alpha_2,$ and α_3 are given by

$$\alpha_1 = \lim_{\kappa \to 0} \eta(\kappa),$$

$$\alpha_2 = \lim_{\kappa \to 0} \sigma_1(\kappa)/\kappa^2 \qquad \text{(A.54)}$$

$$2\alpha_3 = \lim_{\kappa \to 0} (\sigma_2(\kappa) - \sigma_1(\kappa))/\kappa^2.$$

If we consider the stress perturbation $\mathbf{T}^{(1)}$ for the particular fluid model described by (A.53), we find that the material functions in (A.52) are given as functions of $\kappa^{(0)}$ by

$$\eta_1 = \eta_2 = \eta_3 = \eta_4 = \eta_5 = \alpha_1,$$

$$\sigma_3 = \kappa^{(0)^2}(2\alpha_2 + \alpha_3),$$

$$\sigma_4 = \kappa^{(0)^2}(2\alpha_2 + 3\alpha_3), \qquad \text{(A.55)}$$

$$\sigma_5 = \kappa^{(0)^2}(\alpha_2 + \alpha_3).$$

Since (A.55) should become applicable in the region where the shear rate

$\kappa^{(0)}$ is very small, one is led to the following limiting behavior of the new material functions at low shear rates, $\kappa^{(0)} \to 0$:

$$\eta_1 \to \eta_2 \to \eta_3 \to \eta_4 \to \eta_5 \to \eta \to \eta_0,$$

$$\sigma_3 \to (3\sigma_1 + \sigma_2)/2,$$

$$\sigma_4 \to (\sigma_1 + 3\sigma_2)/2,$$

$$\sigma_5 \to (\sigma_1 + \sigma_2)/2.$$

(A.56)

As can be seen from (A.56), the quantities η_1 through η_5 are, in a manner of speaking, similar to the viscosity η, at least at low shear rates. Likewise, the parameters σ_3, σ_4, and σ_5 may be interpreted as normal stresses, analogous to σ_1 and σ_2. Having all the material functions associated with the perturbed kinematics now in hand, we may turn to a consideration of the dynamics.

The dynamical equations for the perturbed motion, which are found in the usual way from the equations of continuity and momentum for an incompressible fluid, are given by

$$\text{(continuity)} \qquad \frac{1}{r}\frac{\partial(ur)}{\partial r} + \frac{\partial w}{\partial z} = 0, \qquad \text{(A.57)}$$

$$\text{(momentum)} \qquad -2\rho v(V/r) = \frac{\partial T_{rr}^{(1)}}{\partial r} + \frac{\partial T_{rz}^{(1)}}{\partial z} + \frac{T_{rr}^{(1)} - T_{zz}^{(1)}}{r},$$

$$\frac{\rho u}{r}\frac{\partial(Vr)}{\partial r} = \frac{1}{r^2}\frac{\partial(r^2 T_{r\theta}^{(1)})}{\partial r} + \frac{\partial T_{\theta z}^{(1)}}{\partial z}, \qquad \text{(A.58)}$$

$$0 = \frac{1}{r}\frac{\partial(rT_{rz}^{(1)})}{\partial r} + \frac{\partial T_{zz}^{(1)}}{\partial z}.$$

We now make use of the simplifying, "small-gap" assumption, strictly valid in limit where the annular gap width is vanishingly small compared to the radii of the cylinders. Here, one can readily show that the local undisturbed shear rate will approach a unique value independent of gap position, which we henceforth refer to as the "gap shear rate." It is also of interest to note that since in the small-gap approximation the shear rate $\kappa^{(0)}$ is constant, the material parameters in (A.52) are likewise constants independent of position.

If we now substitute Eqs. (A.30), (A.31), and (A.52) into Eqs. (A.57) and (A.58) and employ the small-gap approximation, we obtain the following set of equations:

$$\frac{\partial u}{\partial r} + \frac{\partial w}{\partial z} = 0, \qquad \text{(A.59)}$$

$$-2\rho v\Omega_1[1+\alpha x] = \frac{\partial T_{zz}^{(1)}}{\partial r} + (4\eta_1 - \eta)\frac{\partial^2 u}{\partial r^2} + 2\frac{(\eta_2 - \eta_3)}{r_1}\frac{\partial u}{\partial r} - \frac{4\eta_4}{r_1^2}u$$

$$+\frac{\sigma_2\eta_2}{\kappa^{(0)}}\frac{\partial^2 v}{\partial r^2} + \frac{\sigma_2}{\kappa^{(0)}}\frac{\partial^2 v}{\partial z^2} + \eta_5\frac{\partial^2 u}{\partial z^2} - \frac{n_1\sigma_1}{\kappa^{(0)}}\frac{\partial v}{\partial r}, \tag{A.60a}$$

$$\rho u\kappa^{(0)} = \frac{(\sigma_3 - \sigma_1)}{\kappa^{(0)}}\frac{\partial^2 u}{\partial r^2} + \frac{\sigma_4}{\kappa^{(0)}}\frac{1}{r_1}\frac{\partial u}{\partial r} + \frac{\sigma_5}{\kappa^{(0)}}\frac{\partial^2 u}{\partial z^2} + \eta\left[n_0\frac{\partial^2 v}{\partial z^2} + \frac{\partial^2 v}{\partial z^2}\right], \tag{A.60b}$$

$$0 = \frac{\partial T_{zz}^{(1)}}{\partial z} + \eta\frac{\partial^2 w}{\partial r^2} + \frac{\sigma_2}{\kappa^{(0)}}\frac{\partial^2 v}{\partial z\,\partial r} - \eta_5\frac{\partial^2 w}{\partial z^2}, \tag{A.60c}$$

where the gap shear rate $\kappa^{(0)}$ is given by

$$\kappa^{(0)} = \alpha a\Omega_1 \tag{A.61}$$

with

$$a = r_1/(r_2 - r_1) \tag{A.62}$$

and with α given by Eq. (A.32).

Upon inspection of (A.60), one might be tempted to discard certain terms in these relations on the basis of the small-gap approximation. For example, the expression $[2(\eta_2 - \eta_3)/r_1]/(\partial u/\partial r)$ appears, at first glance, to become negligible at large values of r_1. However, the material functions in (A.30) depend on $\kappa^{(0)}$, with the exact form of the dependence unknown for several of the material functions (namely η_1 through η_5, σ_3, σ_4, and σ_5) except at low shear rates. Note further, from (A.61), that for fixed values of the parameters α and Ω_1, the gap shear rate increases in direct proportion to the radius-to-gap ratio a. Therefore, it is not possible to set bounds on the magnitude of the terms involving the newly defined material functions unless we restrict the magnitude of $\kappa^{(0)}$. In the absence of such a restriction, there is no justification for discarding the terms in question. On the other hand, if such a restriction is imposed, we may consider the shear rate $\kappa^{(0)}$ to be constant, in which case the speed of the inner cylinder Ω_1 must decrease inversely with the radius to gap ratio (for fixed α). This will cause the inertial term in Eq. (A.60a) to become negligible; and, without the inertial term, the relations will not reduce to the appropriate form in the Newtonian limit. More precisely, the Taylor number will not emerge as the key parameter in the analysis, and the equations will become simply those for flow between two flat plates. It is clear then, on the basis of the above discussion, that no additional terms in (A.60) can be discarded *a priori* on the basis of the small-gap approximation.

To reduce Eqs. (A.59) and (A.60) to dimensionless form, let us now define the following quantities:

$$
\left.
\begin{aligned}
&X = (r - r_1)/(r_2 - r_1), && w^* = w/\Omega_1(r_2 - r_1), \\
&Z = z/(r_2 - r_1), && R = 2\rho(r_2 - r_1)^2\Omega_1/\eta, \\
&\Sigma = T_{zz}^{(1)}/\Omega_1\eta, && R' = -\rho\kappa^{(0)}(r_2 - r_1)^2/\eta, \\
&u^* = u/\Omega_1(r_2 - r_1), && T = RR' = -\alpha a R^2/2. \\
&v^* = v/\Omega_1(r_2 - r_1),
\end{aligned}
\right\} \qquad \text{(A.63)}
$$

As can be seen from these relations, the *Reynolds numbers* R and R' and the *Taylor number* T are given in terms of the ordinary shear viscosity $\eta(\kappa)$ evaluated at the gap shear rate $\kappa^{(0)}$. By way of contrast, several other treatments of this same stability problem for the case of various specific fluid models have made use of expressions for R, R', and T that involve a constant viscosity, characteristic of the particular fluid under investigation. Such approaches have resulted in the introduction of an additional and superfluous parameter into the analysis, as will soon become obvious. On the basis of the present disturbance equations, which contain $\eta(\kappa^{(0)})$ explicitly as a material function, the most appropriate forms for R, R', and T are those given by (A.63).

Substituting Eqs. (A.62) and (A.63) into Eqs. (A.59) and (A.60), we can now obtain the disturbance equations in dimensionless form:

$$
\frac{\partial u^*}{\partial X} + \frac{\partial w^*}{\partial Z} = 0, \qquad \text{(A.64)}
$$

$$
-v^*(1 + \alpha X)R = \frac{\partial \Sigma}{\partial X} + \frac{(4\eta_1 - \eta)}{\eta}\frac{\partial^2 u^*}{\partial X^2} + \frac{2(\eta_2 - \eta_3)}{\eta a}\frac{\partial u^*}{\partial X} - \frac{4\eta_4 u^*}{\eta a^2}
$$

$$
+ \frac{\sigma_2 \eta_2}{\eta\kappa^{(0)}}\frac{\partial^2 v^*}{\partial X^2} + \frac{\sigma_2}{\eta\kappa^{(0)}}\frac{\partial^2 v^*}{\partial Z^2} + \frac{\eta_5}{\eta}\frac{\partial^2 u^*}{\partial Z^2} - \frac{\eta_1\sigma_1}{\eta\kappa^{(0)}}\frac{\partial v^*}{\partial X}, \qquad \text{(A.65a)}
$$

$$
-R'u^* \equiv n_0 \frac{\partial^2 v^*}{\partial X^2} + \frac{\partial^2 v^*}{\partial Z^2} + \left[\frac{\sigma_3 - \sigma_1}{\eta\kappa^{(0)}}\right]\frac{\partial^2 u^*}{\partial X^2}
$$

$$
+ \left[\frac{\sigma_4}{\eta\kappa^{(0)}}\right]\frac{1}{a}\frac{\partial u^*}{\partial X} + \left[\frac{\sigma_5}{\eta\kappa^{(0)}}\right]\frac{\partial^2 u^*}{\partial Z^2}, \qquad \text{(A.65b)}
$$

$$
0 = \frac{\partial \Sigma}{\partial Z} + \frac{\partial^2 w^*}{\partial X^2} - \frac{\eta_5}{\eta}\frac{\partial^2 w^*}{\partial Z^2} + \frac{\sigma_2}{\eta\kappa^{(0)}}\frac{\partial^2 v^*}{\partial Z \partial X}. \qquad \text{(A.65c)}
$$

We next resolve the disturbance into normal modes, by making the usual assumption that the disturbance is spatially periodic in the z-direction and writing

$$u^* = k\psi(X)\sin kZ,$$
$$v^* = -k\chi(X)\sin kZ,$$
$$w^* = D\psi \cos kZ, \qquad\qquad\qquad (A.66)$$

and

$$\Sigma = kP(X)\sin kZ,$$

where $D = d/dX$ and where k is the dimensionless wave number or spatial frequency of the disturbance. Substitution of Eqs. (A.66), which satisfy (A.64) identically, into (A.65) yields

$$R[1 + \alpha X]\chi + \left[\frac{\sigma_2}{\eta\kappa^{(0)}}\right][n_2 D^2 - k^2]\chi - \frac{\sigma_1}{\eta\kappa^{(0)}}\frac{n_1}{a}D\chi$$

$$= DP + \left[\frac{4\eta_1 - \eta}{\eta}\right]D^2\psi + \frac{2(\eta_2 - \eta_3)}{\eta a}D\psi - \left[\frac{4\eta_4}{\eta a^2} + \frac{\eta_5}{\eta}k^2\right]\psi, \quad (A.67a)$$

$$[n_0 D^2 - k^2]\chi = \left[R' + \frac{(\sigma_3 - \sigma_1)D^2}{\eta\kappa^{(0)}} + \frac{\sigma_4}{\eta\kappa^{(0)}}\frac{1}{a}D - \frac{\sigma_5}{\eta\kappa^{(0)}}k^2\right]\psi, \quad (A.67b)$$

$$0 = k^2 P + D^3\psi + \frac{\eta_5}{\eta}k^2 D\psi - \frac{\sigma_2}{\eta\kappa^{(0)}}k^2 D\psi. \qquad (A.67c)$$

Eliminating the function P between Eqs. (A.67a) and (A.67c), we find that the disturbance equations now take the form

$$[n_0 D^2 - k^2]\chi = \left[R' + \beta_1 D^2 - \beta_2 k^2 + \frac{\beta_3}{a}D\right]\psi, \qquad (A.68a)$$

$$\left[D^4 - 2\beta_4 D^2 k^2 + \beta_5 k^4 - \frac{\beta_6}{a}k^2 D + 4\beta_7\frac{k^2}{a^2}\right]\psi$$

$$= -k^2\left\{R(1 + \alpha X) + \beta_8[(n_2 - 1)D^2 - k^2] - \frac{\beta_9}{a}n_1 D\right\}\chi, \quad (A.68b)$$

where the β's are the following dimensionless rheological parameters, which depend on the gap shear rate $\kappa^{(0)}$:

$$\beta_1 = (\sigma_3 - \sigma_1)/\eta\kappa^{(0)}, \qquad \beta_2 = \sigma_5/\eta\kappa^{(0)}, \qquad \beta_3 = \sigma_4/\eta\kappa^{(0)},$$
$$\beta_4 = (4\eta_1 - \eta_5 - \eta)/2\eta, \qquad \beta_5 = \eta_5/\eta, \qquad \beta_6 = 2(\eta_2 - \eta_3)/\eta, \quad (A.69)$$
$$\beta_7 = \eta_4/\eta, \qquad \beta_8 = \sigma_2/\eta\kappa^{(0)}, \qquad \beta_9 = \sigma_1/\eta\kappa^{(0)}.$$

As can be seen from Eqs. (A.68) and (A.69), the terms operating on χ in (A.68) are related exclusively to the three viscometric functions η, σ_1, and σ_2.

Conversely, those operating on ψ are given exclusively in terms of the newly defined material functions.

Note further from Eqs. (A.68) and (A.69) that all the β-parameters on the left-hand sides of these equations (β_4 through β_7) are "viscosity" type quantities, and are even functions of $\kappa^{(0)}$.

Equations (A.68) contain a total of seven new rheological parameters (β_1 through β_7), unrelated generally to the viscometric functions of the fluid, except at vanishing shear rates. Now, in the original expressions for the components of the stress perturbation $\mathbf{T}^{(0)}$ (A.52), there were eight newly defined material functions (η_1 to η_5 and σ_3 to σ_5). Therefore, in the course of our analysis, the parameters have combined in such a way as to reduce the total number of such unknown material functions by one.

Equations (A.68) represent two ordinary, homogeneous, differential equations in the two unknowns χ and ψ. Together with the homogeneous boundary conditions

$$\psi = D\psi = \chi = 0 \qquad \text{at } X = 0, 1 \qquad (A.70)$$

they define a characteristic value problem for the gap shear rate $\kappa^{(0)}$ as a function $\kappa^{(0)}(k)$ of the dimensionless wave number k. The minimum value of $|\kappa^{(0)}|$ over all values for k determines the critical conditions $\kappa_c^{(0)}$ and k_c at which instability first sets in.

Each of the terms in (A.68) may be expected to have a different influence on the solution to the disturbance equations. By carrying out a sensitivity analysis on the system (using either analytic or numerical methods to solve the disturbance equations) it is possible to determine which terms affect the critical shear rate most strongly and which ones are weak and can be neglected. The outcome of such an analysis is as follows: The terms in (A.68) that contain the operator D to the first power are negligible, provided the corresponding rheological coefficient is not extremely large. If we assume the latter to be the case, then the disturbance equations reduce to

$$[n_0 D^2 - k^2]\chi = [R' + \beta_1 D^2 - \beta_2 k^2]\psi, \qquad (A.71a)$$

$$[D^4 - 2\beta_4 D^2 k^2 + \beta_5 k^4 + 4\beta_7 k^2/a^2]\psi$$
$$= -k^2[R(1 + \alpha X) + \beta_8[(n_2 - 1)D^2 - k^2]\chi \qquad (A.71b)$$

which contain only five new rheological functions, compared with seven in (A.68) (Miller, 1967).

Previous Studies, Special Fluid Models and Limiting Cases

Lockett and Rivlin (1968) have given a comprehensive analysis of the present stability problem, which employs the same Rivlin–Ericksen model as that of the current treatment (see also Smith and Rivlin, 1972). The disturbance equations they obtain are nearly identical in form to (A.68), except for the term $4\beta_7 k^2/a^2$, which is absent from their results. One will recall that in our analyses, we could not justify the neglect of this term on the basis of the small gap approximation, as Lockett and Rivlin have done. The other major differences between the present findings and those of Lockett and Rivlin are:

The present analysis shows clearly that the three viscometric functions η, N_1, and N_2 play a primary role in influencing the stability behavior, and it elucidates specifically how these functions enter into the disturbance equations. In effect, this reduces the total number of new parameters (beyond those associated with viscometric flow) from *nine*, in Lockett and Rivlin's analysis, to *five*.

Instead of Reynolds numbers based on $\eta(0)$, which Rivlin and Lockett employ in their analysis, the present work indicates that a more meaningful basis for these parameters is the shear viscosity $\eta(\kappa)$ evaluated at the gap shear rate.

We specify the parity of *all* rheological parameters in the disturbance equations (i.e., which are even functions, and which are odd in shear rate) and, where possible, have related them to the viscometric functions.

Datta (1964) has considered the stability in flow between concentric rotating cylinders for a "second-order fluid," defined by (A.53). This model is known to describe the behavior of all simple fluids with fading memory under extremely slow deformations. Consequently, Datta's findings include the results of Graebel (1961), Thomas and Walters (1964a,b), and Rubin *et al.* (1968) as special cases. In this regard, we note that the analysis of Rubin *et al.* appears to contain several algebraic errors rendering their final disturbance equations incorrect, which in fact is immediately obvious from the disagreement of their equations with those of Datta.

If one compares the present findings with those of Datta, the specific form of his material functions are found to be

$$
\left.
\begin{aligned}
&\eta = \alpha_1, \qquad \beta_1 = \beta_2 = \frac{(\alpha_2 + \alpha_3)\kappa}{\alpha_1} = \frac{(\sigma_1 + \sigma_2)}{2\eta\kappa}, \\
&\sigma_1 = \alpha_2 \kappa^2, \\
&\sigma_2 = (\alpha_2 + 2\alpha_3)\kappa^2, \qquad \beta_4 = \beta_5 = 1, \qquad \beta_7 = 0
\end{aligned}
\right\} \quad \text{(A.72)}
$$

with

$$
n_0 = 1, \, n_2 = 2.
$$

According to (A.72), the material parameters β_1 and β_2 both approach the average of the two normal stresses divided by the shear stress as the gap shear rate $\kappa^{(0)}$ becomes very small. Furthermore, the parameters β_4, β_5, and β_7 are found to approach their Newtonian values in this limit.

Giesekus (1966) and Ginn and Denn (1969) have attempted to extrapolate the results of Datta to situations where the gap shear rate is no longer very small, by assuming that the constants α_1, α_2, and α_3 in Data's final disturbance equations are shear dependent. At a glance, this approach would appear to lead to the correct results for the fluid model defined by (A.53), which is equivalent to Datta's expression with shear dependent parameters. Unfortunately, the approach fails, as can readily be verified by employing (A.53) as the starting equation, deriving the disturbance equations, and comparing the findings with those of Giesekus. One thereby observes that the final disturbance equations are nearly identical in the two situations, except that the parameters n_0 and n_2 which are equal to 1 and 2, respectively, in Giesekus' equations. Clearly, for a general fluid, n_0 and n_2 are not necessarily restricted to these values.

A number of other investigations have been carried out to consider the present stability problem for the case of specific viscoelastic fluid models. These may now be regarded as special case of the present general analyses. In Table IIA and B, the specific forms are presented for the key rheological parameters that apply in each of the previous studies. Although the equations for the rheological parameters depend on the specific fluid model, the general form of the disturbance equations, involving a fixed number of rheological functions, is *model independent*.

We note also that the present analysis carries over to several interesting limiting cases. We recall that Giesekus (1966) has carried out an analysis of the present stability problem for the second-grade fluid in two limiting situations of considerable interest. These are (a) *plane Couette flow*, where the radius-to-gap ratio of the system becomes infinite at a fixed value of the gap shear rate and (b) *inertia-free flow*, where the "rheological" terms dominate over the inertial terms in the disturbance equations. We now reconsider Giesekus' findings in terms of the present more general fluid model.

Considering first the case of *inertia-free flow*, we find that the disturbance equations reduce to

$$(n_0 D^2 - k^2)\chi = [\beta_1 D^2 - \beta_2 k^2]\psi,$$

$$[D^4 - 2\beta_4 D^2 k^2 + \beta_5 k^4 + 4\beta_7 k^2/a^2]\psi = -k^2 \beta_8 [(n_2 - 1)D^2 - k^2]\chi. \quad \text{(A.73)}$$

Since these equations no longer contain inertial terms, no simple modification of the Taylor criterion for the onset of instability can be appropriate here. (In the case of a Newtonian fluid, it is of course precisely the inertial terms which provide the driving force for instability!) Instead, an entirely new criterion is called for, in *which the rheological coefficients themselves determine the instability.*

The analysis of Giesekus (1966) shows that the critical wave number in the inertia-free case is infinite, indicating that the disturbance wavelength and the vortex cell size would be much smaller than in the corresponding Newtonian situation. If we take the limit of (A.73) for $k \to \infty$, we immediately obtain the appropriate stability criterion for negligible inertia in the present model:

$$\frac{\beta_2 \beta_8}{\beta_5} = 1. \tag{A.74}$$

Accordingly, the onset of instability does *not* depend, in the limit, on which of the concentric cylinders was rotating and which was held fixed and, indeed, it is immaterial how the uniform shear of the base flow is established. Rather, the onset is determined only by the base shear rate and flow boundaries. (Such results may have serious implications for both cone-and-plate rheometry and Couette rheometry, each of which depends on the maintenance of a shear flow for the study of viscometric behavior.)

We now turn attention to the case of *plane Couette flow*, in which limit the present disturbance equations reduce to

$$[n_0 D^2 - k^2]\chi = [R' + \beta_1 D^2 - \beta_2 k^2]\psi,$$
$$[D^4 - 2\beta_4 D^2 k^2 + \beta_5 k^4]\psi = -k^2 \beta_8[(n_2 - 1)D^2 - k^2]\chi. \tag{A.75}$$

For this situation, Giesekus' (1966) results indicate that the onset of instability will occur for $k \to \infty$ at small values of R', but that beyond a certain value of R', the point of instability will be shifted back toward smaller k_c. Under no circumstances, however, would the cells become as large as in the Newtonian case. It is interesting, further, to note that Giesekus finds that the introduction of inertia has a *destabilizing* influence on the flow.

To conclude here, we show that the case of the "generalized Newtonian fluid" (cf. Bird *et al.*, 1977a) defined by the constitutive model

$$\mathbf{T} + p\mathbf{1} = \eta(\kappa)\mathbf{A}_1,$$

with

$$\kappa^2 = \text{tr}(\mathbf{A}_1^2/2) \tag{A.76}$$

can be dealt with once and for all. This fluid exhibits neither normal stress behavior nor memory effects in general deformations. It includes many classical models, such as the Ostwald–deWaele or "power-law" fluid, the Bingham-plastic material, the Reiner–Philippoff model, the Eyring fluid, and the Ellis fluid, as well as a host of other "viscosity models" as special cases. We now demonstrate that it is possible to specify the stability in Couette flow of any such fluid definitively.

We begin by writing down the disturbance equations for this model:

$$[n_0 D^2 - k^2]\chi = R'\psi,$$
$$[D^2 - \psi^2]^2\psi = -k^2 R(1 + \alpha X)\chi. \tag{A.77}$$

Next, redefining the parameter ψ such that $\psi_0 = R'\psi$, we have

$$[n_0 D^2 - k^2]\chi = \psi_0,$$
$$[D^2 - k^2]^2\psi_0 = -Tk^2(1 + \alpha X)\chi. \tag{A.78}$$

Equations (A.78) reduce exactly to those for a Newtonian fluid if we set the parameter n_0 equal to 1. Therefore one expects that the onset of instability will be characterized by a critical Taylor number which may be regarded as a function of the parameters α and n_0.

Miller (1967) has solved the disturbance equations numerically for various values of n_0, for the most frequently treated case $\alpha = -1$ (inner cylinder rotating, outer one fixed). The results of the calculations are presented in Figs. 4 and 5. As can be seen from the figures, the critical Taylor number increases monotonically with the flow behavior index n_0, as does the critical wave number k_c. Thus, for *"pseudoplastic"* materials, $(n_0 < 1)$ the onset of instability will correspond to lower values of T_c and k_c than in the Newtonian case, while for *"dilatent"* materials $(n_0 > 1)$ the reverse is true.

ACKNOWLEDGMENTS

An expression of gratitude is due to several organizations and individuals. The present work has been supported in part by Grant ENG 7700458 from the National Sscience Foundation and an endowment from the Fluor Corporation, Irvine. The work in Section II was initiated during the author's tenure as a National Science Foundation Senior Fellow in the University of Cambridge Department of Applied Mathematics and Theoretical Physics. I am indebted to Professor G. K. Batchelor, for serving as host, and to Dr. Rex Dark of Kings College, Cambridge, for helping me to extract some of the results in Section II from the mathematics literature. The work in the Appendix, authored by Dr. C. Miller, is based on work largely completed at the University of Michigan and supported by the National Aeronautics and Space Administration through Grant NSG-659.

In addition to the above mentioned parties, I wish to acknowledge the outstanding assistance of Mr. Robert Nicholson in helping to prepare the manuscript.

REFERENCES

ACIERNO, D., LAMANTIA, F. P., MARRUCCI, G., and TITOMANLIO, G. (1976). A non-linear viscoelastic model with structure dependent relaxation times. *J. Non-Newtonian Fluid Mech.* **1**, 125 and 147.

ACIERNO, D., LaMANTIA, F. P., MARRUCCI, G., and TITOMANLIO, C. (1977). *J. Non-Newtonian Fluid Mech.* **2**, 225.

ASTARITA, G. (1979). Three alternate approaches to development of constitutive equations. *J. Non-Newtonian Fluid. Mech.* **5**, 125.

ASTARITA, G., and DENN, M. M. (1974). The effect of non-Newtonian properties of polymer solutions on flow fields. *In* "Theoretical Rheology" (J. F. Hutton, J. R. A. Pearson, and K. Walters, eds.), Chapter 20. Wiley, New York.

ASTARITA, G., and JONGSCHAAP, R. J. J. (1978). The maximum amplitude of strain for the validity of linear viscoelasticity. *J. Non-Newtonian Fluid Mech.* **3**, 381.

ASTARITA, G., and MARRUCCI, G. (1974). "Principles of Non-Newtonian Fluid Mechanics." McGraw-Hill, New York.

BARTHÉS-BIESEL, D., and ACRIVOS, A. (1974). The rheology and suspensions and its relation to phenomenological theories for non-Newtonian fluids. *Int. J. Multiphase Flow* **1**, 1.

BATCHELOR, G. K. (1971). The stress generated in a non-dilute suspension of elongated particles by pure straining motion. *J. Fluid Mech.* **46**, 813.

BATCHELOR, G. K. (1974). Transport properties of two-phase materials with random structure. *Annu. Rev. Fluid Mech.* **6**, 227.

BEARD, D. W., DAVIES, M. H., and WALTERS, K. (1966). The stability of elastico-viscous flow between rotating cylinders. Part 3. Overstability in viscous and Maxwell fluids. *J. Fluid Mech.* **24**, 321.

BEAVERS, G. S., and JOSEPH, D. D. (1975). The rotating rod viscometer. *J. Fluid Mech.* **69**, 475.

BERNSTEIN, B., and MALKUS, D. S. (1978). Finite element analysis of steady non-Newtonian flow of a memory fluid. *49th Annu. Soc. Rheo. Meet. 1978* Paper A3.

BERSTEIN, B., KEARSLEY, E. A., and ZAPAS, L. J. (1964). Thermodynamics of perfect elastic fluids. *J. Res. Natl. Bur. Stand., Sect. B* **68**, 103.

BIRD, R. B. (1976). Useful non-Newtonian models. *Annu. Rev. Fluid Mech.* **8**, 13.

BIRD, R. B. (1977). Rheology and kinetic theory of polymeric liquids. *Annu. Rev. Phys. Chem.* **28**, 185.

BIRD, R. B., ARMSTRONG, R. C., and HASSAGER, O. (1977a). "Dynamics of Polymeric Fluids," Vol. 1. Wiley, New York.

BIRD, R. B., and CARREAU, P. J. (1968). A non-linear viscoelastic model for polymer solutions and melts. *Chem. Eng. Sci.* **23**, 427.

BIRD, R. B., HASSAGER, O., and ABDEL-KHALIK, S. I. (1974). Co-rotational rheological models and the Goddard expansion. *AIChE J.* **20**, 1041.

BIRD, R. B., HASSAGER, O., ARMSTRONG, R. C., and CURTISS, C. C. (1977b). "Dynamics of Polymeric Liquids," Vol. 2. Wiley, New York.

BLACK, J. R., DENN, M. M., and HSIAO, G. C. (1974). Creeping flow of a viscoelastic liquid through a contraction: A numerical perturbation solution. *In* "Theoretical Rheology" (J. F. Hutton, J. R. A. Pearson, and K. Walters, eds.), Chapter 1. Wiley, New York.

BOGUE, D. C. (1966). An explicit constitutive equation based on an integrated strain history. *Ind. Eng. Chem., Fundam.* **5**, 253.

BRENNER, H. (1974). Rheology of a dilute suspension of axisymmetric Brownian particles. *Int. J. Multiphase Flow* **1**, 195.

CABLE, P. J., and BOGER, D. V. (1978). A comprehensive experimental investigation of tubular entry flow. *AIChE J.* **24**, 869 and 992.

CARREAU, P. J. (1972). Rheological equations from molecular network theories. *Trans. Soc. Rheol.* **16**, 99.

CASWELL, B. (1967). Kinematics and stress on a surface at rest. *Arch. Ration. Mech. Anal.* **26**, 385.

CASWELL, B., and VIRIYAYUTHAKORN, M. (1978). A finite element time-marching scheme for the computation of viscoelastic flows. *49th Annu. Soc. Rheol. Meet. 1978* Paper A2.

CHACON, R. V., and FRIEDMAN, N. (1965). Additive functionals. *Arch. Ratio. Mech. Anal.* **18**, 230.

CHAN, L. L-Y. (1972). Experimental observations and numerical simulation of the Weissenberg climbing effect. Ph.D. Dissertation, University of Michigan, Ann Arbor.

CHAN-MAN FONG, C. F. (1965). On the stability of elastic-viscous flow between rotating cylinders. *Rheol. Acta* **4**, 37.

CHRISTIANSEN, R. L., and BIRD, R. B. (1977–1978). Dilute solution rheology: Experimental results and finitely extensible non-linear elastic dumbbell theory. *J. Non-Newtonian Fluid Mech.* **3**, 161.

COLEMAN, B. D. (1962a). Kinematical concepts with applications in the mechanics and thermodynamics of incompressible viscoelastic fluids. *Arch. Ration. Mech. Anal.* **9**, 273.

COLEMAN, B. D. (1962b). Substantially stagnant motions. *Trans. Soc. Rheol.* **6**, 293.

COLEMAN, B. D. (1968). On the use of symmetry to simplify the constitutive equations of isotropic materials with memory. *Proc. R. Soc. London, Ser. A* **306**, 449.

COLEMAN, B. D., MARKOVITZ, H., and NOLL, W. (1966). "Viscometric Flows of Non-Newtonian Fluids." Springer-Verlag, Berlin and New York.

COLEMAN, B. D., and NOLL, W. (1961). Foundations of linear viscoelasticity. *Rev. Mod. Phys.* **33**, 239 and 439.

COLEMAN, B. D., and NOLL, W. (1962). Steady extension of incompressible simple fluids. *Phys. Fluids* **5**, 840.

CROCHET, M. J., and NAGHDI, P. M. (1978). On the thermodynamics of Polymers in the transition and rubber regions. *J. Rheol.* **22**, 73.

CURRIE, P. K. (1977). Viscometric flows of anisotropic fluids. *Rheol. Acta* **16**, 205.

DATTA, S. K. (1964). Note on the stability of an elasticoviscous liquid in Couette flow. *Phys. Fluids* **7**, 1915.

DAVIES, M. H. (1965). A note on the stability of elastico-viscous liquids. *Appl. Sci. Res., Sect. A* **15**, 253.

DENN, M. M., and ROISMAN, J. J. (1969). Rotational stability and measurement of normal stress functions in dilute polymer solutions. *AIChE J.* **15**, 454.

ERICKSEN, J. L., (1960). Anistropic fluids. *Arch. Ration. Mech. Anal.* **4**, 231.

ERICKSEN, J. L. (1961). Conservation laws for liquid crystals. *Trans. Soc. Rheol.* **5**, 23.

FERRY, J. D. (1970). "Viscoelastic Properties of Polymers," 2nd ed. Wiley, New York.

FISHER, R. J., and DENN, M. M. (1976). A theory of isothermal melt spinning and draw resonance. *AIChE J.* **22**, 236.

FRENKIEL, F. N., LANDAHL, M. T., and LUMLEY, J. L., eds. (1977). Proceedings International Symposium on the Structure of Turbulence and Drag Reduction (Washington D.C., June, 7–12, 1976). *Phys. Fluids.* **20**(10), Part II.

GIESEKUS, H. (1962). Stromungen mit konstanten Geschwindigkeit gradienten und die Bewegung von darin suspendierten Teilchen. *Rheol. Acta* **2**, 101.

GIESEKUS, H. (1966). Zur Stabilität von Stromungen viskoelastischer Flüssigkeiten. *Rheol. Acta* **5**, 239.

GINN, R. F., and DENN, M. M. (1969). Rotational stability in viscoelastic liquids: Theory. *AIChE J.* **15**, 450.

GODDARD, J. D. (1967). A modified functional expansion for viscoelastic fluids. *Trans. Soc. Rheol.* **11**, 381.

GODDARD, J. D. (1971). A general class of motions with explicit pattern for simple fluids. *Pap., Br. Soc. Rheol. Conf. 1971.*

GODDARD, J. D. (1976). The stress field of slender particles oriented by a non-Newtonian extensional flow. *J. Fluid Mech.* **78**, 177.

GODDARD, J. D. (1978a). (Book review.) Dynamics of polymeric liquids. *J. Non-Newtonian Fluid Mech.* **4**, 277.

GODDARD, J. D. (1978b). A comment on the material functions for steady circular shear in the orthogonal rheometer. *J. Non-Newtonian Fluid Mech.* **4**, 365.

GODDARD, J. D., and MILLER, C. (1966). An inverse for the Jauman derivative and some applications to the rheology of viscoelastic fluids. *Rheol. Acta* **5**, 177.

GRAEBEL, W. P. (1961). Stability of a Stoksian fluid in Couette flow. *Phys. Fluids* **4**, 362.

GRAEBEL, W. P. (1964). The hydrodynamic stability of a Bingham fluid in Couette flow. *In* "Second-Order Effects in Elasticity, Plasticity, and Fluid Dynamics" (M. Reiner and D. Abir, eds.), p. 636. Pergamon, Oxford.

GRAESSLEY, W. W. (1974). The entanglement concept in polymer rheology. *Adv. Polym. Sci.* **16**, 1.

GRAESSLEY, W. W., PARK, W. S., and CRAWLEY, R. L. (1977). Experimental tests of constitutive relations for polymers undergoing uniaxial shear flows. *Rheol. Acta* **16**, 291.

GREEN, A. E., and ZERNA, W. (1975). "Theoretical Elasticity," 2nd ed. Oxford Univ. Press, London and New York.

HAND, G. L. (1962). A theory of anisotropic fluids. *J. Fluid Mech.* **13**, 33.

HIGASHITANI, K., and LODGE, A. S. (1975). Hole pressure error measurement in pressure generated flow. *Trans. Soc. Rheol.* **19**, 307.

HINCH, E. J. (1974). The mechanics of fluid suspensions. *In* "Theoretical Rheology" (J. F. Hutton, J. R. A. Pearson, and K. Walters, eds), Chapter 13. Wiley, New York.

HINCH, E. J. (1975a). *Colloq. Int. C.N.R.S.* **233**.

HINCH, E. J. (1975b). Application of the Langevin equation to fluid suspension. *J. Fluid. Mech.* **72**, 499.

HINCH, E. J. (1976). The deformation of a nearly straight thread in a shearing flow with weak Brownian motions. **75**, 765.

HINCH, E. J. (1977). Mechanical models of dilute polymer solutions in strong flows. *Phys. Fluids* **20**, S22.

HINCH, E. J., and LEAL, L. G. (1976). Constitutive equations in suspension mechanics. Part 2. Approximate forms for suspensions of rigid particles affected by Brownian rotations. *J. Fluid Mech.* **76**, 187.

HOU, T-H., TONG, P. P., and DEVARGAS, L. (1977). On the origin of the hole pressure. *Rheol. Acta* **16**, 544.

HUILGOL, R. R. (1969). On the properties of the motion with constant stretch history occurring in the Maxwell orthogonal rheometer. *Trans. Soc. Rheol.* **13**, 513.

HUILGOL, R. R. (1971). A class of motions with constant stretch history. *Q. Appl. Math.* **29**, 1.

HUILGOL, R. R. (1978). On corotational and other rate-type constitutive equations. *J. Non-Newtonian Fluid Mech.* **4**, 269.

HUTTON, J. F., PEARSON, J. R. A., and WALTERS, K., eds. (1974). "Theoretical Rheology," Wiley, New York.

JEFFREY, D. J., and ACRIVOS, A. (1976). The rheological properties of suspensions of rigid particles. *AIChE J.* **22**, 417.

JOHNSON, M. W., Jr., and SEGALMAN, D. (1977). A model for viscoelastic fluid behavior which allows for non-affine deformation. *J. Non-Newtonian Fluid Mech.* **2**, 255.

JOSEPH, D. D., and BEAVERS, G. S. (1976, 1977a). The free surface on a simple fluid between cylinders undergoing torsional oscillations. *Arch. Ration. Mech. Anal.* **62**, 323, **64**, 245.

JOSEPH, D. D., and BEAVERS, G. S. (1977b). Free surface problems in rheological fluid mechanics. *Rheol. Acta* **16**, 169.

KEARSLEY, E. A. (1970). Intrinsic errors for pressure measurement in a slot along a flow. *Trans. Soc. Rheol.* **14**, 419.

KLASON, C., and KUBAT, J., eds. (1976). "Proceedings of the VIIth International Congress on Rheology." Tages-Anzeiger/Regina-Druck, Zürich.

LEROY, P., and PIERRARD, J. M. (1973). Fluides viscoélastiques non-linéaires satisfaisant a un principe de superposition: Etude théorique et expérimentale. *Rheol. Acta* **12**, 449.

LESLIE, F. M. (1964). The stability in Couette flow of certain anistropic fluids. *Proc. Cambridge Philos. Soc.* **60**, 949.

LOCKETT, F. J., and RIVLIN, R. S. (1968). Stability in Couette flow of a viscoelastic fluid. Part I. *J. Mec.* **7**, 475.

LODGE, A. S. (1964). "Elastic Liquids." Academic Press, New York.

LODGE, A. S. (1974). "Body Tensor Fields in Continuum Mechanics." Academic Press, New York.

MARKOVITZ, H. (1977). Boltzmann and the beginnings of linear viscoelasticity. *Trans. Soc. Rheol.* **21**, 381.

MARRUCCI, G., and ASTARITA, G. (1974). Comments on the validity of a common category of constitutive relations. *Rheol. Acta* **13**, 754.

MARTIN, A. D., and MIZEL, A. D. (1964). A representation theorem for certain non-linear functionals. *Arch. Ration. Mech. Anal.* **15**, 353.

METZNER, A. B., WHITE, J. L., and DENN, M. M. (1966). Behavior of viscoelastic materials in short-time process. *Chem. Eng. Prog.* **62**, 81.

MIDDLEMAN, S. (1978). (Book review.) Dynamics of polymeric liquids. *Phys. Today* **31**, 54.

MILLER, C. (1967). A study of the Taylor-Couette stability of viscoelastic fluids. Ph.D. Thesis, University of Michigan, Ann Arbor (unpublished).

MILLER, C., and GODDARD, J. D. (1967). A study of the Taylor-Couette stability of viscoelastic fluids. O.R.A. Rep. 06673-8-T to the National Aeronautics and Space Administration. University of Michigan, Ann Arbor.

MORLAND, L. W., and LEE, E. H. (1960). Stress analysis for linear viscoelastic materials with temperature variation. *Trans. Soc. Rheol.* **4**, 233.

NOLL, W. (1958). A mathematical theory of the mechanical behavior of continuous media. *Arch. Ration. Mech. Anal.* **2**, 197.

NOLL, W. (1962). Motions with constant stretch history. *Arch. Ration. Mech. Anal.* **11**, 97.

OLDROYD, J. G. (1965). Some steady flows of the general viscoelastic liquid. *Proc. R. Soc. London, Ser. A* **283**, 115.

PEARSON, J. R. A. (1967). The lubrication approximation applied to non-Newtonian flow problems: A perturbation approach. *In* "Non-Linear Partial Differential Equations" (W. F. Ames, ed.), p. 73. Academic Press, New York.

PEARSON, J. R. A., and MATOVICH, M. A. (1969). Spinning a molten threadline: Stability. *Ind. Eng. Chem., Fundam.* **8**, 605.

PETRIE, C. J. S., and DENN, M. M. (1976). Instabilities in polymer processing (Journal review). *AIChE J.* **22**, 209.

PHAN-THIEN, N., ATKINSON, J. D., and TANNER, R. I. (1978). The Langevin approach to the dumbbell kinetic theory problem. *J. Non-Newtonian Fluid Mech.* **3**, 309.

PHILLIPS, M. C. (1977). The prediction of time-independent non-linear stresses in viscoelastic materials. *J. Non-Newtonian Fluid Mech.* **2**, 109, 123, and 139.

PIPKIN, A. C. (1972). "Lectures on Viscoelasticity Theory." Springer-Verlag, Berlin and New York.

PIPKIN, A. C., and OWEN, D. R. (1967). Nearly viscometric flows. *Phys. Fluids* **10**, 836.

PIPKIN, A. C., and TANNER, R. L. (1972). A survey of theory and experiment in viscometric flows of viscoelastic liquids. *Mech. Today* **1**, 262.

PIPKIN, A. C., and TANNER, R. I. (1977). Steady non-viscometric flows of viscoelastic liquids. *Annu. Rev. Fluid Mech.* **9**, 13.

RALLISON, J. M. (1978). The effects of Brownian rotations in a dilute suspension of rigid particles of arbitrary shape. *J. Fluid Mech.* **84**, 237.

RIVLIN, R. S. (1976). A note on secondary flow of a non-Newtonian fluid in a non-circular pipe. *J. Non-Newtonian Fluid Mech.* **1**, 391.

RIVLIN, R. S., and SAWYER, K. N. (1971). Nonlinear continuum mechanics of viscoelastic fluids. *Annu. Rev. Fluid Mech.* **3**, 117.

RUBIN, H., ELATA, C., and POREH, M. (1968). On the stability of elasticoviscous flows. *Rheol. Acta* **7**, 340..

SARTI, G. C. (1977). Thermodynamics of polymeric liquids: Simple fluids with entropic elasticity obeying the time-temperature superposition principle. *Rheol. Acta* **16**, 516.

SCHOWALTER, W. R. (1978). "Mechanics of Non-Newtonian Fluids." Pergamon, Oxford.

SMITH, G. F. (1965). On isotropic integrity bases. *Arch. Ration. Mech. Anal.* **18**, 282.

SMITH, M. M., and RIVLIN, R. S. (1972). Stability in Couette flow of a viscoelastic fluid. *J. Mec.* **11**, 69.

SPENCER, A. J. M., and RIVLIN, R. S. (1959). The theory of matrix polynomials and its application to the mechanics of isotropic continuum. *Arch. Ration. Mech. Anal.* **2**, 309 and 435.

SPENCER, A. J. M., and RIVLIN, R. S. (1960).

SUPRUNENKO, D. A., and TYSHKEVICH, R. I. (1968). "Commutative Matrices," Academic Press, New York.

TANNER, R. I. (1975a). Progress in experimental rheology. *In* "Theoretical Rheology" (J. F. Hutton, J. R. A. Pearson, and K. Walters, eds.), p. 235. Wiley, New York.

TANNER, R. I. (1975b). Stresses in dilute solutions of bead-non-linear spring macromolecules. III. Friction coefficients varying with dumbbell extension. *Trans. Soc. Rheol.* **19**, 557.

TANNER, R. I. (1979). Some useful constitutive models with a kinematic slip hypothesis. *J. Non-Newtonian Fluid Mech.* **5**, 103.

TANNER, R. I., and HUILGOL, R. R. (1975). On a classification scheme for flow fields. *Rheol. Acta* **14**, 959.

TANNER, R. I., and PIPKIN, A. C. (1969). Intrinsic errors in pressure hole measurement. *Trans. Soc. Rheol.* **13**, 471.

TAYLOR, G. I. (1923). Stability of a viscous liquid contained between two rotating cylinders. *Philos. Trans. R. Soc. London, Ser. A* **223**, 289.

THIEN, N. P., and TANNER, R. I. (1977). A new constitutive equation derived from network theory. *J. Non-Newtonian Fluid Mech.* **2**, 353.

THOMAS, R. H., and WALTERS, K. (1964a). The stability of elastico-viscous flow between rotating cylinders, Part I. *J. Fluid Mech.* **18**, 33.

THOMAS, R. H., and WALTERS, K. (1964b). The stability of elastico-viscous flow between rotating cylinders, Part II. *J. Fluid Mech.* **19**, 557.

TOWNSEND, P., WALTERS, K., and WATERHOUSE, W. W. (1976). Secondary flows in pipes of square cross-section and the measurement of the second normal stress difference. *J. Non-Newtonian Fluid Mech.* **1**, 107.

TRUESDELL, C., (1966). "The Elements of Continuum Mechanic." Springer-Verlag, Berlin and New York.

TRUESDELL, C., and TOUPIN, R. (1960). The classical field theories. *In* "Hand buch der Physik" (S. Flügge, ed.), Vol. III, Part 1, pp. 268ff. Springer-Verlag, Berlin and New York.

TSAI, C. F., and DARBY, R. (1978). Non-linear viscoelastic properties of very dilute drag reducing polymer solutions. *J. Rheol.* **22**, 219.

VIRK, P. S. (1975). Drag reduction fundamentals (Journal review). *AIChE J.* **21**, 625.

WAGNER, M. H. (1977). Prediction of primary normal stress difference from shear vsicosity data using a single integral constitutive equation. *Rheol. Acta* **16**, 43.

WAGNER, M. H. (1978). A constitutive analysis of uniaxial elongational flow data of a low-density polyethylene melt. *J. Non-Newtonian Fluid Mech.* **4**, 39.

WALTERS, K. (1975). "Rheometry." Chapman & Hall, London.

WEINBERGER, C. B., CRUZ-SAENZ, G. F., and DONNELLY, G. J. (1976). Onset of draw resonance during isothermal melt spinning: A comparison between measurements and predictions. *AIChE J.* **22**, 441.

WHITE, J. L. (1975). Considerations of unconstrained recovery in viscoelastic fluids. *Trans. Soc. Rheol.* **19**, 271.

WINEMAN, A. S., and PIPKIN, A. C. (1964). Material symmetry restrictions on constitutive equations *Arch. Ration. Mech. Anal.* **17**, 184.

ZAHORSKI, S. (1972a). Flows with proportional stretch history. *Arch. Mech. Stosow.* **24**, 681.

ZAHORSKI, S. (1972b). Certain non-viscometric flows of viscoelastic fluids. *Mech. Teor. Stosow.* **10**, 29.

ZAHORSKI, S. (1973). Motions with superposed proportional stretch histories as applied to combined steady and oscillatory flows of simple fluids. *Arch. Mech. Stosow.* **24**, 681.

ADVANCES IN APPLIED MECHANICS, VOLUME 19

Relaminarization of Fluid Flows

R. NARASIMHA and K. R. SREENIVASAN*

Indian Institute of Science
Bangalore, India

* Present address: Department of Engineering and Applied Science, Yale University, New Haven, Connecticut 06520.

I. Introduction

A. GENERAL REMARKS

It is so often said that turbulence is the natural state of fluid motion that reports purporting to have observed relaminarization of a turbulent flow were until not long ago greeted with varying degrees of disbelief. Indeed, a common reaction when the subject was mentioned used to be that the implied transition from disorder to order was thermodynamically impossible! But these reverting flows that we shall discuss below are not closed

FIG. 1. Reversion in a coiled tube. Flow enters at top left, leaves at bottom right. Dye injected continuously at the fourth coil does not diffuse, indicating laminar flow. Dye injected at entry diffuses rapidly, indicating turbulence (photograph is taken just before this dye reaches the fourth coil). Inner diameter of tube about 8 mm; radius of curvature about 55 mm. (From Viswanath *et al.*, 1978.)

systems; if crystallization is possible out of liquids, so too is relaminarization from a turbulent state.

Indeed, should there still be any doubters, it is easy to set up simple experiments showing relaminarization (variously also called reversion, and inverse or reverse transition). We may begin with two examples of a few such experiments described by Viswanath *et al.* (1978). The first is similar to one made by G. I. Taylor in 1929. Turbulent flow is established in a tube of about 8 mm diameter, made of some flexible transparent material like tygon. The tube is then wrapped around a cylinder (say about 100 mm diameter) in a few coils, and comes out straight again. As is easily confirmed by injecting dye, the flow can be made to come in and go out turbulent in the straight sections, while remaining apparently laminar within the coil! Thus, in Fig. 1, the dye continuously injected at the fourth coil does not diffuse at all, indicating laminar flow there; but the dye injected into the straight section just upstream of where the tube bends into a coil always diffuses very rapidly, indicating turbulence. The flow similarly becomes turbulent again in the straight section downstream, although this is not shown in the photograph.

In the second experiment (Fig. 2), we inject a jet of dye vertically upward from the bottom into a tank of water. If the jet velocity is sufficiently high

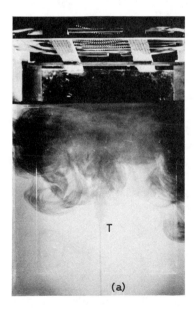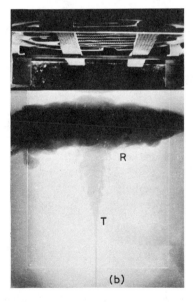

FIG. 2. Reversion in stably stratified flow. (a) Cold flow. (b) Flow with top fluid layers heated. Note the point of transition to turbulence in the jet (marked T), followed by reversion (marked R) in (b). (From Viswanath *et al.*, 1978.)

it breaks into turbulence as usual. If the upper layers of water are heated now, say with an ordinary heating coil placed on top of the tank, the turbulence in the jet dies out and a smooth layer of dye cloud collects near the top; we can observe here both transition and reversion in the jet.

In fact, once the idea of possible reversion is accepted, its presence is seen everywhere. When a laminar jet is seen issuing from a water faucet, such reversion must have occurred between the mains and the faucet, as the flow in the mains will invariably be turbulent. When we are breathing hard, say after exertion, the flow in the wind pipe is turbulent (especially because the flow detaches from the vocal cords in the glottic aperture), but deep inside the bronchioles in the lungs the flow must be laminar—relaminarization has therefore been occurring under our very noses all along! Inversions in the atmosphere (involving an *increase* in the temperature with height) can lead to fairly sudden suppression of turbulence as in the second experiment mentioned above, and were studied more than 50 years ago by Richardson (1920); Prandtl and Reichardt (1934) investigated the effect of stable stratification on turbulent shear flows by heating the top floor of a wind tunnel. Evidence of reversion of some kind has also been found when the flow is subjected to acceleration (Sternberg, 1954; Sergienko and Gretsov, 1959), magnetic fields (e.g., Murgatroyd, 1953), rotation (e.g., Halleen and Johnston, 1967), curvature (e.g., So and Mellor, 1973), heating (e.g., Perkins and Worsoe-Schmidt, 1965), suction (Dutton, 1960), blowing (e.g., Eckert and Rodi, 1968), and so on. Indeed, if turbulence *is* the natural state of fluid motion, the chances are that any flow that is observed to be laminar must have been the consequence of a reversion somewhere (often having passed through the trivial intermediate state of rest!).

Nevertheless, there is also evidence that the pendulum of opinion has perhaps swung too far in the other direction now; there is a strong temptation to attribute any observed departure from some presumed turbulent law to relaminarization. But there are few universally valid turbulent laws; those that pass as such (e.g., the log-law in wall flows) are at best asymptotic limits, and may not be valid in flows that, while being unusual or abnormal, are nonetheless turbulent. Choosing criteria for relaminarization based on such presumed laws is therefore a risky venture.

B. Scope of the Article

At the present time it seems opportune to examine the mechanisms that cause such reversion, and to see if there are any similarities or general principles that govern them. In carrying out this examination, we adopt a rather pragmatic definition of reversion, according to which a flow (or a part of it) has relaminarized if its development can be understood without recourse to any model for turbulent shear flow. This implies, in particular, that the turbulent fluctuations need not necessarily have completely vanished in the relaminarized state; but that, if present, their contribution to mean flow dynamics (e.g., to momentum transport) is negligible. Under these circumstances the flow may be called quasi-laminar; it usually carries a residual turbulence that is inherited from the previous history of the flow, but has been rendered passive.

We begin by considering (in Sections III, IV, and V) three reverting flows in some detail, as we believe them to be basic archetypes (Narasimha, 1977). In the first, turbulent energy is dissipated through the action of a molecular transport property like the viscosity or conductivity, and the governing parameter is typified by the Reynolds number. In the second class, turbulence energy is destroyed or absorbed by work done against an external agency like buoyancy forces or flow curvature; the typical parameter is a Richardson number. In both cases experimental evidence indicates that the suppression of turbulence may go beyond the mere decay of energy to an actual de-correlation of the velocity components contributing to the crucial Reynolds shear stress that governs the mean flow.

The third class of reverting flows is exemplified by a turbulent boundary layer subjected to severe acceleration. Here a two-layer model is suggested. In the outer layer turbulence is fairly rapidly distorted and the Reynolds shear stress is nearly frozen; the inner viscous layer exhibits random oscillations in response to the forcing provided by the residue of the original turbulence. Reversion here is not so much the result of dissipation or destruction of energy (although these mechanisms are also operating), but rather of the domination of pressure forces over slowly responding Reynolds stresses in the outer region, accompanied by the generation of a new laminar subboundary-layer stabilized by the acceleration.

A number of other different reverting flows are then considered in the light of the analysis of these archetypes (Sections VI to XI). We believe that this examination sheds considerable light on certain instances of turbulence suppression which have hitherto been regarded in isolation, and shows how a combination of different mechanisms is often operating. Nevertheless, it will not be possible to come to definite conclusions about the primary mechanism (if there is one) in all cases, either because they have not yet been studied thoroughly or because they are too complex. It is hoped, however, that the attempt to take an overall view will reveal where further work needs to be done. In particular, in some instances of turbulence suppression (e.g., curvature), the observed effect is much larger than what might be expected from naive estimates of energy balance. The explanation lies presumably in the stabilization of the flow; perhaps here one is interfering with the organization of motion in the coherent structures now believed to be present in turbulent shear flows—the destruction of phase differences may be as important as that of amplitudes. A fuller understanding of the nature of such organized motion is likely to provide more insight into relaminarization as well.

In addition to being an attempt at a fairly comprehensive survey of relaminarizing flows, the present article includes the results of some previously unpublished analyses.

We begin with a brief summary of the flow equations to which we will need to make constant reference in the rest of the paper.

II. The Governing Equations

It is convenient from the outset to split the flow into mean and fluctuating components (the latter denoted by a prime, with a mean that is necessarily zero). For an incompressible fluid of constant density, all stresses and forces may be expressed in kinematic units by dividing through by the density. Vector or tensor notation will be used as convenient; in the latter case the convention is adopted that all terms with a repeated Greek suffix are to be summed over the values 1, 2, 3 for the suffix.

The equations governing the mean flow of an incompressible fluid may be written (see, e.g., Monin and Yaglom, 1971, Chapter 6) as

$$\frac{\partial u_\alpha}{\partial x_\alpha} = 0, \tag{2.1}$$

$$\frac{du_i}{dt} \equiv \frac{\partial u_i}{\partial t} + u_\alpha \frac{\partial u_i}{\partial x_\alpha} = -\frac{\partial p}{\partial x_i} + v\nabla^2 u_i - \frac{\partial}{\partial x_\alpha}\langle u'_\alpha u'_i \rangle + X_i. \tag{2.2}$$

These equations respectively express the laws of conservation of mass and momentum, with $\mathbf{u} = (u_\alpha)$ being the mean velocity, p the mean pressure, $\mathbf{X} = (X_i)$ the mean body force (per unit mass), and v the kinematic viscosity; $\nabla^2 \equiv \partial^2/\partial x_\alpha \partial x_\alpha$. The quantity $-\langle u'_i u'_\alpha \rangle$, where the angular brackets denote the mean value, represents the (kinematic) Reynolds stress tensor. An equation for the components of this tensor may be easily derived from the momentum equation for the total velocity $\mathbf{u} + \mathbf{u}'$ (see, e.g., Monin and Yaglom, 1971, Chapter 6):

$$\frac{d}{dt}\langle u'_i u'_j \rangle = -\langle u'_i u'_\alpha \rangle \frac{\partial u_j}{\partial x_\alpha} + \langle u'_i X'_j \rangle$$

$$+ v\langle u'_i \nabla^2 u'_j \rangle - \left\langle u'_i \frac{\partial p'}{\partial x_j} \right\rangle$$

$$- \frac{1}{2}\frac{\partial}{\partial x_\alpha}\langle u'_i u'_j u'_\alpha \rangle + [i \leftrightarrow j]. \tag{2.3}$$

Here $[i \longleftrightarrow j]$ indicates additional terms obtained by switching i and j in all the other terms on the right.

Considerable insight is often gained by examining the turbulent energy

$$q \equiv \tfrac{1}{2}\langle u'_\alpha u'_\alpha \rangle = \tfrac{1}{2}\hat{u}_\alpha \hat{u}_\alpha, \hat{u}_\alpha = \langle u'^2_\alpha \rangle^{1/2}; \tag{2.4}$$

the governing equation is obtained by putting $i = j$ in (2.3) and summing:

$$\frac{dq}{dt} = -\langle u'_\alpha u'_\beta \rangle \frac{\partial u_\beta}{\partial x_\alpha} + \langle u'_\alpha X'_\alpha \rangle - \varepsilon$$

(i)   (ii)   (iii)

$$+ v\nabla^2 q + v \frac{\partial^2}{\partial x_\alpha \partial x_\beta} \langle u'_\alpha u'_\beta \rangle$$

(iva)   (ivb)

$$- \frac{\partial}{\partial x_\alpha} [\langle p' u'_\alpha \rangle + \langle q' u'_\alpha \rangle], \tag{2.5}$$

(va)   (vb)

where

$$q' = u'_\alpha u'_\alpha - \langle u'_\alpha u'_\alpha \rangle, \tag{2.6}$$

and

$$\varepsilon = \frac{1}{2} v \left\langle \left(\frac{\partial u'_\alpha}{\partial x_\beta} + \frac{\partial u'_\beta}{\partial x_\alpha} \right) \left(\frac{\partial u'_\alpha}{\partial x_\beta} + \frac{\partial u'_\beta}{\partial x_\alpha} \right) \right\rangle \tag{2.7}$$

represents viscous dissipation, i.e., loss of kinetic energy into heat through the action of viscosity. Terms (iva) and (ivb) in (2.5) are diffusion-like viscous terms; the gradient terms (va) and (vb) may be called diffusion by pressure and turbulent transport, respectively. Terms (i) and (ii) may together be said to represent "generation": term (i) is the energy production by the action of the Reynolds stresses on the mean shear, and term (ii), which we shall call "absorption" to distinguish it from the mean flow term (i), represents the work done against the fluctuating body forces. The nomenclature must not be interpreted to mean that (i) invariably tends to enhance the energy, or (ii) to diminish it—under appropriate conditions the contrary is indeed possible. If X' is a conservative body force (like gravity), the absorption term represents conversion of turbulent kinetic energy to (e.g., gravitational) potential energy. However, it is possible for the absorption term to represent irreversible dissipation also; one instance is when X' is the magnetic force on an electrically conducting fluid in a magnetic field. In this case X' depends on the current and therefore the conductivity of the fluid; the absorption, in the limit of low magnetic Reynolds numbers that will be discussed in Section X, represents ohmic dissipation. We will use the same nomenclature for the corresponding terms in (2.3).

As is well known, equations like (2.3) for the second-order quantities $\langle u_i' u_j' \rangle$ are incomplete, as they contain third-order quantities like $\langle u_\alpha' u_\beta' u_\beta' \rangle$; equations for these can also be written down, but they would contain terms of the fourth order, and so on. Equations for the evolution of moments up to any given order are not closed, because the basic nonlinearity in the equations of motion always introduces higher order moments. No complete theory of turbulent or reverting flows can therefore be constructed solely from equations of the above type, but much insight can nevertheless be obtained by their examination.

It is clear from (2.5) that turbulent energy could be altered by the action of (i) dissipation, (ii) production against mean shear, (iii) absorption by fluctuating external forces, or (iv) diffusion (in a generalized sense). It appears that observations in reverting flows can usually be traced to the action of one or more of the first three, but there is some evidence (in curved flows, for example, see Section VI) that changes in the diffusion terms can be important.

It is of course possible that the turbulent energy transport described by Eq. (2.5) is affected by a basic change in the mean flow, governed by (2.1) and (2.2); e.g., the imposition of a pressure gradient or a new body force may alter the mean flow sufficiently to cause (as a *consequence*) significant changes in the turbulence structure. In principle, the mean flow equations (2.1), (2.2), and the turbulent transport equations (2.3) and (2.5) are closely coupled to each other, so either set cannot be considered in isolation. However, there are situations where changes are faster in some variables than in others, so it becomes possible to identify rate-controlling mechanisms. We shall illustrate this in the following sections.

III. Reversion by Dissipation

A. INTRODUCTION

A mechanism that is comparatively easy to understand is the relative increase in dissipation that occurs when the Reynolds number goes down in a flow. Consider a two-dimensional parallel or nearly parallel flow with a mean velocity along the x-axis given by $u = u(y)$ and fluctuating velocity components u', v' along the x- and y-axes, respectively. Comparing terms (ii) and (iv) in (2.5), we find

$$\frac{\text{production of turbulent energy}}{\text{dissipation}} = \frac{-\langle u'v' \rangle \, \partial u/\partial y}{\varepsilon}; \qquad (3.1)$$

we may in general expect this ratio to decrease when the mean flow Reynolds number Re becomes sufficiently low. If therefore Re decreases downstream in a flow, we may expect reversion for the same reason that a flow whose Reynolds number decreases in time will revert.

B. Enlargement in Ducts

Experimental investigations of such reverting flows have been reported by Laufer (1962), Sibulkin (1962), and Badri Narayanan (1968). The general situation in these experiments involves the gradual enlargement of a pipe or channel from one diameter or width to another, as illustrated in Fig. 3 (for a channel): the angle of divergence is kept sufficiently small to ensure that no flow separation occurs. In this case the Reynolds number (based on the section-average flow velocity $\bar{U}$ and channel half-height or pipe radius a) goes down from say Re_1 upstream of the divergence to Re_2 downstream. If, therefore, $Re_1 > Re_{cr}$ and $Re_2 < Re_{cr}$, where Re_{cr} is an appropriate critical Reynolds number, it may be expected that an approaching turbulent flow will revert to the laminar state.

This is indeed found to be the case: the turbulence decays, and the mean flow asymptotically approaches the classical laminar (Poiseuille) solution. However, there are several interesting features in the flow that merit attention. Experiments show, for example, that the skin friction reaches the laminar value well before the velocity distribution in the middle reaches the value

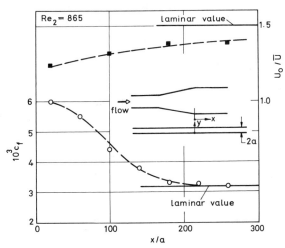

FIG. 3. Variation of center-line velocity U_0 and skin friction coefficient c_f in reverting channel flow. Experimental data from Badri Narayanan (1968).

characteristic of the well-known parabolic profile (see Fig. 3). An inner layer thickness, defined by the location of the maximum of the rms value of the fluctuating longitudinal velocity component, grows downstream like $x^{1/2}$, in a manner typical of *laminar* boundary layers (Fig. 4): there is here an assumption that the distribution of $\hat{u}$ is related to that of the mean velocity, and we shall return to this point in Section V.

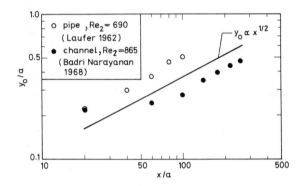

FIG. 4.   Growth of the inner layer thickness during relaminarization due to duct enlargement: y_0 is the distance from the surface at which the rms value of the fluctuating longitudinal velocity component u' is a maximum. The $y_0 \propto x^{1/2}$ line shows the growth rate characteristic of a laminar boundary layer.

The picture that emerges, of relatively rapid adjustment near the wall and a much slower process in the outer layer, is consistent with the general findings in turbulent flow. There is also the suggestion that the process of reversion is associated with the conversion of the viscous sublayer into an effectively new laminar boundary layer near the wall, rather like the entry region in a pipe, except that the core in the reverting flow is sheared and carries the residue of an originally turbulent flow.

We shall now discuss some of these features of the flow in greater detail.

1. *Approach to the Laminar Mean Velocity Profile*

The state of the flow in an advanced stage of reversion must be capable of being described by a perturbation expansion on the final asymptotic state. Downstream of the enlargement in the duct, the effectively new laminar layer referred to above grows until it eventually fills the duct and establishes an entirely laminar profile all across. In an advanced stage of reversion where this has not quite happened, the Reynolds number $\bar{U}\delta/\nu$ based on the thickness δ of this laminar layer is fairly large, and its inverse $\gamma = \nu/\bar{U}\delta$ serves as a convenient small parameter. Outside the layer is a core where departures from the laminar profile can be represented by a power series in γ, say

$$u(x, y) = u_0(y) + \gamma u_1(x, y) + O(\gamma^2), \tag{3.2a}$$

$$v(x, y) = -\gamma v_0(x, y) + O(\gamma^2), \tag{3.2b}$$

$$u'(x, y) = \gamma u_1'(x, y) + O(\gamma^2), \tag{3.2c}$$

$$v'(x, y) = \gamma v_1'(x, y) + O(\gamma^2), \tag{3.2d}$$

where $u_0(y)$ represents the asymptotic laminar state. (This is of course the reason why the expansions for the fluctuations u', v' begin with terms of order γ.) As we shall see later, the correlation coefficient

$$C_\tau \equiv \tau/\hat{u}\hat{v}, \tag{3.3}$$

where $\tau = -\langle u'v' \rangle$, decreases rapidly with increasing x, so that it is possible to assume it to be small for sufficiently large x. Thus,

$$\tau = o(\gamma^2). \tag{3.4}$$

Finally, assuming dp/dx as given, substituting the expansions (3.2) and (3.4) in the momentum equation (2.2) for $i = 1$, we have, to order unity,

$$dp/dx = v(d^2u_0/dy^2),$$

with the well-known Poiseuille solution

$$\dot{u}_0 = \frac{1}{2v}\frac{dp}{dx}(\xi^2 - a^2), \qquad \xi = y - a. \tag{3.5}$$

To order γ we have

$$\left(u_0 - \xi\frac{du_0}{d\xi}\right)\frac{\partial u_1}{\partial x} = v\frac{\partial^2 u_1}{\partial \xi^2},$$

or, from (3.5),

$$-\frac{1}{2}\frac{dp}{dx}(\xi^2 + a^2)\frac{\partial u_1}{\partial x} = v\frac{\partial^2 u_1}{\partial \xi^2}. \tag{3.6}$$

Writing

$$u_1(x, \xi) = u_{1x}(x)u_{1y}(\xi),$$

we get, from (3.6),

$$\frac{du_{1x}}{u_{1x}} = C\frac{dx}{(dp/dx)},$$

where C is a positive constant. If dp/dx is a constant, as it indeed appears to be the case (as shown by the constancy of c_f, see Fig. 3) in the later stages of reversion, we have

$$u_{1x} \sim \exp(-\alpha x/2a), \tag{3.7}$$

where $\alpha = -2aC/(dp/dx)$ is another positive constant. According to (3.7), the deviation of the velocity profile from the Poiseuille value, at any given y in the core of the flow, decreases exponentially with x. Along the centerline, for both pipe and channel flows undergoing reversion, the data of Fig. 5 show this to be indeed the case—perhaps surprisingly from as small a downstream distance as $20a$!

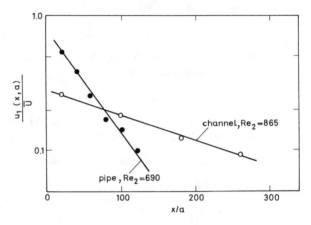

Fig. 5.   Departure of measured mean velocity at the center-line from the Poiseuille value in relaminarizing duct flow.

The equation for u_{1y} can be written as

$$\frac{d^2 u_{1y}}{d\zeta^2} + u_{1y}\left(\frac{1}{4}\zeta^2 - \beta\right) = 0, \tag{3.8}$$

where

$$\zeta \equiv \xi\left(\frac{2C}{v^2}\right)^{1/4}, \qquad \beta \equiv -\frac{a^2}{2v}\left(\frac{C}{2}\right)^{1/2}. \tag{3.9}$$

Equation (3.8) should therefore be solved with the boundary conditions

$$u_{1y} = 0 \qquad \text{at} \quad \xi = \pm a. \tag{3.10}$$

Because the equation as well as the boundary conditions are homogeneous, it is clear that nontrivial solutions can exist only for certain values β_n of the separation constant β. In the later stages of reversion, the laminar boundary layers growing on the two walls will fill a substantial part of the channel width. As a consequence, u_{1y} is exponentially small in magnitude in the two laminar boundary layers. It is then clear that to represent u_{1y} reasonably accurately inside the laminar layers, we require a large number of eigenfunctions which oscillate rather rapidly. To avoid the need for evaluating these eigenfunctions, we may assume *a priori* that u_{1y} is identically zero in these regions and, instead of (3.10), apply the modified boundary condition

$$u_{1y} = 0 \qquad \text{at} \quad \xi = \pm(a - \delta). \tag{3.11}$$

This cannot be strictly correct because it demands an implicit dependence of u_{1y} on x (since δ is a function of x), but should be satisfactory if $d\delta/dx \ll 1$. In spite of a slight inconsistency, the advantage of using (3.11) instead of (3.10) is that there is hope that the first eigenfunction represents u_{1y} reasonably accurately. In any case, it will be sufficient for the purpose of illustrating the principle of the method of calculation.

An approximate solution to (3.8) can be written (see, e.g., Abramowitz and Stegun, 1965, p. 688) as

$$u_{1y}(\zeta) = 1 + \beta \frac{\zeta^2}{2!} + \left(\beta^2 - \frac{1}{2}\right)\frac{\zeta^4}{4!} + \left(\beta^3 - \frac{7}{2}\beta\right)\frac{\zeta^6}{6!}$$
$$+ \left(\beta^4 - 11\beta^2 + \frac{15}{4}\right)\frac{\zeta^8}{8!} + \cdots, \tag{3.12}$$

where β is to be determined by imposing the boundary condition (3.11). Then (3.12) gives the theoretical velocity distribution $u_1(x, y)/u_1(x, a)$.

Figure 6 shows a comparison of the experimental data with the above calculations for the reverting channel flow at $x/a = 260$. For this station, choosing (from measurement) $\delta \simeq 0.4a$, we require $\beta \simeq -1.18$; the corresponding solution agrees well with the measurement.

As an aside we note that the value of α in (3.7), obtained from Fig. 5 implies that the separation constant β is about -1.15 for the reverting channel flow.

FIG. 6. Distribution of mean velocity in the core in reverting channel flow: departure from Poiseuille solution compared with present theory, Eq. (3.12). Experimental data from Badri Narayanan (1968).

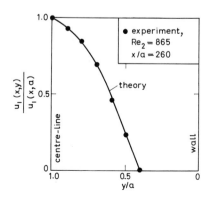

We recall that this value of β is obtained from a consideration of $u_1(x, 0)$ at several x/a. That this agrees well with the β evaluated from a consideration of the velocity profile $u_1(x, y)$ at a single x/a in a good indication of the consistency of the calculations.

2. The Decay of Turbulence

The precise manner in which turbulence decays in such a reverting flow raises many interesting questions. Again, in the parallel (or nearly parallel) two-dimensional shear flow we are considering, with $x_1 = x$, $x_2 = y$, $x_3 = z$ and $u_1 = u$ (y only), $u_2 = v = 0$, $u_3 = w = 0$, Eq. (2.3) shows that the Reynolds stress components have the following production terms:

$$\frac{1}{2} \hat{u}^2 : -\langle u'v' \rangle \frac{\partial u}{\partial y},$$

$$\hat{v}^2, \hat{w}^2 : 0,$$

$$-\langle u'v' \rangle : \hat{v}^2 \frac{\partial u}{\partial y}. \tag{3.13}$$

We shall examine experimental observations in the light of these equations.

Figure 7 shows some representative measurements of turbulence quantities in a reverting flow in a channel, from the experiments of Badri Narayanan (1968). It is seen that both $\hat{u}$ and $\hat{v}$ decay exponentially with distance, and that $\hat{v}$ decays more rapidly than $\hat{u}$. At $Re_2 = 865$, it takes a streamwise distance of $115a$ for a halving in $\hat{u}^2$ at the centerline, but only $75a$ for $\hat{v}^2$. The ratio $\hat{u}/\hat{v}$ is around 4 at $x = 20a$, and increases to 6 at $x = 220a$; over the same distance, the ratio has been found to increase to nearly 15 at $Re_2 = 625$! Thus, the decaying turbulent flow is strongly anisotropic. From (3.13) the production of turbulence occurs only in the $\hat{u}^2$ component; clearly, inter-component energy transfer is too slow and ineffective to correct the anisotropy.

It is further seen from Fig. 7 that the correlation coefficient C_r [defined by Eq. (3.3)] goes down, from a maximum value of 0.36 at $x = 20a$ to 0.13 at $x = 180a$. This shows that a decorrelation mechanism must be at work, thus implying a weakening of the nonlinear effects. It is then clear from Eq. (3.13) that $\hat{u}^2$ must decay because of dissipation overtaking the production. Note that the Reynolds shear stress decreases nearly linearly in x unlike the turbulent energy components $\hat{u}^2$ and $\hat{v}^2$, which decay nearly exponentially in x.

Finally, we may surmise that the decorrelation of u' and v' fluctuations is connected with the destruction of the coherent motion in the wall-layer,

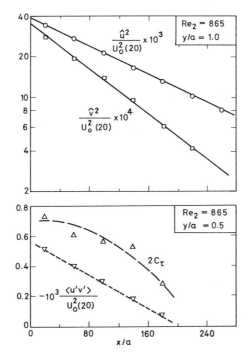

FIG. 7. Decay of turbulence intensity and the shear correlation coefficient in reverting channel flow. U_0 (20) = center-line velocity at $x/a = 20$. (Experimental data from Badri Narayanan, 1968.)

which recent work suggests is a key to the production of turbulent energy (e.g., Laufer, 1975).

3. *The Decay Rate*

Owen (1969) has presented an interesting argument to estimate the rate of decay of the turbulence intensity for given downstream Reynolds number Re_2. He assumes that in the turbulent energy balance equation (2.5) only the production and dissipation terms are significant. This is however not in general valid; Laufer's (1962) measurements, analyzed by Coles (1962), show in fact that on the axis of the pipe diffusion is comparable to or greater than dissipation, at both the first and the last measurement station ($x/a = 20, 100$).

A better procedure is to consider the integral of the energy balance equation (2.5) over (say) the half-height of the channel; the diffusion terms then drop out and we get

$$\int_0^a u \frac{\partial q}{\partial x} \, dy = -\int_0^a \langle u'v' \rangle \frac{\partial u}{\partial y} \, dy - \int_0^a \varepsilon \, dy. \tag{3.14}$$

Following Owen, one may put

$$\frac{\partial u}{\partial y} \sim \frac{U_*}{a}, \quad -\langle u'v' \rangle = b_0 q, \tag{3.15}$$

where $U_* = \tau_0^{1/2}$ is the friction velocity, τ_0 is the wall stress and b_0 is taken as a constant of order 10^{-1}. (Note that this ignores the decorrelation effect mentioned above.) The dissipation ε is often estimated as

$$\varepsilon = 15v \left\langle \left(\frac{\partial u'}{\partial x}\right)^2 \right\rangle = 15v \frac{\hat{u}^2}{\lambda^2}, \tag{3.16}$$

where λ is the Taylor microscale; the expression assumes that the small eddies responsible for dissipation are isotropic, following Kolmogorov's famous postulate. Owen replaces (3.16) by

$$\varepsilon \simeq b_2 vq/a^2, \tag{3.17}$$

where b_2 is taken to be a constant of order 10; the crucial assumption is made that λ is proportional to a. If $q*$ is a characteristic value of q, the energy equation (3.14) then takes the form

$$\frac{dq*}{dx} = q*\left(A_1 - \frac{A_2}{\text{Re}}\right) = A_3 q*\left(\frac{1}{\text{Re}_{cr}} - \frac{1}{\text{Re}}\right),$$

whose solution is

$$q* \sim \exp\left[-\frac{A(x - x_0)}{a}\left(\frac{1}{\text{Re}} - \frac{1}{\text{Re}_{cr}}\right)\right], \tag{3.18}$$

where A, A_1, A_2, A_3, and x_0 are all some constants. Owen finds that the experimental evidence is not inconsistent with the above expression.

This analysis is however not entirely satisfactory. The decorrelation mentioned earlier shows that $\langle u'v' \rangle$ does not decay in proportion to q but much faster. Because of the large anisotropy the usual estimate (3.16) for dissipation is likely to be in error; furthermore, experiments show that λ does not remain constant during decay (at any rate as estimated from the u' data), but rather tends to increase like $\hat{u}^{-2}$ (Laufer, 1962), so that ε is proportional to $\hat{u}^6$ rather than $\hat{u}^2$. (It is however possible that the dissipation will be governed by the microscale in the normal direction, which could well be limited by the dimension of the duct.)

Figure 8 shows the channel data of Badri Narayanan (1968) plotted against $1/\text{Re}$. The data do not lie too convincingly on a straight line, as they should if (3.18) were valid.

On the other hand, a good fit to the data is a law of the type

$$\hat{u}^2 \sim \exp[-B(\text{Re}_{cr} - \text{Re})^3(x - x_0)/a], \qquad \text{Re} < \text{Re}_{cr}, \tag{3.19}$$

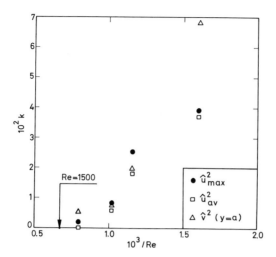

FIG. 8. Turbulence decay rates in relaminarizing channel flows. k is defined by $\hat{u}^2 \propto \exp(-kx/2a)$.

where B is some positive constant (Fig. 9). Indeed the diagram suggests, by extrapolation, $Re_{cr} \simeq 1500$. Note that measurements in reverting pipe flow are also included with the channel data in Fig. 9. Based on the (only) two available data points, it does not seem necessary to infer a different critical Reynolds number for reversion in pipe flow. It may be of interest to note, however, that the value $Re_{cr} = 1500$ (equivalent to 3000, if based on the pipe diameter), corresponds to the *end* of the transitional regime and the maximum skin friction coefficient, rather than to the beginning of transition and the minimum skin friction (both of which occur at a diameter-Reynolds number of about 2300).

A remarkable feature of this kind of dissipative reversion is its slowness. In a channel, even at $Re_2 = 865$, which is about 60% of the critical Reynolds number, it takes a distance of $200a$ for $\hat{u}$ to fall to a third of its initial value.

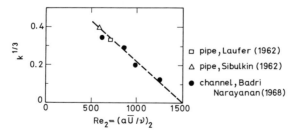

FIG. 9. Turbulence decay rates at different downstream Reynolds numbers in reverting duct flows.

The reason for this is that there is apparently only a slight difference between production and dissipation in the flow even at Reynolds numbers well below the critical. Consequently, the flow is nearly in a state of equilibrium; and both spectra and intensity distributions show a universal similarity during decay (Laufer, 1962). We next discuss this briefly.

4. *The Turbulent Spectrum*

In all the experiments reported (Laufer, 1962; Sibulkin, 1962; Badri Narayanan, 1968), the measured spectral density of u' at the centerline exhibits an approximate similarity over the entire wave-number range, when normalized by the centerline velocity and the turbulence integral scale l. This is an unusual simplicity, considering that even in homogeneous turbulence no such similarity over the entire wave-number region is found (see, e.g., Batchelor, 1953) during the major part of the decay process. Furthermore, the observed spectral similarity* in reverting flows is of the type $e^{-k_1 l/2}$, where k_1 is the component of the wave-number vector in the direction of the man flow. This suggests considerable nonlinear interaction among wave numbers, and in this sense the decay process is not entirely viscous-dominated: if it were, the situation would be diffusion-controlled as in the "final period" of decay in homogeneous turbulence, where the spectral density exhibits a similarity of the type $e^{-\beta k_1^2}$ (Batchelor and Townsend, 1948). We have however already seen in earlier sections that intercomponent correlations become increasingly less important with increasing downstream distance.

Another pointer to the importance of the nonlinear mechanism in spectral energy transfer is that, during a major part of the process of decay, the Reynolds number $\mathrm{Re}_\lambda (\equiv \hat{u}\lambda/\nu)$ *increases* downstream. For example, in Laufer's (1962) reverting pipe flow, Re_λ increases by a factor of nearly 2.5 from $x/a \simeq 25$ to $x/a \simeq 100$. However, the quantitative significance of this observation is doubtful because of the increasing anisotropy of the turbulence field.

C. OTHER INSTANCES OF REVERSION BY DISSIPATION

There are a large number of situations where the variation of flow parameters increases dissipation and so suppresses turbulence. The decrease in

---

* There are two consequences of this similarity. First the two-point double-velocity correlation $\langle u'(x_0)\, u'(x_0 + x) \rangle$ decreases rather slowly with distance (like x^{-2} at large x). No measurements of this correlation have been made, however, in this class of reverting flows. Second, the spectral similarity implies that the ratio λ/l is a constant during decay. This constant is about $\frac{1}{3}$ in Laufer's experiment and about $\frac{1}{2}$ in Sibulkin's and Badri Narayanan's experiments.

Reynolds number can occur either due to an increase in the cross-sectional area of a single duct carrying the fluid, as in the flows studied above, or by branching. An interesting example of the latter is in the human lungs, where Reynolds numbers (based on diameter) vary from about 3100 in the trachea to 1230 in the segmental bronchi (Owen, 1969). Although the bronchial tubes in the upper airways are too short for developed flow to be obtained, the decrease in Reynolds numbers is such that well before the very fine bronchial tubes are reached, the flow will have reverted to the laminar state.

Another example of dissipative reversion occurs in a pipe of uniform diameter containing an orifice, say for metering fluid flow. There is a certain Reynolds number range in which turbulence arises just downstream of the orifice even when the approach flow is laminar, as the free shear layers springing from the orifice lip are highly unstable. Further downstream of the orifice, however, this turbulence must be suppressed as the flow reverts to its original laminar state. Measurements by Alvi and Sridharan (see Alvi, 1975) show how the orifice characteristics are affected by the sequence of events going through transition in the shear layer followed by reversion downstream to fully turbulent flow everywhere (Fig. 10). In particular, the so-called settling length, defined as a measure of the distance downstream of the orifice required for the reestablishment of the pressure gradient characteristic of fully developed flow, increases rapidly to more than 20 diameters as the pipe critical Reynolds number of about 2300 (based on diameter) is reached; this is a simple consequence of the slowness of the dissipative reversion process that we have already noted above. Once the critical Reynolds number is exceeded the settling length drops steeply to six to seven diameters, characteristic of fully developed turbulent flow.

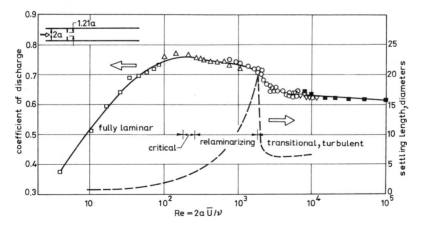

FIG. 10. Variation of selected flow characteristics for a sharp-edged orifice with approach flow pipe Reynolds number. (Data from Alvi, 1975.)

Other characteristics of the orifice, like the discharge coefficient, also show the effects of reversion at the appropriate Reynolds numbers, but these are much less pronounced.

Other forms of dissipation arising from different molecular transport properties (e.g., thermal or electrical conductivity) may also bring about reversion. Magnetohydrodynamic flows provide some interesting examples, but they are sufficiently complex to demand separate attention, and so are discussed in Section X.

IV. Reversion in Stably Stratified Flows

A. INTRODUCTION

As noted in Section I, the possibility of the suppression of turbulence in the presence of a stabilizing density gradient has been known for a long time. However, although it is almost 60 years since Richardson proposed a criterion for the phenomenon in 1920, no full-fledged theory exists yet. Physically, the presence of a lighter fluid on top (i.e., lighter than when the fluid is in hydrostatic equilibrium) means that rising fluid has to work against gravity, and so turbulent energy could be converted into gravitational potential energy. It is such energy absorption that leads to the reversion shown in Fig. 2.

Such stable stratification can be observed frequently in the atmosphere. For example, following sunset on a day during which sunshine has produced much convection, there can be sudden cooling of the ground by radiation (see, e.g., Scorer, 1958, p. 209), producing stable density gradients and a suppression of turbulence; this phenomenon can be recognized by the appearance of clouds with flat, smooth tops, similar to that in Fig. 2. Similar effects may take place following a sharp cooling shower of rain (Fig. 11), and may also occur during a solar eclipse, which we shall discuss in Section IV,C below.

B. ANALYSIS

The effects of stratification can be approximately analyzed by making what is known as the Boussinesq assumption, according to which the only effect of density changes is to provide an additional body force, the fluid being otherwise incompressible—effectively therefore its mass remains constant but weight changes. (See a discussion of this assumption in Monin and Yaglom, 1971, Chapter 6.) There is now therefore a body force in Eq. (2.3), represented by

$$\mathbf{X}' = -\mathbf{k}\rho g b',\tag{4.1}$$

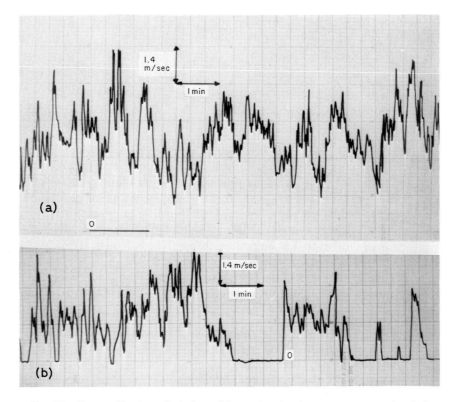

FIG. 11.   Traces of horizontal wind speed from a low inertia cup anemometer, just before
(a) and after (b) a cooling rain shower: note the patches of laminar flow in (b). (From A. Prabhu,
unpublished, 1977.)

where **k** is a unit vector along the vertical, g is the acceleration due to gravity,
and b' is the fluctuating part of the specific volume ρ^{-1}.

Consider now a steady mean flow $u(z)$ along the x-axis, with the z-axis
being vertical and the flow assumed homogeneous in all horizontal planes.
The relevant generation terms in (2.3) are then respectively as follows (see
also Stewart, 1959):

$$\tfrac{1}{2}\hat{u}^2 : -\langle u'w'\rangle \frac{\partial u}{\partial z}, \tag{4.2a}$$

$$\tfrac{1}{2}\hat{v}^2 : 0, \tag{4.2b}$$

$$\tfrac{1}{2}\hat{w}^2 : -g\rho\langle b'w'\rangle, \tag{4.2c}$$

$$-\langle u'w'\rangle : g\rho\langle b'u'\rangle + \hat{w}^2\frac{\partial u}{\partial z}. \tag{4.2d}$$

In addition, we may derive from the energy equation the generation terms for the specific volume fluctuation $\hat{b}$ and the flux $\langle b'w' \rangle$:

$$\hat{b}^2 : -\langle b'w' \rangle \frac{d}{dz}\left(\frac{1}{\rho}\right), \tag{4.3a}$$

$$-\langle b'w' \rangle : \rho g \hat{b}^2 + \hat{w}^2 \frac{d}{dz}\left(\frac{1}{\rho}\right). \tag{4.3b}$$

It is interesting to note that there is a production from interaction with the mean flow only for the component $\hat{u}$, but this production depends also on $\langle u'w' \rangle$. In turn, $\langle u'w' \rangle$ can be produced through the interaction of $\hat{w}$ and du/dz; however $\hat{w}$, not being directly produced by mean flow interaction, must be sustained by transfer of energy from the other components through the action of pressure (see discussion in Batchelor, 1953, p. 84, about the mechanism). If, however, $\hat{w}$ is suppressed through the buoyancy term $\langle b'w' \rangle$, then the production of $\hat{u}$ and $\langle u'w' \rangle$ will also diminish.

A measure of the ratio of the energy so absorbed by buoyancy forces to the production through the mean shear is the so-called flux Richardson number,

$$Rf \equiv \frac{\rho g \langle b'w' \rangle}{\langle u'w' \rangle \, du/dz}$$

$$= \frac{g}{T} \frac{\langle T'w' \rangle}{\langle u'w' \rangle \, du/dz}, \tag{4.4}$$

where $T' = T\rho'/\rho$ is the temperature fluctuation in a perfect gas, making the reasonable assumption that in low-speed flows the relative pressure fluctuation p'/p is negligible.

If mean flow production and viscous dissipation are approximately equal, as they are in many shear flows (see, e.g., the energy balance in a turbulent boundary layer, presented below in Section V,B,1), then a relatively small loss to buoyancy can affect turbulence intensities appreciably; suppression of $\hat{w}^2$ quickly diminishes the other components as well. Townsend (1958) has made a detailed analysis of the energy balance in such stratified flows, and shows that the critical flux Richardson number Rf_{cr} must be less than $\frac{1}{2}$, as otherwise it is impossible to find a real turbulence intensity that will satisfy the equation for turbulent energy as well as that for temperature fluctuations. Experimental observations, to be discussed in the next section, show that Rf_{cr} is in fact much less than Townsend's upper bound, being only 0.1 or less. Ellison (1957) has presented an analysis that suggests $Rf_{cr} \simeq 0.15$, and further that density and velocity fluctuations are decoupled rather than destroyed under stable conditions.

Before examining the experimental evidence, it is worth noting that, if the energy and momentum fluxes in (4.4) are respectively proportional to the gradients of $1/\rho$ and u (through appropriate eddy diffusivities of heat and momentum, say K_h and K_m), we can write

$$Rf = \frac{K_h}{K_m} \frac{g \, d(1/\rho)/dz}{(du/dz)^2}$$

$$= \frac{K_h}{K_m} Ri, \qquad (4.5)$$

where Ri is known as the gradient Richardson number. In an atmosphere at rest, hydrostatic equilibrium requires $dp/dz = -\rho g$, and if entropy were constant the temperature would decrease with altitude at the so-called dry adiabatic lapse rate $dT/dz = -g/c_p$, where c_p is the specific heat at constant pressure. (The temperature of a parcel of dry air would change at this rate if it were to move vertically at constant entropy.) The temperature gradient that is relevant for stability is therefore the excess over the adiabatic lapse rate, i.e., $(dT/dz) + (g/c_p)$, or the gradient of the potential temperature $T + gz/c_p$. The appropriate definition of Ri in this case is thus

$$Ri = -\frac{g}{T} \frac{(dT/dz + g/c_p)}{(du/dz)^2}. \qquad (4.6)$$

Note that if the gradients in these definitions are estimated in terms of characteristic velocity and length scales U and L, Ri is proportional to $gL \, \Delta\rho/\rho U^2$, which is often called a bulk Richardson number and resembles an inverse Froude number.

A large number of experimental observations quote a critical Ri rather than a critical Rf, presumably because the former is easier to measure. The significance of Ri_{cr} is, however, less certain because of the variability of the ratio K_m/K_h [see (4.4)] with stability conditions. Even so, the observed values of Ri_{cr} also seem to be fairly low.

We may note in passing that Miles (1961) showed that the sufficient condition for an inviscid stratified flow to be stable is that $Ri > \frac{1}{4}$ everywhere in the flow.

C. Experimental Observations

Apart from the early work of Prandtl and Reichardt (1934) and the interesting measurements recently reported by Nicholl (1970), there have not been many detailed turbulence measurements in stably stratified turbulent shear flows. By using a hot wire at two widely different temperatures, Nicholl was able to measure both the turbulent intensities and the fluctuating

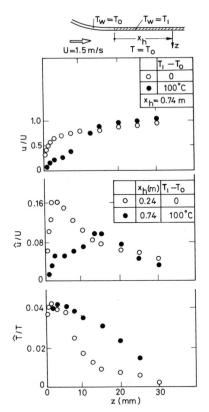

FIG. 12. Suppression of turbulence in the boundary layer on the heated roof of a wind tunnel; experimental data from Nicholl (1970). Note that while velocity fluctuations are much lower on the heated surface, temperature fluctuations are not.

temperature T'. The measurements were made in the boundary layer on the roof of a wind tunnel, a step change in the surface temperature, making it about 100°C hotter than the free stream, subjected an approaching "cold" turbulent boundary (at a momentum thickness Reynolds number of about 600) to a stable stratification beyond a certain streamwise station. Measurements 74 cm (about 25δ) downstream of this station show a large deceleration of the mean flow near the wall, an appreciable drop in $\hat{u}$, and an *increase* in $\hat{T}$ in the outer layer (Fig. 12). The increase in $\hat{T}$ is presumably due to strong vertical motions induced by the deceleration near the surface and the resulting increase in the displacement of the outer streamlines. Interestingly, Nicholl found no significant changes in either the shear stress or the heat flux correlation coefficients,*

$$-\frac{\langle u'w'\rangle}{\hat{u}\hat{w}}, \qquad \frac{\langle w'T'\rangle}{\hat{w}\hat{T}}.$$

* Measurements by Webster (1964), in a flow with uniform velocity and density gradients at a Reynolds number that was perhaps not high enough, show on the other hand an appreciable drop in these correlation coefficients (in particular the heat flux one) with increasing Ri.

Townsend (1958), making an analysis of these (then still unpublished) measurements, plotted distributions of the local gradient Richardson number (4.6) across the boundary layer. The scatter in estimates of Ri was so large that Townsend felt "only the most docile reader would agree ... that the mean values [of Ri] in the outer parts of the layers are all less than 0.1 in." But using mean gradients, Townsend concluded that the Richardson number just before collapse of turbulence was less than 0.1; indeed, in one of the experiments, at Ri $\simeq$ 0.022 the collapse of turbulence was nearly complete.

In an interesting study of the wake of a circular cylinder towed in stably stratified salt water (Ri $\simeq$ 0.15), Pao (1969) observed that the turbulence created in the near-wake region died away sufficiently far from the cylinder. At a distance of 110 diameters, for example, both large- and small-scale turbulence had collapsed, and the residual motion (as observed by shadowgraph pictures) consisted exclusively of periodic internal waves.

Ellison and Turner (1960) examined the behavior of a layer of dense salt solution introduced through a slot in the floor of a sloping rectangular channel in which there was a main turbulent flow. When the main flow was uphill, the stabilizing effect of the density gradient caused a suppression of the spread of the salt solution. Ellison and Turner showed that their observations on the transfer coefficients of salt and momentum were consistent with a critical value of about 0.15 for the flux Richardson number.

Similar interesting observations have also been made in the atmosphere. Measurements of velocity and temperature distributions across inversions, explicitly showing the disappearance of turbulence, have been reported by Businger and Arya (1974). They estimated Ri_{cr} to be about 0.21. Lyons *et al.* (1964) have reported atmospheric measurements in which turbulence always existed for Ri $\simeq$ 0.15, but was completely absent when Ri > 0.5; presumably the critical value is somewhere in-between.

Another illustrative case is that of a total (or nearly total) solar eclipse. As a result of the sudden cessation of the incoming solar radiation, the ground cools off faster than the air above it, and a stable temperature gradient gets established relatively quickly near the ground. If this cooling of the surface continues, the depth of the stably stratified layer grows, and reaches a height where, presumably, a critical value of Ri is reached. This then results in a sudden decay of turbulence. Measurements made recently during a total (R. Luxton, private communication, 1976) or near-total solar eclipse (Antonia *et al.*, 1979) essentially confirm these expectations (Fig. 13).* An interesting feature of these measurements is the nearly constant time lag between the decay of temperature and velocity fluctuations. The reason probably is that

---

* Unfortunately, the final stages of the eclipse (which lasted roughly from 1500 to 1700 hr) gradually merged into the normal sunset period, so the return to normal sunny conditions were not obtained.

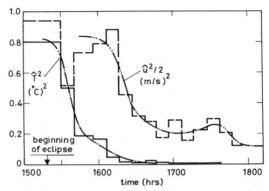

FIG. 13.   Decay of velocity and temperature fluctuations during a nearly total solar eclipse (data from Antonia *et al.* 1979). Note how the velocity fluctuations lag behind the temperature fluctuations by a nearly constant time interval (about 50 min).

the temperature fluctuation starts decaying almost instantaneously in response to the cessation of solar radiation, while velocity fluctuations start decaying only after the stably stratified layer has been established and has grown to the height above the ground of the measurement station ($\simeq 4\,\mathrm{m}$ in this case).

V. Reversion in Highly Accelerated Flows

A. INTRODUCTION

We now examine a class of reverting flows that do not seem to belong entirely to either of the other two types we have described until now, namely, a turbulent boundary layer subjected to a large favorable pressure gradient or acceleration. In their well-known work on transition from laminar to turbulent flow, Schubauer and Skramstad (1947) and Liepmann (1943, 1945) had already observed how a favorable pressure gradient could suppress incipient turbulence. But the first evidence that a fully turbulent flow subjected to large accelerations may revert to a laminar state appears to have come from experiments at high speeds, such as those of Sternberg (1954) at the shoulder of a cone-cylinder junction and of Sergienko and Gretsov (1959) in an axisymmetric supersonic nozzle. Indeed, Sternberg's remarkable report must have inspired many of the studies that followed in later years. Although we shall defer discussion of high speed flows until Section V,C, it is worthwhile to take a quick look at Figs. 14(a) and (b) which show

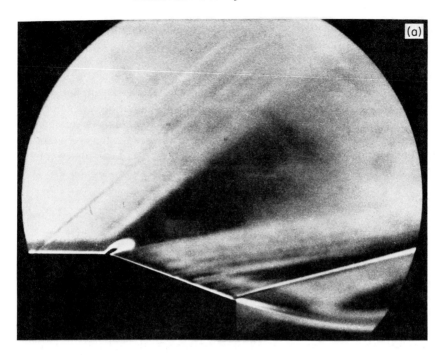

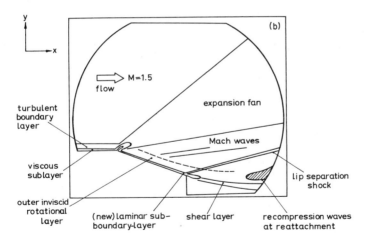

FIG. 14.   (a) Schlieren photograph of flow past a boat-tailed step. Note how, downstream of the Prandtl–Meyer corner, a thin streak representing a laminar subboundary-layer grows underneath the thick remnant of the original turbulent boundary layer. (From Viswanath and Narasimha, 1975.) (b) Key to Fig. 14(a).

(among other things) the behavior of an initially turbulent boundary layer as it negotiates a Prandtl–Meyer corner in supersonic flow. The photograph, which is a *schlieren* from Viswanath and Narasimha (1975), shows clearly the generation of a thin new boundary layer downstream of the corner—a layer that measurements of quantities such as wall temperature and surface heat flux show to be essentially laminar. This new laminar layer is embedded underneath the remnant of the original turbulent boundary layer.

It is, however, the flow at low speeds that has been extensively studied in recent years and we examine these first. (For a critical review of the experimental data, see Sreenivasan, 1972.)

B. Low Speed Flows

1. *General Remarks*

To fix our thoughts, consider the situation in which a fully turbulent boundary layer develops at constant pressure up to the point x_0, beyond which a steep favorable pressure gradient is imposed. Observation shows that downstream of x_0 the boundary layer thins down; eventually the velocity profile departs from the well-known law of the wall, the shape factor (H) increases, the skin friction coefficient (c_f) drops, the relative turbulence intensity goes down, and the flow becomes effectively laminar. Figure 15 demonstrates how complete such reversion can be, in terms of the velocity profile.

Many detailed studies of such flows have been made in the last 15 years, but different workers have used different methods for recognizing the onset of relaminarization, and proposed different parameters as criteria for the

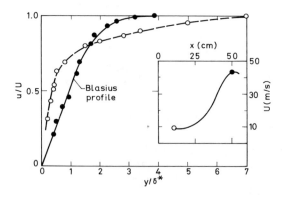

FIG. 15.   Relaminarization in accelerated boundary layer flow: velocity profiles just before and after acceleration, shown in inset. (Data from Ramjee, 1968.)

occurrence of reversion (see Table I: some of the symbols used here will be defined as we proceed). The large majority of these proposals involve the viscosity of the fluid, and so can be interpreted as some kind of Reynolds number: the differences therefore lie in the choice of length and velocity scales. The original flow being turbulent, it offers both the free stream velocity U and the friction velocity U_* as velocity scales, and because the perturbation is a *function* describing the pressure distribution $p(x)$, many choices are possible for the scales that may form the Reynolds number. One of the most widely used parameters, namely

$$K = \nu U'/U^2, \qquad U' \equiv dU/dx, \qquad (5.1)$$

where $U(x)$ is the free stream velocity, carries no information at all about the boundary layer whose reversion is the object of study: it is purely a free stream parameter (and, for that reason, a very convenient one to use). Various combinations of K and the skin friction coefficient c_f, of the form Kc_f^{-n} with n varying between $\frac{1}{2}$ and $\frac{3}{2}$, have also been suggested (Back *et al.*, 1964; Launder and Stinchcombe, 1967), but it is clear that the general approach has been, "Seek the Reynolds number."

The difficulties with some of these proposals have been discussed by Narasimha and Sreenivasan (1973). It is eminently reasonable that departure from the standard log-law in the wall layer depends on the pressure gradient non-dimensionalized with respect to the wall variables,

$$\Delta p = \nu p_x/U_*^3, \qquad p_x \equiv dp/dx, \qquad (5.2)$$

as suggested by Patel (1965). However, as pointed out by Narasimha and Sreenivasan (1973), this parameter reaches a minimum *upstream* of where Patel and Head (1968) infer reversion to have occurred—so that the suggested critical value for $-\Delta_p$ has already been encountered once, and exceeded, before reversion apparently occurs. (Note that in favorable pressure gradients $p_x < 0$.) The same objection applies also to the later proposal of Patel and Head (1968), which replaces dp/dx in (5.2) by $\partial \tau/\partial y$:

$$\Delta_\tau = \frac{\nu}{U_*^3} \frac{\partial \tau}{\partial y}. \qquad (5.3)$$

There is in this case the further difficulty of measuring or estimating $\partial \tau/\partial y$ in relaminarizing flows.

While these parameters are certainly useful in indicating when the flow begins to depart from the standard constant-pressure laws, we must recognize that such departures may not necessarily imply reversion, although they must certainly precede it. Such "standard" laws are often based on assumptions (e.g., matching similarity solutions) that are only asymptotically valid at large Reynolds numbers, if not on fairly specific models (e.g., mixing length)

TABLE I

STUDIES OF HIGHLY ACCELERATED LOW-SPEED BOUNDARY LAYERS

Reference	Flow studied	Method of identification of reversion	Proposed criterion for reversion
Wilson and Pope (1954)	Turbine nozzle cascade	Reduction in heat transfer coefficient	—
Senoo (1957)	Turbine nozzle cascade	Reduction of high frequency part of turbulent signal in throat region of nozzle	—
Launder (1963, 1964)	Two-dimensional nozzle flow	—	$K = \nu U'/U^2 \gtrsim 2 \times 10^{-6}$ for about 20δ
Schraub and Kline (1965)	Two-dimensional boundary layer flow in a water-channel	Cessation of bursting in wall region	$K/c_f^{1/2}; K \gtrsim 3.5 \times 10^{-6}$
Moretti and Kays (1965)	Two-dimensional nozzle flow	Failure of a Stanton number prediction procedure	$K \gtrsim 3.5 \times 10^{-6}$
Patel (1965)	Accelerated wall layer in pipe flow	Breakdown of log-law in wall layer	$-\Delta_p = -\nu p_x/U_*^3 = K(2/c_f)^{3/2} \gtrsim 0.025$
Patel and Head (1968)	Accelerated wall layer in pipe flow	Departure of wall-layer flow from a modified wall law	$-\Delta_t = -\dfrac{\nu}{U_*^3}\dfrac{\partial \tau}{\partial y} \gtrsim 0.009$

250

Reference	Flow type	Characteristic	Parameter
Fielder and Head (1966)	Accelerated turbulent boundary layer	Spreading of intermittency to wall region	$Kc_f^{-3/2}$
Launder and Stinchcombe (1967)	Sink-flow boundary layer	Departure from normal turbulent boundary layer characteristics	
Back and Seban (1967)	Highly accelerated boundary layer	Reduction of heat transfer and skin-friction below that given by a standard prediction procedure	$K \gtrsim 3 \times 10^{-6}$
Bradshaw (1969a)	Highly accelerated boundary layer	Overlapping of energy containing and dissipating eddy scales	$\tau^2/\nu\varepsilon \lesssim 12$
Badri Narayanan and Ramjee (1969)	Highly accelerated boundary layer	Decrease in $(\tilde{u}/U)$ sub	$\mathrm{Re}_\theta \lesssim 300$
Narasimha and Sreenivasan (1973)	Highly accelerated boundary layer	Flow described by quasi-laminar limit	$\Lambda = -p_x\delta/\tau_0 \gtrsim 50$
Okamoto and Misu (1977)	Boundary layer in contraction	Departure from log-law	—
Simpson and Shackleton (1977)	Nozzle flow	Cessation of entrainment	—

whose limitations are well known. It is possible that such departures merely indicate that these assumptions are not valid or that the flow is abnormal in some sense, rather than that there is reversion.

If the criterion for reversion is a Reynolds number, it is implied from our discussion in Section III that viscous dissipation exceeds production. Measurements of the turbulent energy balance in an accelerated boundary layer, reported by Badri Narayanan, *et al.* (1974), show however that the dissipation always remains smaller. Figure 16 compares the energy budget at a station where $K \simeq 2.8 \times 10^{-6}$ with that at zero pressure gradient, $K = 0$. All contributions to the budget are nondimensionalized here by the outer scales U and δ, it is seen that while both production and dissipation are reduced in these units, their distributions across the boundary layer remain similar, and dissipation remains smaller. Certain changes in advection and diffusion may be noticed, but, except possibly in the region $y/\delta < 0.1$ which is not covered by the measurements, the observed phenomena cannot be attributed to dissipation.

We need to consider also whether energy production decreases because of a favorable pressure gradient. From (2.5), it can be shown that the relevant terms in two-dimensional flow are

$$q: -\langle u'v' \rangle \frac{\partial u}{\partial y} - (\hat{u}^2 - \hat{v}^2) \frac{\partial u}{\partial x}; \qquad (5.4)$$

the last term is negative if $\hat{u} > \hat{v}$, $\partial u/\partial x > 0$. In the energy balance measurements cited above, this term was found to be negligible. More generally,

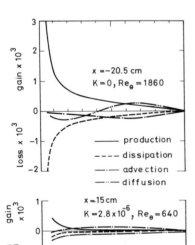

FIG. 16. Turbulent energy balance in constant pressure and accelerated boundary layers; all terms scaled on outer variables U, δ. (Data from Badri Narayanan *et al.*, 1974) $K = 0$ at $x = -20.5$ cm, $K = 2.8 \times 10^{-6}$ at $x = 15$ cm.)

one may define, in analogy with the flux Richardson number (4.4), a parameter that measures the ratio of the "absorption" to the production:

$$\chi = \frac{(\hat{u}^2 - \hat{v}^2)\,\partial u/\partial x}{-\langle u'v'\rangle\,\partial u/\partial y}. \tag{5.5a}$$

At $y_+ \simeq 15$, where the production in a constant-pressure boundary layer is largest, the value of this parameter is estimated by Back *et al.* (1964) to be

$$\chi \simeq \frac{22v}{U_*^2}\frac{\partial U}{dx} = 44\,\frac{K}{c_{\mathrm{f}}}. \tag{5.5b}$$

This shows that for χ to be of order unity, K would have to be about 10^{-4}; in the many flows where reversion has been observed at $K \simeq 3 \times 10^{-6}$, the absorption could clearly have been no more than a few percent where the production is largest. (This conclusion may not be valid in the outer layer; see Section V,B,3.)

Finally, we may note that in accelerated flows there is no evidence of any significant decorrelation among the components contributing to the Reynolds shear stress; Fig. 17 shows that over much of the boundary layer C_τ remains constant at about 0.5 in zero as well as favorable pressure gradients.

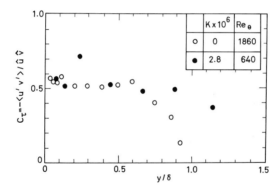

FIG. 17.   Shear stress correlation coefficient at two stations (same as in Fig. 16) in a reverting accelerated boundary layer. (Data from Badri Narayanan *et al.*, 1974.)

From considerations such as these, Narasimha and Sreenivasan (1973) were led to propose that reversion in such flows is primarily the result of the domination of pressure forces over nearly frozen Reynolds stresses, rather than of absorption or dissipation, although these could contribute (especially the latter near the wall). They noted first that the completion of the process of reversion has a fairly definite meaning, for it certainly occurs, for the mean flow, when the net effect of the Reynolds stresses is negligible. Random fluctuations inherited from previous history will remain, but they

are no longer relevant to the dynamics of the flow, which under these circumstances may be said to have reached a quasi-laminar state.

2. The Mean Flow

In this approach, we first formulate such a quasi-laminar limit for sufficiently large values of a pressure gradient parameter

$$\Lambda = -\frac{\delta}{\tau_0}\frac{dp}{dx},\tag{5.6}$$

where δ is the boundary layer thickness and τ_0 the wall stress in the boundary
layer just before the pressure gradient is applied. The two-dimensional turbulent boundary layer equations

$$\frac{\partial u}{\partial x} + \frac{\partial v}{\partial y} = 0,$$

$$u\frac{\partial u}{\partial x} + v\frac{\partial u}{\partial y} = -\frac{dp}{dx} + v\frac{\partial^2 u}{\partial y^2} + \frac{\partial \tau}{\partial y}\tag{5.7}$$

then split, in the limit $\Lambda \to \infty$, to an outer inviscid but rotational flow

$$u\frac{\partial u}{\partial x} + v\frac{\partial u}{\partial y} = -\frac{dp}{dx},\tag{5.8a}$$

and an inner viscous flow governed by

$$u\frac{\partial u}{\partial x} + v\frac{\partial u}{\partial y} = -\frac{dp}{dx} + v\frac{\partial^2 u}{\partial y^2},\tag{5.8b}$$

i.e., there is a laminar subboundary-layer underneath a sheared inviscid
flow. The boundary conditions for the two equations have to be obtained
by the method of matched asymptotic expansions. Solutions are quite easily
worked out for any given initial flow (Narasimha and Sreenivasan, 1973).
    These solutions turn out to be remarkably effective in describing the flow
parameters, but before showing this, let us note two very simple consequences
of (5.8a), which implies the conservation of total head and vorticity along
each streamline in the outer flow. From the former, it follows that the velocity
at the edge of the inner layer, say $U_s(x)$ [the inner limit of the solution of
(5.8a)], increases downstream sufficiently rapidly that the difference $U(x) -
U_s(x)$ decreases. To conserve vorticity, therefore, the boundary layer must
thin down. Correspondingly, also the outer edge of the (whole) boundary
layer must be a streamline, and the entrainment must vanish (to the lowest
order in Λ^{-1}, of course). Thus the total mass flux in the boundary layer,
proportional to $U(\delta - \delta^*)$ where δ^* is the displacement thickness, must also
remain constant; as δ^* is quite small compared to δ in flows of this kind,

it follows that the boundary layer Reynolds number $U\delta/\nu$ should not show large streamwise variations. Thus we have a flow in which different Reynolds numbers behave in different ways: Ux/ν increases, $U\theta/\nu$ (θ = momentum thickness) decreases, $U\delta/\nu$ varies little, and Δ_p^{-1} first decreases and then increases!

All of these deductions are largely verified by experiment: Fig. 18a,b show representative comparisons for various boundary layer mean flow parameters. The good agreement seen here demonstrates the value of looking at

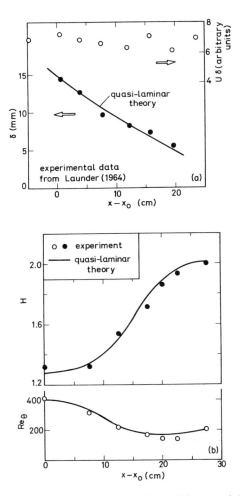

FIG. 18. (a) Streamwise variation of boundary layer thickness and the Reynolds number $U\delta/\nu$ (arbitrary units) in reverting accelerated boundary layer. (b) Streamwise variation of boundary layer Reynolds number and shape factor: comparison between experiment (Badri Narayanan and Ramjee, 1968, run 3) and theory. (From Narasimha and Sreenivasan, 1973.)

reversion as the generation of a new laminar boundary layer, rather as in the extreme case depicted in Fig. 14.

The position regarding wall variables, like the skin friction and heat flux, is slightly different. Figure 19 shows how the skin friction coefficient rises at first as the free stream accelerates, exactly as any reasonable turbulence model suggests [in the present case, the method of Spence (1956) has been used to provide estimates]. However, it is observed that thereafter c_f reaches a maximum and plunges steeply down, in accordance with the quasi-laminar theory, whose predictions are in excellent agreement with measurement especially beyond the point where $\Lambda \simeq 50$. Still further downstream, as the pressure gradient falls, c_f rises once more as the flow goes back to turbulence—note the last experimental point in the diagram! A detailed stability analysis (Narasimha and Sreenivasan, 1973) shows that this retransition occurs very near the point where instability might be expected to set in in the inner laminar boundary layer. Clearly, the maintenance of an effectively laminar inner layer in spite of the highly disturbed state of the flow above it must be attributed to the strong stabilizing influence of the favorable pressure gradient. Corresponding, the slightest sign of instability in the inner layer is all that is required to trigger sufficient energy production to cause quick retransition.

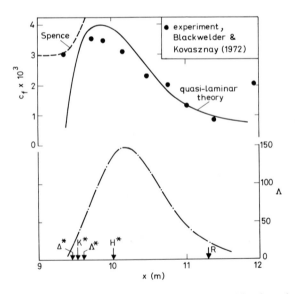

FIG. 19.   Variation of skin friction coefficient in a relaminarizing boundary layer: comparison with quasi-laminar theory. Markers on the abscissa denote location of following events. $\Delta^*:\Delta_p = -0.025$; $K^*:K = 3 \times 10^{-6}$; $\Lambda^*:\Lambda = 50$; H^*: minimum in H; R: retransition to turbulence. (From Sreenivasan, 1974.)

3. *The Turbulence Quantities*

An interesting part of this quasi-laminar theory is the ability to predict the turbulence quantities during decay. Consider first the inner layer. Very close to the surface, there is a spanwise structure in the turbulent flow; the length scale of this structure is of order $250v/U_*$ during reversion, according to the measurements of Schraub and Kline (1965). Although this length could be small compared to the boundary layer thickness, it is about 10 times larger than the height where the intensity $\hat{u}$ is a maximum. It is therefore conceivable that the fluctuating motion in the inner layer is approximately two-dimensional.

We assume further that it is quasi-steady, which is again plausible as the high-frequency components of the motion are known to decay fast (Launder, 1964). Now the development of *steady* perturbations on a laminar boundary layer has been studied by Chen and Libby (1968); by solving the appropriate eigenvalue problem, they find that, if the basic flow belongs to the Falkner–Skan family $U_s \sim (x - x_0)^m$, the maximum value of the perturbation velocity decays like

$$U_s(x - x_0)^{-\frac{1}{2}(1+m)\lambda_1}, \tag{5.9}$$

where $\lambda_1 = \lambda_1(m)$ is the lowest eigenvalue for the flow. Higher modes decay faster. We may now expand the velocity fluctuation $u'(x_0, t; y)$ at some initial x_0 in a series of the Chen–Libby eigenfunctions. The first term in this expansion, corresponding to the lowest mode, will eventually dominate the higher modes decaying faster, so that the fluctuation intensity $\hat{u}$ finally obeys the inverse power law suggested by (5.9). In those experiments where the above Falkner–Skan assumption is a reasonable approximation for the inner layer, the decay of $\hat{u}$ is indeed found to follow (5.9); Figure 20 shows a comparison.*

In other words, the fluctuating motion in the inner layer appears to be a random low frequency oscillation excited by the ambient turbulence.

Although the basic assumptions of this section cannot all be rigorously justified, experimental data show that it contains a large measure of truth.

In the outer layer, conditions are closer to what is known as rapid distortion theory (e.g., Batchelor and Proudman, 1954), in which both inertial and viscous forces are ignored. Sreenivasan (1974) has estimated that the time of flight of a particle through the region of pressure gradient in reverting flows (e.g., those reported by Badri Narayanan and Ramjee, 1969; Blackwelder and Kovasznay, 1972) is a fifth to a tenth of a characteristic time scale

---

* The same theory also predicts that the location of the maximum $\hat{u}$ scales on the thickness of the inner laminar layer; this is also borne out roughly by experiment (see Narasimha and Sreenivasan, 1973).

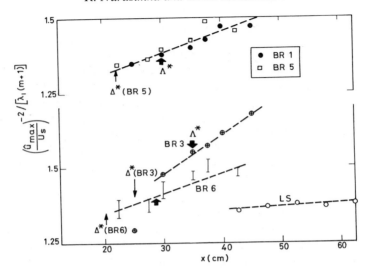

FIG. 20.   Decay of velocity fluctuations in inner layer during relaminarization. Points are data from various experiments; theory suggests the points should lie on straight lines. (From Narasimha and Sreenivasan, 1973.) BRn indicates the nth run of Badri Narayanan and Ramjee (1969).

TABLE II

RELATIVE RAPIDITY OF DISTORTION IN VARIOUS FLOWS

Source	Type	Rapidity parameter[a]
MacPhail (1944)	Grid turbulence	0.29–0.83
Dryden-Schubauer (1947)	Grid turbulence	0.67
Hall (1938)	Grid turbulence	0.83
Townsend (1954)	Grid turbulence	1.0
Uberoi (1956)	Grid turbulence	3.3
Ramjee et al. (1972)	Channel flow	5.9
Badri Narayanan and Ramjee (1969)	Boundary layer flow	5.5
Blackwelder and Kovasznay (1972)	Boundary layer flow	10.0
Crow (1969)	Sonic boom propagating in the atmosphere	10^2 or more

[a] Defined as $l/\hat{u}t_f$, where l is a characteristic size of energy-bearing eddies, t_f is the time of flight of a fluid particle through distortion. Estimates of parameter from Sreenivasan (1974).

for the energy containing eddies; the distortion is therefore relatively rapid, certainly more so than in many grid turbulence experiments set up to study rapid distortion, as the accompanying Table II shows. Figure 21 shows a comparison of the observed variation of $\hat{u}$ in a rapidly accelerated flow with that predicted by a straight application of the rapid distortion theory of Batchelor and Proudman (1954) for initially isotropic turbulence. The assumption of isotropy in calculating the effect of rapid distortion on each component of the energy yields reasonable results even when the turbulence is not strictly isotropic, as Sreenivasan and Narasimha (1978) have shown.

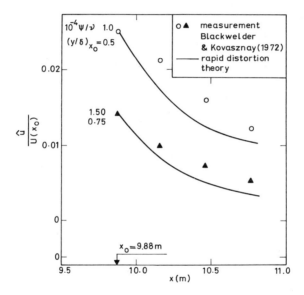

FIG. 21. Variation of turbulence intensity in outer layer. Measurements of Blackwelder and Kovasznay (1972) compared with rapid distortion theory; ψ = stream function.

A simple approximation to the dynamics of such rapid distortion of a *sheared* flow is obtained by ignoring all viscous and nonlinear terms (including in the latter the pressure diffusion term also) in the stress transport equation (2.3), and considering the limit when the strain ratio $(U - U_s)/U'\delta$ is small. (This implies that in the outer flow $\partial u/\partial x \gg \partial u/\partial y$.) We then obtain (Sreenivasan and Narasimha, 1974)

$$\frac{d}{dt}\,\hat{u}^2 = -2\hat{u}^2\,\frac{\partial u}{\partial x} + \left[2\tau\,\frac{\partial u}{\partial y}\right], \qquad (5.10a)$$

$$\frac{d}{dt}\,\hat{v}^2 = 2\hat{v}^2\,\frac{\partial u}{\partial x}, \qquad (5.10b)$$

$$\frac{d}{dt}\,\hat{w}^2 = 0, \tag{5.10c}$$

$$\frac{d}{dt}\,q = 0, \tag{5.10d}$$

$$-\frac{d}{dt}\langle u'v' \rangle = \left[\hat{v}^2 \frac{\partial u}{\partial y}\right], \tag{5.10e}$$

where the terms in square brackets are one order higher in the strain ratio. To the lowest order, therefore, $\hat{u}$ goes down by the working of the pressure gradient against the normal Reynolds stress, $\hat{v}$ goes up for the same reason; and $\hat{w}, q$, and the shear stress are frozen along each streamline. To the next order, there is a small production of both $\hat{u}$ and τ. Experiment shows reasonable agreement with these predictions. In fact, recent measurements in such a reverting flow (Badri Narayanan et al., 1974) show, consistently, a small drop in $\hat{u}$, a small increase in $\hat{v}$ and hardly any change in either the stress $-\langle u'v' \rangle$ (Fig. 22) or (as we have already noted) the correlation coefficient C_τ.

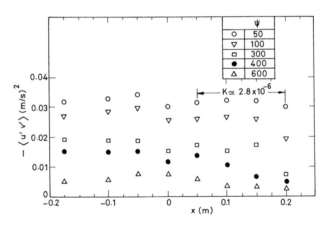

FIG. 22. Variation of Reynolds shear stress in accelerated boundary layer (measurements by Badri Narayanan et al., 1974). Stream function ψ in arbitrary units, in which the edge of the boundary layer is at approximately $\psi = 700$. (Note that in these accelerated flows the edge of the boundary layer is itself a stream line to a good approximation.)

A combination of such a rapid distortion theory in the outer layer, with an eigenfunction theory in the inner layer, produces turbulence intensity predictions in good agreement with observation for the longitudinal intensity (Fig. 23). The theory is not quite as successful for the normal component, in this as in other rapidly distorted flows.

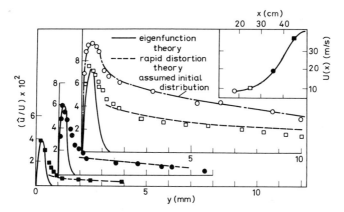

FIG. 23. Comparison of measured $\hat{u}$ distribution (Badri Narayanan and Ramjee (1969), run 3), with calculations using rapid distortion theory in the outer layer and eigenfunction theory in the inner layer (From Narasimha and Sreenivasan, 1973.)

We may in summary divide the flow in highly accelerated boundary layers into different regions as shown in Fig. 24. Region I is fully turbulent; region III is quasi-laminar—in the outer layer it is valid almost from the beginning of the pressure gradient at x_0. Thus, in a narrow but justifiable sense reversion in the outer flow can be said to occur almost immediately after the acceleration at x_0. (It is this kind of behavior that makes the proposals listed in Table

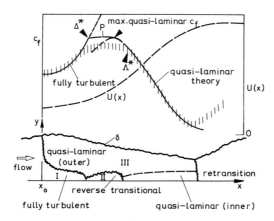

FIG. 24. Flow regimes in reverting boundary layer subjected to acceleration. The variation of true c_f (shown hatched) is predicted closely by fully turbulent theory up to the point Δ^* (corresponding to $\Delta p = -0.025$), and by quasi-laminar theory after the point Λ^*, corresponding to $\Lambda = 50$. The gap between the two theories is patched by a region P of constant c_f. (From Narasimha and Sreenivasan, 1973.)

I seem like an excessive preoccupation with the Reynolds number.) There is only a small bubble-shaped region (numbered II) near the wall where neither the fully turbulent nor the quasi-laminar solution is valid. It is only this region that needs to be modeled more elaborately; but it does not strongly affect many important mean flow characteristics (e.g., the various thicknesses), which can be determined without invoking specific models for turbulence. The reason for such success is chiefly that the turbulence, especially in the outer region, is distorted rather than destroyed by the acceleration, and the Reynolds shear stress is nearly frozen. The reversion observed is thus essentially due to the domination of pressure forces over the slowly responding Reynolds stresses in an originally turbulent flow, accompanied by the generation of a new laminar boundary layer stabilized by the favorable pressure gradient.

4. Burst Parameters

Kline et al. (1967) have suggested that relaminarization in accelerated flows is associated with the cessation of turbulent bursting. They have correlated the rate of occurrence of bursts per unit span, F [measured by Schraub and Kline (1965) in a boundary layer subjected to a "strongly favorable pressure gradient"], with the acceleration parameter K, and deduced by extrapolation a critical value $K_{cr} \simeq 3.5 \times 10^{-6}$ at which bursting ceases. In coming to such conclusions, the burst parameter has to be assessed against an appropriate scale; Kline et al. (1967) suggest a wall scaling

$$F_+ = Fv^2U_*^{-3},$$

but Narasimha and Sreenivasan (1973) have shown that an equally appropriate scaling is

$$\tilde{F} = Fv\delta^*/UU_*,$$

based on the argument that while the lateral spacing of bursts scales on wall variables, the temporal rate scales on outer variables (as shown by Rao et al., 1971). Further, Narasimha and Sreenivasan (1973) found that, whatever scaling is used, the burst parameter falls exponentially in Λ over the whole range covered by the available data (Fig. 25); no critical station in the flow can be identified, nor can an obvious extrapolation to $F = 0$ made. This, of course, is consistent with our earlier description of relaminarization in accelerated flows as an asymptotic process. In particular, in the experiment reported by Schraub and Kline (1965), the skin friction coefficient was still rising at the last measurement station in the flow, and the parameter Λ had not exceeded 30, suggesting that reversion was not complete in the sense of Section V,B,2. It would therefore appear that there is an exponential decrease in the bursting rate in an accelerated turbulent boundary layer before the wall variables begin to assume quasi-laminar values.

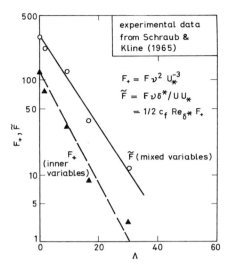

FIG. 25.   Burst parameters in accelerated flow: F = number of bursts per unit time, unit span. (From Narasimha and Sreenivasan, 1973.)

There is unfortunately no direct evidence on the effect of a favorable pressure gradient on the time rate of occurrence N_b of bursts at a given point in the flow. Badri Narayanan *et al.* (1974, 1977) report an average frequency (say N_p) for the occurrence of high frequency pulses in filtered u' signals, obtained according to a technique developed by Rao *et al.* (1971). These measurements show that the parameter $N_p \delta / U$ decreases with downstream distance in the region of strong acceleration. This is consistent with the conclusions of Kline *et al.* if $N_p \simeq N_b$, as Rao *et al.* found in the constant-pressure boundary layer. However, as no estimate of the spacing of the wall-layer streaks was obtained by Badri Narayanan *et al.*, F^+ cannot be estimated and hence no direct comparisons can be made.

A rapid (although not catastrophic) drop in the bursting rate is also significant from another point of view. With the cessation of eruptions from the inner layer, and also of the sweeps from the outer layer, there is little or no interaction between the two layers. It thus appears quite justified to ignore the usual transfer processes between the two layers, as is implicitly assumed in the quasi-laminar theory of Narasimha and Sreenivasan (1973).

Simpson and Wallace (1975) and Simpson and Shackleton (1977) have recently reported certain measurements that must be mentioned here. These authors evaluated the frequency (say N_s) corresponding to the peak in the first moment $nF_u(n)$ of the spectral density $F_u(n)$ of the streamwise velocity fluctuation u', and assume that $N_s \simeq N_b$. This assumption was based on the work of Strickland and Simpson (1975) in a constant pressure boundary layer, where N_b itself had in turn been inferred from an assumed correspondence with the second peak in the short-time autocorrelation function of the wall shear stress. Simpson and Wallace and Simpson and Shackleton also obtained an estimate of the average spacing of the wall-layer streaks. Based on these assumptions it is possible to estimate a parameter equivalent to F mentioned above. In the accelerated flow, it was found that this quantity actually *increases* downstream, contrary to the trend that one expects from the other measurements.

Perhaps these measurements are too indirect to provide a convincing indication of the behavior of the burst rate in accelerated flows; there is need for further work before any firm conclusion can be drawn.

C. SUPERSONIC FLOWS

These ideas that have proved so successful at low speeds appear to work well at high speeds too, although here detailed turbulent structure measurements are rarely available: the one exception is Morkovin's (1955) work on the interaction between an expansion fan and a turbulent boundary layer on a flat plate. In the Prandtl–Meyer flow past a corner, Sternberg (1954) found a significant drop in the recovery factor downstream; others have found marked changes in boundary layer velocity profiles (Vivekanandan, 1963), in base pressure in sharply boat-tailed bases (Viswanath and Narasimha, 1972), and in heat transfer rates (Zakkay et al., 1964).

The basic idea in Section V,B,1, namely, that during high acceleration the Reynolds stresses over a large part of the boundary layer have little influence on mean flow development, should be valid with even greater force in supersonic flow past an expansion corner, where the pressure gradients are extremely large. Indeed the parameter suggested there as relevant, namely $\Lambda = -p_x \delta/\tau_0$, takes on a simpler form in supersonic corner flow. For, the interaction of the expansion fan with the boundary layer spreads the pressure drop Δp (here considered positive in an expansion, for convenience) over a few boundary layer thicknesses. The extent of this region appears to be relatively insensitive to the total corner flow deflection, from the work of Ananda Murthy and Hammitt (1958). Thus, p_x is of order $\Delta p/\delta$, and Λ is of order $\Delta p/\tau_0$. This reasoning suggests that reversion will occur if $\Delta p/\tau_0$ is sufficiently large. (An alternative approach is to realize that in Prandtl–Meyer flow p_x is nearly a Dirac delta function, and a mean value of Λ would be proportional to $\Delta p/\tau_0$.)

To test this argument, all available experimental data were examined by Viswanath and Narasimha (1975), taking reversion to have occurred if there is evidence either of the growth of a thin new shear layer from the corner, or of an appreciable change in the associated flow characteristic downstream (e.g., turbulence intensity in boundary layer or base pressure). This definition is not as precise as one may desire, but nothing better is possible at present, and it certainly seems appropriate for engineering calculations.

Using estimates of τ_0 from Tetervin (1967), Narasimha and Viswanath (1975) have plotted these data as shown in Fig. 26 (which is their Fig. 1 with some additional data). Here points are shown (i) as filled symbols where authors reported indication of reversion, (ii) as filled, flagged symbols when reversion is inferred by us, and (ii) as open symbols when no evidence of reversion can be found. It is seen that all points above the line $\Delta p/\tau_0 = 75$ show reversion, those below $\Delta p/\tau_0 = 60$ show no evidence of it, and in-between there are both types. It thus appears that reversion occurs if Δp is more than about $70\tau_0$. Once again, this is clearly a case of reversion by domination of pressure forces over the Reynolds shear stress.

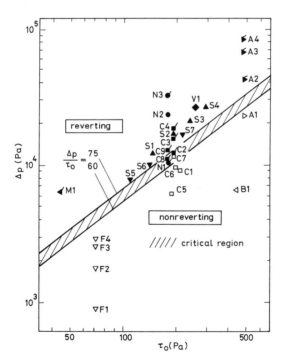

FIG. 26. Pressure drop and wall stress just upstream of corner in supersonic flow past an expansion corner. Filled symbols indicate reversion reported by authors; filled, flagged symbols reversion inferred by us; open symbols no evidence of reversion. Code for experimental points: A, Ananda Murthy and Hammitt (1958); B, Bloy (1975); C, Chapman *et al.* (1952); F, Fuller and Reid (1958); M, Morkovin (1955); N, Viswanath and Narasimha (1972); S, Sternberg (1954); V, Vivekanandan (1963). (After Narasimha and Viswanath, 1975.) The numerical suffix in each case indicates the experimental run.

As pointed out by Sternberg (1954), we may expect that turbulence passing through the expansion fan will be rapidly distorted. Interestingly, as the streamlines spread *apart* during acceleration in supersonic flow, the turbulence is stretched along as well as normal to the flow, leading to a reduction in the turbulent intensity. Such reductions have recently been measured by Gaviglio et al. (1977) in their study of a supersonic near-wake; their observations are consistent with the criterion proposed by Narasimha and Viswanath (1975).

D. Concluding Remarks

One striking feature of reversion in highly accelerated flows is that it is not catastrophic, but rather gradual and asymptotic. No unique "point of reversion" can be identified, as it can in direct transition (Narasimha, 1957; Dhawan and Narasimha, 1958). Nevertheless, the process is more rapid than dissipative reversion; typically, in many flows, the shape factor H may reach a minimum in a streamwise distance of the order of 10–20 initial boundary layer thicknesses from the commencement of acceleration, and a maximum in an additional distance of a similar order. There can, of course, be considerable variation in these distances.

We also note that the two pressure gradient parameters K, Λ introduced above are related to each other through the Reynolds number $\mathrm{Re}_\delta \equiv U\,\delta/\nu$:

$$\Lambda = 2K\,\mathrm{Re}_\delta/c_f. \tag{5.11}$$

Most laboratory experiments are carried out at about the same Re_δ (of order 10^4); it is thus difficult to discriminate between the various criteria that have been listed in Table I. If Λ governs the completion of reversion, as argued by Narasimha and Sreenivasan (1973), then according to (5.11) the value of K required for reversion should decrease as Re_δ increases. Although there is as yet no convincing evidence for this prediction, some support is available from certain experiments reported by Nash–Webber and Oates (1972) on nozzle boundary layers. They found that an increase in blowing pressure from 5 to 20 in. Hg abs. (which roughly doubled the Reynolds numbers) brought down the K at the beginning of reversion (identified as the point of minimum H) from about 5.7×10^{-6} to about 2.3×10^{-6} (see Fig. 7 of their paper). This is the trend predicted by the "domination" theory.*

---

* The final diagram presented by Nash-Webber and Oates appears to contradict this, as the boundary of the laminarizing region in the K–Re plane shows K increasing with Re. However, this region is inclusive, in the sense that reversion does not necessarily occur at the boundary, but somewhere *within* it; hence a relation between K and Re *at* reversion cannot be inferred from the diagram.

This seems to imply that relaminarization should be easier to achieve at higher Reynolds numbers, but note that it would be harder to maintain it, because the Reynolds number of the laminar subboundary-layer would be correspondingly higher and so retransition would also be quicker!

VI. Curved Flows

A. Radial Poiseuille Flow

Consider the flow between two parallel disks, with fluid coming in at the center and moving radially out, as shown in the inset to Fig. 27. If $\bar{U}$ is the mean velocity at radius r, $2a$ is the separation between the disks and $Q = 4\pi ra\bar{U}$ the total volume flow of the fluid, the local Reynolds number of the flow varies inversely with the radius:

$$\text{Re}(r) \equiv a\bar{U}(r)/\nu = Q^*a/r, \qquad Q^* = Q/4\pi\nu a, \qquad (6.1)$$

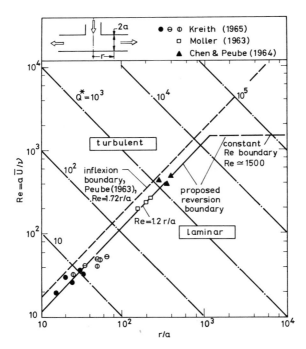

Fig. 27. Transition/reversion boundaries in radial Poiseuille flow. $-\cdot-$ indicates the Re $-\ r/a$ relation for a given Q^*.

where Q^* is a dimensionless flow-rate number. It may thus be expected that an incoming turbulent flow will relaminarize as its Reynolds number drops downstream.

Reversion has indeed been observed in such flows, but the phenomenon does not appear to occur at a fixed value of the Reynolds number, and so is rather more complex than may be imagined at first sight. The observed critical Reynolds numbers can in fact be very low: they vary from around 400 in the experiments of Moller (1963) to as low as around 15 in those of Kreith (1965). (Kreith does not quote this value, which is inferred by us from the data presented by him.)

Moller judged reversion by the agreement of measured pressure loss with laminar theory; when this is obtained reversion is complete, and so must have been initiated at a location far upstream where the Reynolds number would have been higher. This argument is plausible because of the slowness of dissipative reversion noted in Section III. Moller in fact concluded that the observed critical Reynolds numbers were not far different from those in channel or pipe flows. However this is contradicted by the very low values observed by Kreith (1965).

An alternative explanation is suggested by Chen and Peube (1964). They note that in laminar flow, according to the calculation of Peube (1963), an inflexion point is present in the velocity profile up to a critical radius r_i given by

$$r_i/a = 0.762Q^{*1/2} \qquad \text{or} \qquad \text{Re}_i = 1.72r_i/a. \qquad (6.2)$$

Flow at $r < r_i$ is therefore highly unstable, and could become turbulent even at very low values of the Reynolds number. Correspondingly, the value of Re at reversion may also be expected to be relatively low. Chen and Peube therefore suggested a criterion for reversion in terms of r_i/a.

To test the Chen–Peube hypothesis, Kreith (1965) made experiments under different flow conditions in the following way. He placed a hot wire at various radial distances halfway between the disks and slowly varied the volume flow rate until, at a given location of the hot wire, the flow underwent transition from *laminar* to a *turbulent* state. A good fit to his measurements shows that the radius r_{cr} at which this transition occurs is related (to within experimental error) to the corresponding mass flow rate Q by the equation

$$r_{cr}/a = 0.26Q^{*1/2}. \qquad (6.3)$$

This relation, although obtained by examining the laminar to turbulent transition at a given radius, holds also for reversion; if the hot wire is moved radially *outward*, (6.3) gives the critical radius at which turbulent fluctuations just cease to exist (or, in some sense, reversion is complete). Kreith's criterion

(6.3) implies that $r_{cr} \sim Q^{1/2}$, whereas a simple constant-Reynolds number criterion implies [from (6.1)] that $r_{cr} \sim Q$.

To see more clearly how the criteria differ we note that, putting (6.1) in (6.3), we get

$$r_{cr}/a \simeq 0.26(4\pi)^{1/2}\bar{U}a/v \simeq 0.91\,\text{Re} \qquad \text{or} \qquad \text{Re}_{cr} = 1.2r_{cr}/a. \qquad (6.4)$$

In the $\text{Re} - (r/a)$ plane shown in Fig. 27, this plots as a straight line; the same diagram also shows how Re varies with r for given Q^*. According to Kreith's criterion, reversion should occur when the curve for given Q^* intersects the line (6.4); furthermore, as Q^* increases the corresponding critical Reynolds number increases without limit.

However, there should be a critical Reynolds number above which the flow will be turbulent even in the absence of an inflexion point, as in plane Poiseuille flow. The value of this critical Reynolds number for radial Poiseuille flow is not known, but it is reasonable to take it to be of order 1500, as in plane flow (Section III), as the velocity distribution assumes the parabolic form asymptotically.

Based on this argument, we suggest that the region in which the flow will revert is bounded by (6.4) on one side, and by a constant critical Reynolds number above, as shown in Fig. 27.

Experimental data from Moller (1963), Kreith (1965), and Chen and Peube (1964) are also plotted in Fig. 27. Unfortunately none of these points corresponds to sufficiently high Q^*, which is therefore an area that needs further investigation.

To summarize, the radial geometry could produce an unstable mean flow whose critical Reynolds number is lower than in plane Poiseuille flow. For sufficiently large dimensionless flow rate Q^*, however, the critical Reynolds number should reach a constant value independent of Q^*. Reversion in this flow is therefore essentially dissipative, with a Reynolds number criterion that is a function of the flow rate parameter.

B. CONVEX BOUNDARY LAYERS

In the last 10 years many detailed studies have been made of boundary layer flow on surfaces with streamwise curvature. One measure of the curvature is the parameter $\kappa\delta$, where κ^{-1} is the radius of curvature (considered positive when convex) and δ is the boundary layer thickness. When $\kappa\delta$ is small, the mean flow may be represented by writing Eqs. (2.1) and (2.2) in a locally Cartesian coordinate system with x along the (curved) surface and y normal to it; the major effects of curvature then appear in these equations as additional terms, comprising a source of strength $-\partial(\kappa v y)/\partial y$ and body force $(-\kappa uv, \kappa u^2)$ along the streamwise and normal directions respectively (So and Mellor, 1972). As a consequence of the centrifugal component κu^2, there is a normal pressure gradient, the pressure being higher toward the edge in a convex boundary layer. This gradient affects the stability of the flow significantly. For example, a slow lump of turbulence thrown outward from a convex surface possesses a lower centrifugal acceleration than the faster ambient fluid, and so is driven back by the normal pressure gradient that balances the mean flow. Similarly, a faster lump moving toward the surface is thrown out. Consequently, there is a tendency to suppression of turbulence, very much as in an inversion. Indeed, as long ago as 1884, Reynolds had listed "curvature with the velocity greatest on the outside" as one of the factors "conducive to direct or steady ($=$ laminar) motion," and had noted how a small curvature may have a large effect. An analogy between buoyancy and curvature effects in the convection/cylindrical Couette flow stability problems was demonstrated by Jeffreys (1928); Prandtl (1929) drew a similar analogy for turbulent flows.

Although we will speak here only of convex boundary layers, the same effects will be encountered on *concave* surfaces if the flow velocity *decreases* away from the surface. The important thing is that the sense of flow curvature and flow vorticity must be the same; the effects we discuss below for convex boundary layers must be present in other instances of what we may call "co-curving" flows.

Another way of appreciating the effect of curvature is to examine the Reynolds stress transport equations for curved boundary layers. When (2.3) is written in the appropriate coordinate system, the following production terms appear in the respective transport equations (Rotta, 1967; So and Mellor, 1972):

$$\frac{1}{2} u^2 : -\langle u'v' \rangle \frac{\partial}{\kappa \, \partial y} (\kappa u) - 2\kappa u \langle u'v' \rangle$$

$$= -\langle u'v' \rangle \frac{\partial u}{\partial y} - \kappa u \langle u'v' \rangle, \tag{6.5a}$$

$$\frac{1}{2} \hat{v}^2 : 2\kappa u \langle u'v' \rangle, \tag{6.5b}$$

$$q : -\langle u'v' \rangle \frac{\partial u}{\partial y} + \kappa u \langle u'v' \rangle, \tag{6.5c}$$

$$-\langle u'v' \rangle : \hat{v}^2 \frac{\partial u}{\partial y} - \kappa u (2\hat{u}^2 - \hat{v}^2). \tag{6.5d}$$

[These are the leading terms in the limit of small $\kappa \delta$ for a two-dimensional boundary layer, and have been extracted from the complete equations given by So and Mellor (1972), Appendix E.] The tendency of these terms, for *given* $\langle u'v' \rangle$ <0, is to enhance $\hat{u}$ but diminish $\hat{v}$ and the total energy q in a convex boundary layer. However (6.5d) shows that $-\langle u'v' \rangle$ tends to go down as well if $\hat{u} > \hat{v}/2^{1/2}$. The experiments of So and Mellor (1973), at $\kappa\delta \simeq 0.074$, show in fact that all three components of the energy, $\hat{u}^2$, $\hat{v}^2$, $\hat{w}^2$, as well as the stress $-\langle u'v' \rangle$, are lower in a convex boundary layer than on a flat one. Figure 28 shows some of these results; note the spectacular drop in the correlation coefficient C_τ [defined in (3.3)], and the implied disappearance of the shear stress at $y/\delta \simeq 0.4$! The curvature terms in these experiments are comparable to the others; for example, the curvature "absorption" term in (6.5c) is about half the other production term.

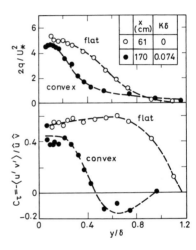

FIG. 28. Effect of convex curvature in turbulence energy and the shear stress correlation coefficient in a boundary layer: experimental data of So and Mellor (1973).

Bradshaw (1969b) has recently made extensive use of the analogy with buoyancy effects to calculate curved turbulent flows; in particular he defines a flux Richardson number for curvature,

$$\text{Rf}_c = \frac{2u}{\partial(u\kappa^{-1})/\partial y} \qquad (6.6)$$

which is approximately equal to $\kappa u/(\partial u/\partial y)$ for very small curvatures, and provides a measure of the ratio of the absorption term to the production term in, say, (6.5c). This Richardson number tends to be very small near the wall, where the flow is therefore hardly ever affected.

As in buoyant flows, however, even a relatively small value of the curvature Richardson number (6.6), i.e., even mild surface curvature, can cause significant changes in the turbulence structure. The experiments of Thomann (1968) in supersonic flow, which were so arranged that the external Mach number remained constant as the flow curved, revealed changes in wall heat flux that would be 10 times more than in laminar flow at the same value of $\kappa\delta$ (Bradshaw, 1973). Furthermore, Bradshaw finds a 10% change in the mixing length when the boundary layer thickness is still only a three-hundredth part of the radius of curvature! Thus, even when the magnitude of the terms *explicitly* involving curvature in the stress transport equations is very small, the effects of curvature are not negligible, and clearly spread to the other terms as well. We now examine the effects more closely.

As Fig. 28 shows, there is not only a reduction in turbulence intensity q on a convex surface, but a significant decorrelation of the velocity components contributing to $\langle u'v'\rangle$: the correlation coefficient C_τ drops from about 0.6 in the flat boundary layer to zero in the convex one at $y/\delta = 0.4$. Clearly, curvature is not only affecting the amplitudes of the fluctuating motion but their relative phases as well.

That these effects reflect basic changes in turbulence structure are shown by the detailed measurements recently reported by Ramaprian and Shivaprasad (1978) at $\kappa\delta \simeq 0.013$. They find that, compared to a flat boundary layer, the production and dissipation in a convex boundary layer are not significantly different (the wall stress drops only slightly), but the diffusion [represented by the terms (va) and (vb) in Eq. (2.5)] is: There is a small gain by diffusion over a large part of the boundary layer ($y/\delta \gtrsim 0.1$), from a supply apparently very close to the wall. (In these measurements, the diffusion term does not show the expected change in sign across the boundary layer; there must therefore be a peak on the loss side for the convex flow, but this apparently occurs too close to the surface to be detected by the probe used.)

Equally interesting are the spectacular changes found in the large scale motions even with $\kappa\delta$ as low as 0.01. Fig. 29 shows the integral time scale

$$T_u \equiv \int_0^{T_0} \frac{\langle u'(t)u'(t+\tau)\rangle}{\hat{u}^2}\, d\tau,$$

where T_0 is the time delay up to the first zero of the autocorrelation function $\langle u'(t)u'(t+\tau)\rangle$. Figure 30 shows the spectrum E_{22} of the v' fluctuations, normalized so that

$$\int_0^\infty E_{22}(k_1\delta)\, d(k_1\delta) = 1,$$

k_1 being the longitudinal wave number. Both diagrams show the large changes caused by convex curvature, especially at low wave numbers (integral scale is given by this end of the spectrum). Clearly the stabilizing influence of curvature is first destroying the organization of the motion in the large-scale structures in the boundary layer: the effects must later spread to the high wave number end and presumably also to the bursting process through

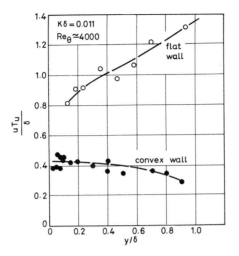

FIG. 29. Integral time scale T_u in convex boundary layer; measurements of Ramaprian and Shivaprasad (1978).

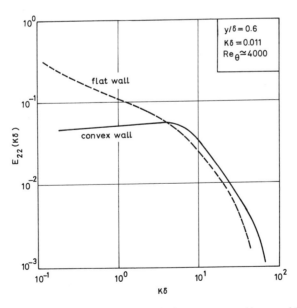

FIG. 30. Power spectral density of normal velocity component v' in curved flow; measurements of Ramaprian and Shivaprasad (1978).

the couplings that have recently been revealed by experimental studies (Rao et al., 1971; Badri Narayanan et al., 1977; Brown and Thomas, 1977).

With the Reynolds shear stresses suppressed, and the viscous stresses anyway negligible, it follows that the flow in the outer layer must be effectively inviscid and rotational, as in the rapidly accelerating flows of Section V. This is indeed found to be the case; Fig. 31 shows how the measured velocity profile in one of Ramaprian and Shivaprasad's experiments compares very well with an inviscid calculation, in which the vorticity was assumed conserved along each stream line from a station about 14δ upstream.

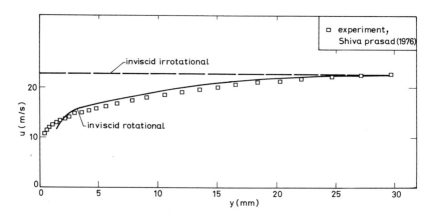

FIG. 31. Velocity profile in convex boundary layer; comparison between measurement at station 35 (Shivaprasad, 1976) and inviscid calculations.

In the inner layer, on the other hand, the effective curvature Richardson number usually remains small (because $\partial u/\partial y$ is very large), and a turbulent wall flow of the classical type remains, although even this corresponds to a slightly reduced skin friction. In analogy again with the discussion of Section V, we may expect to observe departures in the wall flow from the standard law of the wall when the curvature scaled on wall variables,

$$\kappa_+ = \kappa v/U_*,$$

is sufficiently large. In So and Mellor's experiments, with $\kappa\delta \simeq 0.074$, κ_+ is only of order 5×10^{-6}—still far too small to be significant.

To summarize, therefore, the outer flow in a convex boundary layer reverts to a quasi-laminar state, as the turbulence energy is "absorbed" by curvature forces, the organization of the motion is destroyed and the Reynolds shear stresses are suppressed.

VII. **Relaminarization by Rotation**

A. DYNAMICS OF MEAN AND FLUCTUATING VELOCITY

Like curvature, rotation can also produce reversion from turbulence to a laminar state. Experiments in an effectively two-dimensional channel rotating about a spanwise axis (Halleen and Johnston, 1967; Lezius and Johnston, 1971; Johnston *et al.*, 1972) have shown that reversion occurs on what these authors call the trailing side; as this is the side on which the imposed rotation and the basic flow vorticity have the same sense, it is convenient to call it the corotating side. For small rotation rates, the flow remains essentially turbulent although it is slightly modified, but for rotation rates higher than a certain value, flow visualization studies have shown that the following changes occur progressively on the corotating side:

(a) appearance of intermittency in the streaky wall-layer;
(b) basic laminar flow interspersed with turbulent spotlike characteristics;
(c) quiet, purely laminar layer close to the wall.

Furthermore, turbulent bursting ceases, intensities decrease, the velocity profile departs from the standard log-law form, and the skin-friction goes down considerably.

Rotation introduces two additional forces in the rotating frame of reference. The centrifugal force, which is $\frac{1}{2}\Omega^2 r^2$ per unit mass (where Ω is the angular velocity of rotation and r is the radial distance of the field point from the axis of rotation), can be combined with the static pressure and does not then explicitly enter the equations of motion. The Coriolis force, on the other hand, will have to be considered explicitly. For a rotating channel with no flow variation in the streamwise coordinate x, we have the momentum equations (y being parallel to the smallest dimension of the channel)

$$dp/dx = \partial\tau/\partial y \tag{7.1}$$

$$2\Omega u = -\frac{\partial P}{\partial y} - \frac{\partial \hat{v}^2}{\partial y}, \tag{7.2}$$

where $P \equiv p - \frac{1}{2}\Omega^2 r^2$. Usual order of magnitude estimates in (7.2) show that

$$\partial\hat{v}^2/\partial y = o(2\Omega u)$$

everywhere in the flow, so that to a first approximation

$$2\Omega u = -\partial P/\partial y. \tag{7.3}$$

Thus, an important effect of the Corolis force is to produce a normal pressure gradient proportional to Ω. As in co-curving flows (Section VI,B), this pressure gradient has a stabilizing effect: if a fluid particle from layer (1) moves accidentally outwards to layer (2) and preserves its momentum and (thus) its Coriolis force, the inward-acting pressure gradient in position (2), being larger than the Coriolis force acting on this fluid particle, will push it back to its original position. Similarly, a fluid particle moving from position (2) to position (1) is pushed back to position (2). The same argument shows that on the opposite (antirotating) side, the Coriolis force is destabilizing.

Consider now the transport equation for the components of the Reynolds stress. The relevant generation terms are:

$$\tfrac{1}{2}\hat{u}^2 : 2\Omega\langle u'v'\rangle - \langle u'v'\rangle(\partial u/\partial y), \tag{7.4}$$

$$\tfrac{1}{2}\hat{v}^2 : -2\Omega\langle u'v'\rangle, \tag{7.5}$$

$$\tfrac{1}{2}\hat{w}^2 : 0, \tag{7.6}$$

$$q : -\langle u'v'\rangle\,\partial u/\partial y, \tag{7.7}$$

$$-\langle u'v'\rangle : 2\Omega(\hat{u}^2 - \hat{v}^2) + \hat{v}^2\,\partial u/\partial y, \tag{7.8}$$

where the sign convention is such that on the co-rotating side $\Omega < 0$. The tendency of these terms in co-rotating flows is to enhance $\hat{u}$, diminish $\hat{v}$, and leave $\hat{w}$, q unaffected, for *given* $\langle u'v'\rangle$, $\partial u/\partial y$, etc. However, as in curved flows, $-\langle u'v'\rangle$ is itself reduced, and hence also $\partial u/\partial y$, so that we may expect all the Reynolds stress components to be altered.

B. Criterion for Reversion

The ratio of the additional 'absorption' term [in Eq. (7.4), for example] to the conventional production again defines an appropriate Richardson number for rotating flows,

$$\mathrm{Rf}_r = -\frac{2\Omega}{\partial u/\partial y}, \tag{7.9}$$

which we may expect to play a role similar to that in curved flows*: thus, the flow should be more stable for $\mathrm{Rf}_r > 0$.

---

* From an analogy with curved and stratified flows, Bradshaw (1969b) concluded that the appropriate Richardson number governing the stability of the flow is the parameter

$$-2\Omega\left(\frac{\partial u}{\partial y} - 2\Omega\right)\Big/\left(\frac{\partial u}{\partial y}\right)^2,$$

but as noted by Johnston *et al.* this parameter is nearly equal in magnitude to (7.9) in most boundary layer situations where $\Omega \ll \partial u/\partial y$.

Johnston *et al.* (1972) used an inverse Rossby number $\text{Ro}^{-1} = 4\Omega a/u$ as a convenient measure of the rotation, and identified, in a Re–Ro plot, regions of the fully turbulent, reverse-transitional and laminar states (as schematically indicated in Fig. 32). This classification is based on the flow visualization studies in the wall region: the laminar state is defined by the absence of spotlike individual turbulent structures in hydrogen bubble and dye visualizations, and fully developed turbulent flow by the absence of isolated laminar-like regions. Figure 32 shows that, with increasing Reynolds number, reversion occurs at an increasingly higher value of rotation, and that no unique value of Ro describes the boundaries between the different flow regimes.

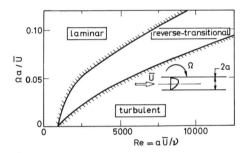

FIG. 32. Boundaries between different flow regimes in corotating channel layer; classification based on flow visualization using hydrogen bubble and dye techniques. (Data of Johnston *et al.*, 1972.)

However, close to the wall, an appropriate measure of $\partial u/\partial y$ in (7.9) is not U/a but U_*^2/v, so that we expect the "wall flow" Richardson number $\Omega_+ \equiv \Omega v/U_*^2$ to be a logical parameter with which to attempt a correlation of the *onset* of reversion. (Note the analogy here to the parameters Δ_p, Δ_τ of Section V,A.) We have therefore plotted the data of Johnston *et al.* as Ω_+ vs. Re in Fig. 33.* From our analysis of Section III, we know that reversion occurs even in the zero-rotation case when the Reynolds number Re < 1500. From Fig. 33, the rotation required to cause reversion sharply increases with Re for $1500 < \text{Re} < 3000$, and thereafter attains a nearly constant value of $\Omega_{+\text{cr}} \simeq 2.5 \times 10^{-3}$. No unique value of Ω_+ can be identified along the boundary between the laminar and reverse-transitional regimes; this again

---

* Here, we have used U_{*0}, the friction velocity corresponding to the nonrotating case at any given Reynolds number, because of its greater predictive convenience. Rotation modifies the friction velocity, as we shall discuss shortly, and so a more logical choice may appear to be the friction velocity U_* in the rotating case. However, it turns out that U_*/U_{*0} depends chiefly on Ω_+, so that the critical value of $\Omega v/U_*^2 = (U_{*0}/U_*)^2\Omega_+$ is proportional to $\Omega_{+\text{cr}}$ $[(\Omega v/U_*^2)_{\text{cr}} \simeq 4.3 \times 10^{-3}]$.

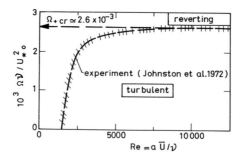

FIG. 33. A replot of the turbulent/reverse transitional boundary of Fig. 32, in terms of a wall-scaled rotation Richardson number Ω_+.

is consistent with the discussion in Section V of how wall-scaled parameters do not govern the completion of reversion.

It is useful to ascertain independently the validity of this argument by considering some other more easily measurable parameter. Figure 34, which is essentially a replot of the data of Johnston *et al.* converted to the present coordinate Ω_+ using estimated values of c_f, shows the ratio of the friction velocity U_* in the rotating flow to that in the nonrotating flow (U_{*0}) at the same Reynolds number, as a function of Ω_+. For $\Omega_+ \lesssim 2.5 \times 10^{-3}$, there is only a small and gradual reduction in the wall friction as a result of rotation; presumably, this line represents a turbulent state. Beyond $\Omega_+ \simeq 2.5 \times 10^{-3}$ a sharper change in U_*/U_{*0} seems to occur, thus possibly marking the onset of reversion.

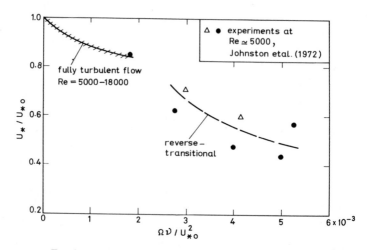

FIG. 34.   Friction velocity in corotating channel layer.

VIII. **Thermal Effects**

A. HEATED HORIZONTAL GAS FLOWS

There is considerable evidence that an initially turbulent internal gas flow reverts to a laminar state when the pipe carrying the gas is heated sufficiently (Perkins and Worsoe-Schmidt, 1965; Magee and McEligot, 1968; and especially Coon and Perkins, 1970; Bankston, 1970; Perkins *et al.*, 1973). An important effect of laminarization in this case is a sharp reduction in energy transfer to the gas, with a consequent large rise in the wall temperature (which can be high enough to be of practical concern in gas-cooled nuclear reactors).

The most usual experimental configuration used in these studies is a tube heated either resistively or by an external heat exchanger. The gas entering the tube is often pre-cooled to achieve higher values of wall-to-bulk gas temperature ratio without operating at excessively high tube temperatures; in this manner, the experiments of Perkins and Worsoe-Schmidt achieved a ratio as high as 7.

In principle, it appears possible that an internal gas flow at any entry Reynolds number can be made to revert to a laminar state by heating it enough so that, due to the increase in kinematic viscosity, the local Reynolds number falls below the critical value (as in the flows discussed in Section III). Flows with entry bulk Reynolds numbers of the order of 10^4 are known to have reverted to an effectively laminar state on heating (the Reynolds number being based on pipe radius a, average flow velocity $\bar{U}$ and kinematic viscosity $\bar{v}$ corresponding to the average flow temperature at a given section). However, reversion has been observed to be *complete* (i.e., the heat transfer characteristics, for example, have reached the appropriate laminar values) at bulk Reynolds numbers of about 2500, and is possibly initiated at Reynolds numbers of about 4000–5000 (see, for example, Coon and Perkins, 1970; Bankston, 1970). These Reynolds numbers are substantially higher than the critical value of 1500 determined in Section III for reversion in adiabatic pipe or channel flows.

The reason could well be fluid acceleration: with heating, the gas density goes down and velocity increases (to conserve mass flow) with distance downstream. The (bulk) acceleration parameter $\bar{K} = \bar{v}(d\bar{U}/dx)/\bar{U}^2$ attains values of the order of 2×10^{-6} in the experiments of Bankston (1970), Coon and Perkins (1970), and Perkins *et al.* (1973). Accelerations of this magnitude are not far from values thought sufficient to cause reversion in external (initially turbulent) boundary layers (see Section V). Although the induced acceleration in these internal flows is not an independent parameter, being

controlled by the retarding effects of the wall friction, its magnitude is sufficient to warrant a further investigation to demarcate the acceleration and
Reynolds number effects.

If the primary mechanism causing reversion is the viscous dissipation,
we expect a Reynolds number criterion to hold. In the heated flow we may
expect the critical Reynolds number to depend on a parameter like $\bar{K}$
characterizing the acceleration (which itself is related to the amount of
heating). Unfortunately, obtaining such a functional relation is not easy in
the experiments cited above. The reasons are that reconstruction of various
details of the experimental data is rarely possible from the published information, and that no consistent definition of reversion has been used in the
various experiments; also no quantitative data on turbulent fluctuations are
available. The data collected by McEligot et al. (1970) suggest that the departure of a measured-wall parameter (such as the Stanton number) from the
empirically established "wall-laws" occurs at a bulk inlet Reynolds number

$$\mathrm{Re}_i \simeq 6.1 \times 10^4 (q_i^*)^{1/2}, \tag{8.1}$$

where the surface heat flux is

$$q_w = q_i^* \bar{\rho} \bar{U} \bar{C}_p \bar{T};$$

the suffix i indicates the inlet conditions, and overbars denote sectional
averages. Further, if one defines reversion to be complete when the local
Stanton number approaches the laminar value corresponding to the local
Reynolds number, the data collected by McEligot et al. show that the
corresponding bulk inlet Reynolds number is

$$\mathrm{Re}_i = 1.72 \times 10^4 (q_i^*)^{1/2}. \tag{8.2}$$

Both (8.1) and (8.2) hold for $q_i^* > 6 \times 10^{-4}$. For square ducts, Perkins et al.
(1973) propose, instead of (8.2), the relation

$$\mathrm{Re}_i = 2.5 \times 10^3 (q_i^*)^{3/4},$$

but we shall not consider this further because of the uncertain effects of duct
geometry.

From energy balance across a short section of pipe of radius a,

$$2\pi a q_w = \frac{\pi a^2}{4} \frac{d}{dx} (\bar{\rho} \bar{C}_p \bar{T} \bar{U});$$

if $\bar{\rho}$, $\bar{C}_p$, $\bar{T}$ and p do not vary much, and taking $\bar{\mu}/\bar{\mu}_i$ as approximately $\bar{T}/\bar{T}_i$,
we get $\bar{K} = 2q^*/\mathrm{Re}$. From (8.1) and (8.2), therefore

$$\mathrm{Re}_i = 1.85 \times 10^9 \, \bar{K} \, (\bar{T}_i/\bar{T})$$

for "initiation of reversion," and

$$\mathrm{Re}_i = 1.48 \times 10^8 \, \bar{K} \, (\bar{T}_i/\bar{T})$$

for "completion of reversion." Both of these equations express the *local* critical Reynolds numbers in terms of the *local* acceleration parameter and the inlet-to-local temperature ratio. The result is shown in Fig. 35.

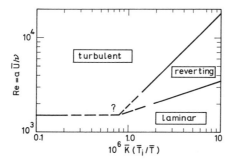

FIG. 35. Proposed variation of critical Reynolds number with acceleration parameter in heated pipe flows, based on McEligot *et al.* (1970).

Finally, we note that the dynamic viscosity of a gas increases with temperature while that of a liquid decreases. So, reversion would perhaps occur also when liquid flows are cooled sufficiently, although the flow in this case would not undergo an acceleration.

B. Heated Vertical Gas Flows

Steiner (1971) has observed reversion in the airflow passing upward through an insulated vertical pipe subjected to a constant heat flux. Measurements were made at initial Reynolds numbers Re ($\equiv \bar{U}a/\nu$) of about 2000, 3000, 4000, and 6000. It was observed that, in the first three cases, the Nusselt number based on the difference between wall and mean temperatures decreased to half the values appropriate to the turbulent flow; reversion to a laminar state may thus be suspected. In the experiment at the highest Reynolds number, on the other hand, no such reduction in the Nusselt number was observed; this and other corroborating mean velocity and mean temperature data showed that reversion had not occurred here.

Steiner attributed the observed reversion to the acceleration associated with the decrease in density due to the heating of the air. But two other mechanisms are also possible in this flow: absorption of turbulence by buoyancy effects, and decrease in Reynolds number below the critical value because of the increase in kinematic viscosity associated with the heating. Accurate demarcation of these effects cannot be made from the data available, but a rough estimate will be attempted below.

Consider first the acceleration effects. Steiner quotes the maximum value K_{max} of the acceleration parameter K in the pipe to be in the range 0.3×10^{-6} ($Re \simeq 6000$) to 3×10^{-6} ($Re \simeq 2000$). However the parameter K_{max} does not have the same significance in internal flows as K does in external flows: a more useful indicator is perhaps the bulk acceleration parameter $\bar{K}$, defined using average values over a cross section (as in Section VIII,A). However, $\bar{K}$ is rather lower than K_{max}; for instance, $\bar{K} \simeq (3/5)K_{max}$ in the flow with $Re \simeq 2000$. Thus, in Steiner's experiments $\bar{K} \simeq 2 \times 10^{-6}$ in the flow with $Re \simeq 2000$, and less than 10^{-6} in the other three flows. The acceleration effects thus cannot be very important except in the flow with $Re \simeq 2000$.

We now use the arguments of the previous section to examine whether because of heating, the Reynolds numbers in these flows could have reached critical values. Although the mean temperature rise in these flows is not known directly, other data seem to suggest that it is of the order of a couple of hundred degrees centigrade. Noting that

$$\bar{K}(\bar{T}_i/\bar{T}) \simeq (3/5)\, K_{max}\, (\bar{T}_i/\bar{T}) \simeq 0.3\, K_{max} \simeq 10^{-6},$$

we find from Fig. 35 that Re_{cr} would not be very different in these flows from the adiabatic value of about 1500 (see Section III). As Reynolds numbers in all the flows are above 1500, it is unlikely that reversion occurred by dissipation.

Consider now possible buoyancy effects. We first note that the flux Richardson number

$$Rf \simeq Gr/(Re^3\, c_f\, Pr),$$

where

$$Gr = gl\,\Delta T/(v^2 T)$$

is the Grashoff number, Pr is the fluid Prandtl number ($\simeq 0.72$ for air), l is the vertical distance between two measuring stations along the pipe, and ΔT the mean temperature difference between them. From the values of Gr and Re quoted by Steiner, using estimated values of c_f, it is easy to show that

$$Gr/(Re^3\, c_f\, Pr) \simeq 0.0125$$

for the $Re \simeq 6000$ experiment, whereas for the other three flows it is about 0.09 ($Re \simeq 4000$), 0.16 ($Re \simeq 3000$), and 0.32 ($Re \simeq 2000$). The last three values are comparable to the critical Richardson number in buoyant flows (see Section IV), thus strongly suggesting that the observed reversion is possibly due to buoyancy effects.

IX. **Surface Mass Transfer**

A. INJECTION

Eckert and Rodi (1968) and Pennell *et al.* (1972) showed that a fully developed turbulent pipe flow exhibits some laminar features when a uniform circumferential injection is applied to the flow. Their experimental configuration consisted essentially of a long pipe, a part of whose length was porous (see inset to Fig. 36). The flow at the entrance to the porous section was turbulent and fully developed, and the fluid was injected over a length of 24 diameters uniformly across the porous surface. The injection velocity V_w was 1–5% of the average flow velocity at the entrance of the porous section, and was low enough to cause no flow separation or vortex formation. As a consequence of fluid injection, the velocity in the porous section increases with the streamwise distance x (measured from the beginning of injection):

$$\bar{U}(x) = \bar{U}_0 + 2V_w(x/a), \qquad (9.1)$$

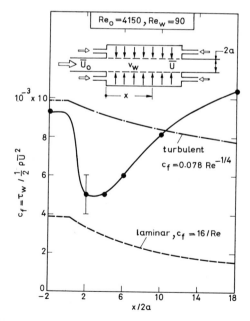

FIG. 36. Skin friction coefficient in pipe flow with injection, based on mean velocity data of Pennell *et al.* (1972); $\mathrm{Re}_0 = 2a\bar{U}_0/\nu$, $\mathrm{Re}_w = 2aV_w/\nu$.

where $\bar{U}(x)$ is the average flow velocity at the station x, and

$$\bar{U}_0 = \bar{U}\,(x = 0).$$

Several changes occur in the flow as a result of fluid injection, but a large number of these are confined to the downstream vicinity of the beginning of injection (i.e., $x/a \lesssim 6\text{--}10$), even though injection continues at the same rate a long way downstream ($0 < x < 48a$). Thus, within the first two diameters, the skin-friction coefficient $c_f\ (\equiv \tau_w/\frac{1}{2}\rho\bar{U}^2)$ drops almost to the local laminar value (Fig. 36); these c_f values have been estimated from the slope of the velocity profiles measured by Pennell et al. (1972). Further downstream, however, c_f quickly catches up with the preinjection level, and, beyond $x/a = 24$, even exceeds the standard turbulent values.

There is also a dramatic reduction in the intensity of the streamwise velocity fluctuation $\hat{u}$ near the wall: at $\mathrm{Re}_0 \equiv 2a\bar{U}_0/\nu \simeq 4250$, $x \simeq a$, $y \simeq 0.036a$, $\hat{u}$ is less than a sixth of its initial value! This reduction increases with the injection velocity, and, at $y_+ = 12$, for example, is approximately

$$\Delta\hat{u} \simeq 3.75\ V_w$$

for $V_w/\bar{U}_0 \lesssim 0.04$. This reduction is again confined in streamwise extent to $x/a < 4$, downstream of which turbulent fluctuations increase fairly rapidly. Along the pipe centerline, there is a smaller and more gradual reduction in the turbulent intensity, extending to $x/a \simeq 10$. In the annular region, there may or may not be any reduction; in fact, $\hat{u}$ may sometimes increase continuously. As a consequence of the large reduction of $\hat{u}$ near $y_+ \simeq 12$ and little or no reduction further away from the wall, there is physically an inward movement—toward the axis of the pipe—of the peak in $\hat{u}$, which now occurs around $y/a \simeq 0.3$ or $y_+ \simeq 25$. The new peak in $\hat{u}$ in its new location is lower by about 40% than the preinjection peak at $y_+ \simeq 12$, and is considerably more spread out.

From this brief description, it is clear that the response of the turbulent flow to fluid injection is quite complex. The catastrophic drop in the mean-square intensity in the wall region is probably a result of the strong interference of fluid injection with the bursting process mainly responsible for the turbulent energy production. In a fully developed turbulent boundary layer, the processes of low speed streak formation, lift-up, oscillatory growth, and breakup, followed by a large scale in-rush, are all in statistical equilibrium. One may speculate that injection upsets this equilibrium by affecting the low-speed streak formation, and thus the complete cycle of events. Coles, on several occasions (see, especially, Coles, 1978), and Brown and Thomas (1977) have proposed that streak formation may be the result of an intermittent Taylor–Görtler instability in the wall-layer, possibly driven by the large eddies of the outer flow. The essential idea is that when the high speed outer fluid moves close to the wall, the sublayer will be thinned, and locally concave (i.e., anticurving) fluid trajectories could be created, thus triggering the Taylor–Görtler instability. For the flow with injection, it is clear that the curvature is strongly convex (cocurving) immediately downstream of $x = 0$; so the instability may be suppressed, streak formation and bursting will be inhibited and turbulent intensities and shear stress will drop. Further downstream, however, the curvature and the turbulence production cycle begin to assume their normal form.

Close to the pipe axis, where injection effects cannot be directly felt, the observed initial reduction in turbulent intensity must be due to the concomitant fluid acceleration. The maximum value of the bulk acceleration parameter $\bar{K}$ (see Section VIII) reached in the different experiments of Pennell *et al.* varies between 4×10^{-6} and 4×10^{-5}; the pressure-gradient parameter Λ (Narasimha and Sreenivasan, 1973; see also Section V) attains values as high as 20. Thus, acceleration effects are significant in the core, especially in the region $x/a \lesssim 2$, where both $\bar{K}$ and Λ reach their maximum values. If the observed reduction in the turbulent intensity is solely due to the effect of sudden acceleration, we should expect rapid distortion theories to be able to predict the observed changes.

Figure 37 shows a comparison of the measured turbulent intensity at several points along the axis of the flow in the region $x/a \lesssim 10$, with results

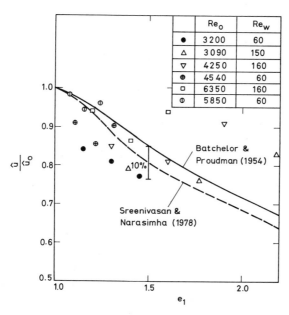

FIG. 37. Turbulence intensity in pipe flow with injection: comparison between measurement (data from Pennell *et al.*, 1972) and rapid distortion theories.

from the rapid-distortion theories of Batchelor and Proudman (1954) for initially isotropic turbulence, and of Sreenivasan and Narasimha (1978) for initially axisymmetric turbulence. (The applicability of these rapid distortion theories in the nonhomogeneous environment has already been discussed in Section V.) It is clear that the rapid-distortion calculations are reasonably successful.

The recovery of the flow downstream may be due partly to turbulent diffusion, and partly to the diminishing importance of fluid injection; when the injection *rate* remains unaltered, and the average flow velocity increases according to (9.1) in the streamwise direction, the ratio $V_w/\bar{U}$ drops. For example, in the experiment of Pennell *et al.* with the highest injection rate, the ratio $V_w/\bar{U}$ at $x/a = 36$ is nearly half of its value at $x = 0$. It would be interesting to observe the flow development in a case where $V_w/\bar{U}$ remains constant.

B. SUCTION

When a uniformly distributed suction is applied to a flat plate laminar boundary layer, it is known that the distribution of velocity in the boundary layer, and the skin friction coefficient, become asymptotically independent of the streamwise direction x and the fluid viscosity, being determined solely by the suction ratio V_s/U, where V_s is the suction velocity (see, for example, Rosenhead, 1963, p. 141). The boundary layer is then said to have reached the laminar asymptotic state.

In the case of a turbulent boundary layer, on the other hand, Dutton (1960) found an asymptotic layer whose form is independent of x only for a certain critical value of the suction ratio. Further, this asymptotic profile depended on such conditions as the initial boundary layer Reynolds number and the precise manner in which suction was applied.

Of particular interest to us here is an observation of Dutton (1960) that for suction ratios higher than a critical value, the boundary layer thins down, the shape factor of the velocity profile increases, the turbulent energy production decreases and, for suction ratios of about 0.01, the initially turbulent boundary layer approaches the *laminar* asymptotic state appropriate to the particular value of the suction ratio (see Fig. 38).

Details of the process of actual relaminarization in this case have not yet been sufficiently well-documented. We can get some insight into the nature of the process by examining energy balance in the asymptotic turbulent boundary layer (corresponding, for one particular set of experimental conditions, to a suction ratio of about 0.007). The Reynolds shear stress distribution in this layer can be evaluated easily from the mean momentum balance.

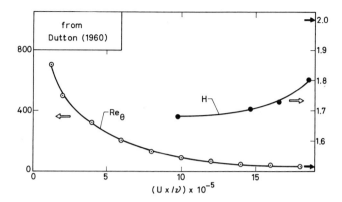

FIG. 38.   Variation of the momentum-thickness Reynolds number Re_θ and the shape factor H for an initially turbulent boundary layer undergoing relaminarization. Suction ratio = 0.0125. Arrows on the right ordinate scale are values of Re_θ and H corresponding to the appropriate asymptotic laminar profile having the same suction ratio.

Dutton (1960) showed that the Reynolds stress so calculated is small all across the boundary layer, its maximum value being only about $0.2U_*^2$. These calculations further showed that 75% of the energy lost by the mean flow is dissipated directly by viscosity acting on the mean velocity gradient, only the remaining 25% being converted to turbulent energy. The reduced turbulent energy production comes about because of the reduced Reynolds shear stress, which more than offsets the increased mean velocity gradient near the wall. It is plausible that for higher suction ratios, essentially the same mechanisms operate with only quantitative differences, eventually resulting in the relaminarization of the flow.

   Reversion has also been observed when there is concentrated suction. Wallis (1950) has studied the problem with suction through a single two-dimensional slot. When the entire turbulent boundary layer is sucked off, a completely new laminar boundary layer grows downstream of the slot, at the rate given by the laminar theory. This useful limiting case is however not interesting in the present context, because the turbulent boundary layer on which we seek to establish the effects of suction has been completely removed form the scene.

   Even in the less extreme case when a sizeable part (but not all) of the turbulent boundary layer is removed by suction, all measures of the boundary layer thickness decrease for a certain distance downstream of the slot; for example, when 3/8 of the mass flux of the approaching turbulent boundary layer is removed, this thinning occurs over a distance of about five times the thickness of the oncoming turbulent boundary layer (Wallis, 1950). Further downstream, however, the boundary layer grows in thickness.

In the region of decreasing thickness, an essentially new laminar sub-boundary-layer develops downstream of the slot, underneath the remnants of the oncoming turbulent boundary layer. Low frequency u' and v' fluctuations are both present in this inner laminar layer, although the Reynolds shear stress is much lower than in the normal turbulent boundary layer. Clearly, a decorrelation mechanism operates here also. (Incidentally, the effect of suction is very much higher on the normal velocity fluctuation v' than on the streamwise velocity fluctuation u'.) Eventually, this inner layer undergoes transition to turbulence like a normal laminar flow.

On the other hand, the local suction thins down the outer layer, and hence also reduces the total boundary layer thickness, even though the laminar subboundary-layer itself grows. Both u' and v' are reduced in the outer layer.

Under these circumstances, the boundary layer just downstream of the suction slot can be said to have reached a quasi-laminar state (Section V). In fact, the similarity to rapidly accelerated boundary layers is so striking that we suspect that a major part of the quasi-laminar theory of Narasimha and Sreenivasan (1973) (see Section V) would in principle be applicable here, although the extent of the relaminarization region is relatively short. Unfortunately, data are not available in enough detail to test this suggestion.

X. Magnetohydrodynamic Duct Flows

A. Introduction

An electrically conducting fluid flowing in a magnetic field experiences a magnetic body force

$$\rho \mathbf{X} = \mathbf{j} \times \mathbf{B} \tag{10.1}$$

per unit volume, where $\mathbf{j}$ is the current density and $\mathbf{B}$ the magnetic field (more precisely the induction); the fluid also suffers an energy loss of j^2/σ per unit volume by "ohmic" dissipation, σ being the electrical conductivity. It has been known for some time that these effects can often relaminarize an originally turbulent flow; we consider here some simple duct shear flows in which such relaminarization has been studied.*

The precise effect of the field depends on its orientation relative to the mean flow. In a channel, for example, the field could be either longitudinal (i.e., aligned with the mean flow) or normal. It may appear necessary to

---

* There are also situations where the magnetic force alters the flow so as to make it unstable, e.g., by creating free shear layers that quickly break down to turbulence; an example is given by Lehnert (1955).

consider two possible normal directions, respectively parallel to the shorter and longer sides of the channel when its cross section is not a square; however, it is convenient to discuss instead the equivalent problem of the effect of a given normal field on channels of different aspect ratio.

We restrict ourselves to a consideration of low magnetic Reynolds numbers,

$$\text{Rm} \equiv Ua/\lambda, \qquad \lambda = 1/\mu\sigma, \tag{10.2}$$

where U and a are characteristic velocity and length scales of the flow, μ is the magnetic permeability and λ, the magnetic diffusivity. The value of Rm provides us a measure of the ratio of convective to (magnetic) diffusion times, or of the induced to the imposed field (see, e.g., the discussion in Shercliff, 1965, Chapter 3). For the fluids used in most laboratory experiments $\nu \ll \lambda$; thus (see, e.g., Kirko, 1965) $\nu/\lambda \simeq 1.5 \times 10^{-7}$ for mercury at $20°C$, and 2.3×10^{-6} for sodium at $300°C$. This means that even when the flow Reynolds number is quite high, the magnetic Reynolds number tends to be low. Whereas at high Rm the field is convected with (or "frozen" into) the fluid, at low Rm the field varies chiefly by diffusion, and is hardly affected by the flow.

The current density is taken here to be given by a simple Ohm's law,

$$\mathbf{j} = \sigma(\mathbf{E} + \mathbf{u} \times \mathbf{B}), \tag{10.3}$$

where $\mathbf{E}$ is the electric field. The field puts energy into the fluid at the rate $\mathbf{E} \cdot \mathbf{j}$ per unit volume and unit time; a part of this, given by j^2/σ, is dissipated irreversibility into heat.

Taking the curl of (2.2) with $\mathbf{X} = \mathbf{j} \times \mathbf{B}/\rho$ we obtain the vorticity equation (with $B^2 = |\mathbf{B}|^2$)

$$\frac{d\boldsymbol{\omega}}{dt} - (\boldsymbol{\omega} \cdot \nabla)\mathbf{u} - \mathbf{F} = \frac{\sigma}{\rho}[-\text{curl } B^2\mathbf{u} + (\mathbf{u} \cdot \mathbf{B})\,\text{curl }\mathbf{B}$$

$$+ (\mathbf{u} \cdot \mathbf{B}) \times \mathbf{B} + \text{curl}(\mathbf{E} \times \mathbf{B})], \tag{10.4}$$

where $\mathbf{F}$ is a term involving the viscous and Reynolds stresses. The effect of the field on the vorticity is thus quite complex; but if the field is normal to the flow and (as at low Rm) hardly affected by it, and the last term in (10.4) involving $\mathbf{E}$ is negligible, it is clear that the first term on the right of (10.4) acts to suppress the vorticity with a characteristic decay time $\tau = \rho/\sigma B^2$.

B. Aligned Fields

If $\mathbf{B}$ is in the direction of the mean flow and $\mathbf{E}$ is negligible, the magnetic force is zero, and the mean momentum balance can be affected only indirectly, through the action of the field on the Reynolds stresses. Many experiments have shown a tendency toward suppression of turbulence (e.g., Globe, 1961, and Fraim and Heiser, 1968, in pipes; Sajben and Fay, 1967, in a jet). It may be shown by a detailed examination of the vorticity equation (Moreau, 1969) that, as above, the field tends to suppress (turbulent) vorticity normal to it in times of the order $\rho/\sigma B^2$. The turbulence should thus tend to become axisymmetric about **B**, and eventually even two-dimensional with vorticity only along **B**.

The experiments of Fraim and Heiser (1968) suggest that reversion occurs for sufficiently small values ($\lesssim 30$) of the ratio of the Reynolds to the Hartmann number

$$\frac{\mathrm{Re}}{\mathrm{Ha}} = \frac{U}{B}\left(\frac{\rho}{\nu\sigma}\right)^{1/2} \tag{10.5}$$

where the Hartmann number

$$\mathrm{Ha} = Ba(\sigma/\rho\nu)^{1/2},$$

when squared, provides a measure of the ratio of the magnetic to the viscous forces. Note that the parameter (10.5) does not involve any length scale characteristic of the reverting flow. Earlier explanations of the role of this parameter (interpreting it as "the square root of the product of the viscous and magnetic forces divided by the inertial forces," as Fraim and Heiser do, for example) are unnecessarily obscure and do not give us an insight into the physical phenomena involved. We offer a simpler interpretation below.

A measure of the ratio of the magnetic to the inertial forces is the Stuart number or interaction parameter, which is given by $S = \mathrm{Ha}^2/\mathrm{Re}, = \sigma B^2 l/\rho u$ for eddies of size l and velocity u. This parameter is large for the larger eddies, which therefore tend to be damped out by *ohmic* dissipation (Moffatt, 1967; Moreau, 1969). The largest eddy that escapes ohmic dissipation is of size $U\tau \sim \rho U/\sigma B^2$, which can be much smaller than the characteristic flow length scale when B is high. Reversion must occur if eddies of this size or smaller are dissipated by viscosity, i.e., if the Reynolds number based on $U\tau$, namely,

$$\frac{U}{\nu}\frac{U\rho}{\sigma B^2} = [\mathrm{Re}/\mathrm{Ha}]^2$$

is sufficiently small. It is thus no surprise to find a critical value for the ratio Re/Ha.

This explanation suggests then that when the field is aligned with the flow, reversion occurs by dissipation—ohmic for the large eddies, viscous for those not so large.

C. Normal Fields

The same two sources of dissipation must affect the turbulence even when the field is normal to the flow, but the $\mathbf{j} \times \mathbf{B}$ force may now act in addition to suppress the *mean* vorticity of the flow normal to $\mathbf{B}$. Using the previous estimate of the decay time, the characteristic streamwise distance should be of order

$$U\rho/\sigma B^2 \sim a/S,$$

where S is now the Stuart number based on U and a. Further downstream the velocity therefore tends to become uniform across the duct, except in thin "Hartmann" layers at the surface that ensure that the no-slip condition is satisfied. At large values of the Hartmann number the thickness δ of this layer is proportional to $a/\text{Ha} = (\rho v/B^2\sigma)^{1/2}$.

The first experiments on the effects of a magnetic field on laminar flow were conducted in a channel by Hartmann and Lazarus (1937); since then there have been many other studies in turbulent flow as well (an interesting survey of earlier Western and Russian work is Branover *et al.*, 1967). Such experiments present many difficulties: measurement techniques are not easy (see, e.g., Branover and Gershon, 1976; Branover *et al.*, 1977), and the flow situations are bedeviled by such effects as the absence of a fully developed state due to insufficient length of duct or field, the fringing of the magnetic field lines near the ends of the magnet, and the finite aspect ratio of the cross section of the duct when it is a channel. Thus, while many interesting experiments have been made (Murgatroyd, 1953; Branover *et al.*, 1977; Gardner and Lykoudis, 1971; Hua and Lykoudis, 1974; Reed and Lykoudis, 1978), and much insight obtained, one can hardly say that the picture is yet complete.

All experiments show that with a sufficiently strong field, the skin friction changes from the initial turbulent value to one characteristic of laminar m.h.d. flow. As an example, we show in Fig. 39 the measurements of Gardner and Lykoudis (1971) in a pipe. Note that the final m.h.d. laminar value of c_f is generally much higher than the non-m.h.d. turbulent value at the same Reynolds number! At the higher Re, the skin friction shows a dip before rising again as the field increases. At lower Re, the recent measurements of Reed and Lykoudis (1978) show an even more involved behavior of c_f than the data of Fig. 39: there is a small but noticeable initial dip, then a hump followed by another dip before c_f rises to the laminar value. The first dip manifests itself because of the damping of turbulence by the magnetic field while the Hartmann flattening has not yet become dominant. The flow is presumably still fully turbulent at this stage. With increasing field strength, the Hartmann effect wins over, accounting for the rise in c_f. The presence

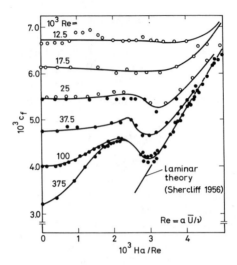

FIG. 39. Typical skin friction coefficient data for a pipe flow in a normal magnetic field (Gardner and Lykoudis, 1971).

of the following hump and the (second) dip is not surprising and are easily explained if we allow that during transition (or reversion) there is a regime in which the flow alternates between the laminar and turbulent states. If at the point of change in flow character the turbulent skin friction is higher than the laminar, a dip must appear (as in a flat plate boundary layer; see Dhawan and Narasimha, 1958); otherwise it will not. The crucial fact is that at low Re, the laminar c_f is comparable to the turbulent c_f near reversion, but at high Re it is much lower. Of course a convincing theory for the variation of turbulent c_f with magnetic field is required before this argument can be considered complete.

It is generally thought that relaminarization—as judged by the agreement between measured c_f (or pressure loss) and laminar theory—occurs when Ha/Re exceeds a critical value. This can by no means be considered to have been established firmly yet; indeed Fig. 39 suggests that the value of Re/Ha at which laminar skin friction is attained decreases with increasing Re, and we shall return to this point shortly. A related point is that the values quoted for $(Re/Ha)_{cr}$ by different workers do not agree. Murgatroyd (1953) suggested $(Re/Ha)_{cr} \simeq 225$ for a channel; the more recent measurements of Hua and Lykoudis (1974) indicate the higher value of nearly 330. Hua and Lykoudis show that it is important to take into account the effect of the aspect ratio on the laminar solution in attempting to judge when reversion is complete. They also observe a departure from the laminar solution at high Ha/Re, and attribute it to magnetic entrance effects that prevent the establishment of the fully developed flow.

Branover and Gershon (1976) suggest the empirical formula

$$(\text{Re}/\text{Ha})_{\text{cr}} = 215 - 85\exp(-0.35\beta),$$

where

$$\beta = \frac{\text{side perpendicular to field}}{\text{side parallel to field}}$$

is the aspect ratio of the channel.

Note that all these values for $(\text{Re}/\text{Ha})_{\text{cr}}$ are appreciably higher than those with aligned fields, presumably reflecting the additional effect on the mean flow due to the magnetic force. The relatively lower value of $(\text{Re}/\text{Ha})_{\text{cr}}$ at lower β is again consistent with this idea, because the Hartmann effect now operates on the shorter sides of the channel and is therefore weaker.

To examine this effect more closely, we note that when Ha is large the Hartmann number based on δ,

$$\text{Ha}_\delta = B\,\delta(\sigma/\rho\nu)^{1/2},$$

is $O(1)$; it follows that

$$\frac{\text{Re}}{\text{Ha}} \sim \frac{U}{B}\left(\frac{\rho}{\nu\sigma}\right)^{1/2} \sim \frac{\text{Re}_\delta}{\text{Ha}_\delta} \sim \text{Re}_\delta.$$

At high Ha, the velocity profile in the Hartmann layer takes the simple form (see, e.g., Shercliff, 1965, Sect. 6.5)

$$u = U\left[1 - \exp\left(-\frac{y}{a\text{Ha}}\right)\right],$$

where y is distance measured from the surface; this velocity distribution is the same as in the asymptotic suction profile (Rosenhead, 1963, p. 241). The critical Reynolds number for this profile is about 45,000 (see Rosenhead, 1963, p. 543) based on the displacement thickness; the Reynolds number at turbulence suppression in m.h.d. channel flow, corresponding to Re_δ of a few hundreds, is thus some orders of magnitude lower than this critical value. The "Hartmann effect" of the normal field therefore produces a highly stable mean velocity distribution. Indeed, the detailed linear stability analysis of *laminar* flow by Lock (1955) finds that the chief reason for the greater stability of channel flow subjected to high normal magnetic fields is the significant alteration of the mean flow.

As the time required for suppression of either the mean or the fluctuating vorticities is apparently of order $\rho/\sigma B^2$, it seems *a priori* possible that the two effects reinforce each other. Interesting experiments on the turbulent quantities in these flows have been conducted by Gardner and Lykoudis (1971) in pipes, and by Branover *et al.* (1977) and Branover and Gershon (1976) in channels. These measurements depict the decay of turbulence at a fixed point in the channel as the field is increased, and do not unfortunately trace the streamwise variation at a fixed field. Nevertheless, the experiments do show that at a station where the surface friction has reached laminar values, turbulent fluctuations have not vanished, although they evidently do not contribute to momentum transport. In fact, Reed and Lykoudis (1978) find that the Reynolds shear stress was completely suppressed even though u' and v' were not; clearly a decorrelating mechanism is again at work here. The flow is therefore quasi-laminar, in the sense of Section V. There has been some discussion whether these fluctuations are a residue of the original turbulent flow, or arise from the creation of unstable velocity profiles in the channel where the (originally turbulent) flow enters the magnetic field: gradients in the field due to the "fringe effect" at the poles used in the experiment can *create* vorticity (see Shercliff, 1965, pp. 94–95), and so lead to the so-called M-shaped profiles possessing inflexion points. Branover and Gershon argue that the occurrence of such profiles depends on the Stuart number, and offer evidence that in those experiments where S was lower (and hence the M-profile effect weaker), the residual turbulence was also weaker.

At the same time, Branover and Gershon also find evidence that the residual turbulence approaches two-dimensionality; as may be seen from Fig. 40, the correlation coefficient of the longitudinal velocity fluctuations at

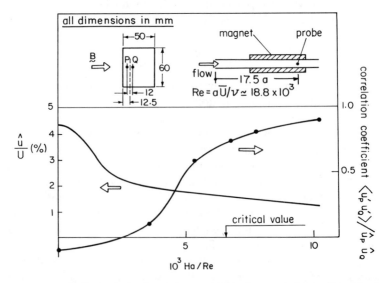

FIG. 40.   Longitudinal velocity fluctuations and field-wise correlation coefficient in channel flow subjected to a normal field. (Data from Fig. 3.3, Branover *et al.*, 1977). Note that at the critical value of Ha/Re $\simeq 6.6 \times 10^{-3}$, when the skin friction has attained the laminar value, the turbulence intensity is still appreciable but the correlation coefficient very large.

two fixed points along the field shows, in one case, an increase from near zero at Re/Ha $\simeq 10^3$ to about 0.9 at Re/Ha $\simeq 10^2$.

On the other hand, Gardner and Lykoudis (1971) found that the magnetic entrance effect extends only a few diameters in their pipe experiments. However, in other respects the situation in a pipe is rather complex, as the current flows in two loops in the cross-sectional plane—across the field in a central core, then around the circumference. This introduces a dependence on the azimuthal angle in the flow parameters; e.g., the turbulence is suppressed first along the field. The Hartmann flattening of the velocity profile is also most pronounced in the same direction. This azimuthal dependence makes interpretation of pipe data difficult. Gardner and Lykoudis point out that there is some evidence, nevertheless, that on the application of the magnetic field the large eddies are damped out first, presumably by ohmic dissipation (Fig. 41). This indicates that the changes in mean velocity profile occur faster, and that these changes, through the creation of stabler profiles, bring viscous dissipation into play. At a later stage of relaminarization, there is however *relatively* greater energy at the lower frequencies (the *total* energy is of course much less).

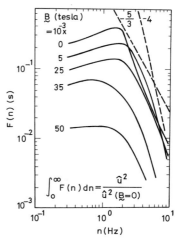

FIG. 41. Spectra of axial velocity fluctuations at center-line in m.h.d. pipe flow: Re = 5150. (From Gardner and Lykoudis, 1971.)

It must be noted that the variation of turbulence intensity in these experiments correlates better with $Ha^2/Re^{0.75}$ than with Ha/Re, especially at low fields.

To summarize, a normal field should directly suppress turbulent vorticity not aligned with it, but the associated Hartmann effect, altering the mean velocity profile in a stabilizing way, must also play a strong role in relaminarizing the flow, if indeed it is not the rate-controlling mechanism. An indication of this is the much higher value of $(Re/Ha)_{cr}$ in normal fields as compared to aligned fields (around 300 vs. about 30). Further support is lent by the lower value of $(Re/Ha)_{cr}$ when the field is along the longer side of the channel; the Hartmann effect, confined now to the shorter sides, is obviously weaker in this case. The presence of significant turbulence even when the skin friction has attained laminar values bears a strong resemblance to several other cases discussed earlier (Sections III and IV) and is of course not peculiar to m.h.d. flows, in spite of the impression sometimes given in the literature of the subject.*

If the primary (or rate-controlling) mechanism of reversion should be the domination of the magnetic force over the Reynolds stress gradient in the originally turbulent flow, we may expect the correlating parameter (following the arguments on accelerating flow in Section V) to be

$$\text{at low fields:} \frac{\sigma U B^2}{\tau_0/D} \sim \frac{Ha^2}{c_f Re}, \tag{10.6a}$$

$$\text{at high fields:} \frac{\sigma U B^2}{\tau_0/\delta} \sim \frac{Ha^2}{c_f Re} \cdot \frac{\delta}{D} \sim \frac{Ha}{c_f Re}. \tag{10.6b}$$

The experiments of Hua and Lykoudis (1974) in a channel show that the decrease in turbulence intensity with increase in field correlates with Ha^2/Re at low field strengths; Gardner and Lykoudis (1971) similarly find the correlating parameter in pipes to be $Ha^2/Re^{0.75}$, which is the same as (10.6a) if c_f is taken to obey the well-known Blasius law in turbulent flow, $c_f \sim Re^{-1/4}$. The Blasius law should be a reasonable approximation at low fields and relatively low Re, but as the field and Re increase it appears from Gardner and Lykoudis's experiments that the c_f varies less with Re than the Blasius law suggests. The correlation of $(Ha/Re)_{cr}$ with Re provided in Fig. 27 of their paper shows two distinct regions, which we may describe by

$$(Ha/Re)_{cr} \sim Re^{-0.05}, \qquad 10^4 < Re < 8 \times 10^4$$
$$\sim Re^{-0.3}, \qquad 10^5 < Re < 5 \times 10^5.$$

The high Re correlation is rather like (10.6b).

Considering the difficulties in these experiments (e.g., the c_f in the relaminarized region shows a systematic departure from the laminar solution of Shercliff (1956), which according to the authors could have been due to roughness, insufficient entry length or the residual turbulence), all we can say is that the criterion for reversion appears to be a parameter of the form Ha/Re^n, where n is definitely less than 1, and probably around $\frac{1}{2}$ for relatively

---

* Branover and Gershon (1976) say that "the problem of remaining disturbances constitutes one of the most paradoxical phenomena of fluid mechanics."

low fields. Recalling that reversion in the non-m.h.d. flow in a channel occurs at Re $\simeq$ 1500 (see Section III,B,2), we may sketch the likely boundary between laminar and turbulent flow as in Fig. 42.

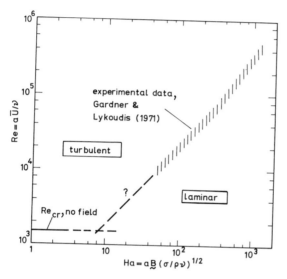

FIG. 42.   Sketch of proposed reversion boundaries in m.h.d. pipe flow.

XI. Other Instances of Relaminarization

It was pointed out by Landau and Lifshitz (1959, p. 137) that an axisymmetric turbulent wake should eventually revert to a laminar state. The wake behind any body of nonzero, finite drag is characterized by a momentum thickness θ which remains constant along the wake. Sufficiently far downstream, the momentum deficit in the wake $U^2\theta^2$ is proportional to $\delta^2 U \Delta U$, where ΔU is the maximum velocity defect, U is the free stream velocity, and δ is the wake thickness. It follows that the wake Reynolds number $\mathrm{Re}_\delta = \delta \Delta U/\nu \sim (U\theta/\nu)(\theta/\delta)$, so that if the wake keeps growing its Reynolds number must keep decreasing. A detailed calculation of the wake, assuming that it is "self-preserving" (Townsend, 1956, Section 7.15), shows that $\delta \sim x^{1/3}, \Delta U \sim x^{-2/3}$, $\mathrm{Re}_\delta \sim x^{-1/3}$. Sufficiently far downstream, therefore, the Reynolds number becomes so small that the wake must relaminarize, by dissipation; standard turbulent behavior cannot therefore be expected. It is probably for similar reasons that Freymuth's (1975) search for the final period of decay in the wake behind a sphere proved fruitless. However, the

Reynolds number variation with x is very weak, and we have already seen in Section III that dissipative reversion is very slow; it is therefore no great surprise that explicit observations of such relaminarization have not been made.

The argument for relaminarization is valid for the wake behind any finite drag-producing body. However, there is no tendency to reversion in the wake behind a *two-dimensional* body, as the corresponding wake Reynolds number remains constant in this case.

The tip vortex produced behind a finite wing also appears to exhibit reversion, presumably as a result of the curvature of the flow. Poppleton (1971) shows very interesting photographs of flow visualization using smoke in the wing-tip vortex generated by a Hercules aircraft in flight. There is a delicate relationship between axial and transverse velocities in such vortices: the axial velocity relative to the wing changes sign across the vortex, with the result that there is co-curving flow in the core; the photographs show that the vortex core is laminar, although the outer part is turbulent.

Laboratory experiments reported by Singh and Uberoi (1976) also suggest a reversion in the wing-tip vortex; they find that, while at a distance of one chord behind the tip the flow is turbulent, at 30 chords the vortex appears laminar, with a periodic instability (see their Fig. 4). At 60 chords, however, no periodicity was evident, but turbulent patches appeared in the velocity signals.

We believe that the relaminarization observed in swirling flames (e.g., Beer et al., 1971) is a related phenomenon.

We have discussed at great length in Section V how spatially accelerated shear flows may relaminarize. It may be expected that when the acceleration is temporal similar effects should be observed. Some preliminary experiments during transient flow in a pipe by Leutheusser and Lam (1977) show that such acceleration certainly delays transition, i.e., the critical Reynolds number increases. In analogy with spatial acceleration, we may also anticipate that the relevant parameters governing departure from the law of the wall and completion of reversion respectively would be (with superscript t standing for the temporal problem)

$$\Lambda^{(t)} = \nu \dot{U}/U_*^2, \quad \Lambda^{(t)} = \dot{U}\delta/U_{*0}^2$$

where $\dot{U} \equiv dU/dt$ is the time derivative of a characteristic flow velocity. When variations in c_f may not be appreciable, we may also expect the free stream parameter

$$K^{(t)} = \nu\dot{U}/U_*^3$$

to play a role similar to that of K in Section V. However, there is not enough experimental data yet to make a detailed analysis.

When the temporal variations are not monotonic, a complex series of transitions and reversions may be expected, depending on ratios of the time scales characterizing the acceleration to those characterizing transition and reversion.

We may briefly return to the reversion observed in coiled pipes, which was mentioned in Section I. There is probably a combination of several mechanisms at work here. There is first of all the suppression of turbulence on a convex wall—in the present instance on the inside of the bend—that was discussed at some length in Section VI. On the outside of the bend, the concave flow curvature should if anything enhance the turbulence, but this effect may be counteracted by the secondary flow that is known to be generated. Lighthill (1970) has pointed out that one effect of the secondary flow is to shift the velocity maximum outward, with the formation on the outside of the bend of a thin layer with an effectively lower Reynolds number. For example, measurements by Rowe (1970, Fig. 1) in a pipe bent to a radius of 12 diameters show that, after a 60° bend, the maximum total pressure has moved out from the centre by about two-thirds of the radius. Such a reduction in the Reynolds number may well dissipate turbulence on the outside of the bend, and allow the development of a laminar subboundary layer there. More detailed experimental investigations are necessary to elucidate the mechanisms operating in this flow.

There is also the possibility of reversion on swept wings in an aircraft—the turbulent boundary layer on the fuselage is first swept down the leading edge of the wing before rolling over to chordwise flow; in the process, it is subjected to a large favorable pressure gradient near the leading edge. Thompson (1973) has shown how it might be possible to exploit the possibilities of reversion in such flows to save a few percent on the drag of the wings of large transport aircraft. It remains to be seen whether, and if so to what extent, the beneficial effect of the favorable pressure gradient is counteracted in this case by the curvature that the boundary layer experiences along the flow.

XII. Conclusion

In the preceding sections we have examined and described a variety of flows in which relaminarization has been observed or may be expected to occur. It is certain that many other reverting flows exist, and more will be discovered in future.

It is instructive however to compare the different types of reversion that we have already described. In both dissipative and absorptive types of reversion, there is a net decrease in turbulence energy: in the first instance, this energy is lost essentially by the action of a molecular transport parameter like the viscosity or the electrical resistivity; in the second, it is destroyed by the work done against a body force like gravity. In the enlarged pipes discussed in Section III, even when the final Reynolds number drops to as low as half of the critical value, the distance required to complete reversion is of the order of a hundred diameters: thus the reversion is very slow. In contrast, in the Richardson type reversions, the destruction of turbulence energy appears to proceed rapidly once the critical value of the parameter is exceeded: turbulence is suppressed in a few jet widths in Fig. 2. In both cases, there is evidence that what happens goes beyond a mere decrease in turbulence energy; in fact some mechanism seems to be at work to decorrelate the velocity components that generate the crucial Reynolds stresses.* In particular, the effects of even mild curvature on a turbulent shear flow seem astonishingly strong. Clearly one is not merely wearing the mechinery of turbulence down in these cases—it is more as if one were throwing a spanner into the works! In other words, these external influences imposed on the flow must be interfering with its organization—with the coherent structures and the bursting cycle that sustain turbulence.

In making this statement, we have come full circle from the thermodynamic objection against the possibility of reversion, mentioned in Section I. For reversion can also be viewed in many cases as the destruction of the large scale quasi-order of turbulent flows into the molecular-scale disorder of laminar flows! A naive order–disorder approach to relaminarization can therefore be misleading for more reasons than one.

The third type of reversion, viz., that which occurs during acceleration, shares some characteristics with the other two types, but has some obviously distinctive features also. For example, the generation of a new laminar layer (inevitably involving dissipation of turbulent energy) is common to reverting flows of different types; and the pressure gradient parameter Λ resembles a "bulk" Richardson number. On the other hand, the shear stresses during acceleration are not really destroyed; rather they are frozen, and everything is over, so to speak, before the machinery of turbulence has time to adjust itself to the external agency (i.e., the imposed pressure gradient). In low speed flows, this agency therefore continues to dominate the dynamics of the flow—not because the turbulence is dissipated or destroyed in absolute magnitude, but rather because the agency is so much larger. One is reminded of the story about Emperor Akbar and Birbal; one way of shortening a stick without wearing it out or breaking it is to place a bigger stick next to it.

* In general, all forms of reversion seem particularly sensitive to interference with the relatively small normal velocity.

We recall our pragmatic definition of reversion (Section I) according to which a flow has laminarized if its development does not demand understanding of the dynamics of turbulence. Thus, it is not necessary for the characteristically random velocity fluctuations of turbulence to be entirely absent in a relaminarized flow, but only that such fluctuations, even if present, should not lead to appreciable momentum or energy transport. We have specifically demonstrated that this is true (although for different reasons) in reverting pipe or channel flows (Section III) and in accelerated reverting boundary layers (Section V), but it is likely to be true at least in a somewhat restricted sense in all the other flows discussed here: perhaps the most restrictive case is that with injection, where the dynamics of turbulence becomes unimportant only fairly locally, close to the beginning of injection. By the same token, it is not clear at present whether our definition of reversion encompasses the technologically important case of drag reduction by dilute polymer addition, even though a phenomenal reduction of turbulent intensities and Reynolds stress has been observed here.

It is sometimes suggested that one ought to decide whether a flow has reverted to the laminar state by examining whether it becomes turbulent again as in direct transition. This seems an attractive idea; in some reverting boundary layers it has been possible to recognize the generation of new turbulent spots leading to a second transition to turbulence. However, as an invariable criterion the idea bristles with difficulties. The chief reason is that unlike direct transition, which is often catastrophic, reversion is often asymptotic. Thus, the birth of a spot in a boundary layer is a dramatic event that heralds turbulence; but during reversion, there is no such event whose occurrence guarantees that the flow is going to become laminar downstream, although, as we have seen in Section V, many flow parameters do vary *rapidly* during reversion. A reverting flow is often in a rather noisy quasi-laminar state, in which even the occurrence of what might otherwise have been a dramatic event (e.g., birth of a spot indicating retransition) tends to get masked.

In any case, it is neither practical nor fruitful to try and decide whether a flow has reverted at any given point by having to investigate whether it could go turbulent again in some standard way. In fact, the concept would be meaningful only if *points* of reversion (as of transition) could be uniquely determined.

In many situations where reversion does occur, a combination of the different basic mechanisms can be at work. For example, in a highly heated duct flow, it is still not certain how much of the resulting changes are due to:

(i) the increase in viscosity, and hence in dissipation, that follows the higher temperatures;

(ii) the acceleration consequent on the reduction in gas density;
(iii) the stratification introduced into the flow by heating;
(iv) the possible domination of the conduction heat flux over the other forms of heat transfer.

It is possible that in any given situation, more than one of these mechanisms contribute; thorough studies are still required to determine which of the different mechanisms are the more significant in any given situation.

A key to understanding these complex situations will surely be an appreciation for the *rate* at which different mechanisms operate. We have seen evidence for the slowness with which dissipative reversion takes place. On the other hand, an external agency can quickly alter the mean flow significantly, and hence either suppress or overwhelm the turbulence; it is also possible that the turbulent energy is either quickly absorbed, or the phase relations promoting momentum exchange destroyed, with *consequent* changes in the mean flow. The possible pathways are schematically illustrated in Fig. 43. Further studies should perhaps concentrate on helping us to decide what the rate-controlling mechanism is in any given flow situation.

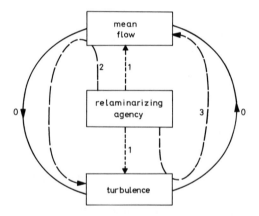

Fig. 43. Pathways to relaminarization. 0, the coupling between mean flow and turbulence in fully developed turbulent flow; 1, relaminarizing agency, working on both mean flow and turbulence at comparable rates; 2, mean flow affected more rapidly than turbulence; 3, turbulence affected more rapidly than mean flow.

We may conclude by remarking on how often reversion apparently occurs, and how easy it appears to be to suppress turbulence—it looks almost as if whether you suck or blow, squeeze or bend, heat or cool, or do any of a vast number of other things to it, turbulence can be destroyed, or at least disabled; *provided* of course the operation is done properly. There are just now beginning to be a few technological examples where turbulence might be controlled by promoting reversion. Thus, we have already noted Thompson's (1973) suggestion that it might be possible to engineer reversion near the leading edge of swept wings on the larger transport aircraft, and so save a few percent on the drag. In many rocket nozzles, wind tunnel contractions, and other such devices, reversion has undoubtedly been occurring without having been explicitly designed in. It would be very useful if high-speed wind tunnel test sections, which tend to be noisy because of sound radiated from the turbulent boundary layers on the wall, could be made "quiet" by the device of relaminarization. Nevertheless, "turbulence control" is still something that remains a bit of a dream. Perhaps when there is greater understanding of the nature of coherent structures and of their role in turbulence, it will be possible to aim spanners better at the machinery of turbulence and so win the ability to control it when necessary.

ACKNOWLEDGMENTS

This paper is an extended version of an invited lecture given by R.N. at the Sixth Canadian Congress of Applied Mechanics, Vancouver 1977. The work described here has been the result of collaboration with several colleagues; we particularly wish to acknowledge out debt to Dr. P. R. Viswanath, Dr. M. A. Badri Narayanan, Dr. A. Prabhu, and Dr. B. G. Shivaprasad.

The preparation of the paper has been supported in part by a grant from the Indian National Science Academy. K.R.S. acknowledges his indebtedness to Professor S. Corrsin for making available financial support from NASA Grant NSG 2303 during the later stages of this work at the Department of Mechanics and Materials Science of The Johns Hopkins University.

REFERENCES

ABRAMOWITZ, M., and STEGUN, I. A., (eds.) (1965). "Handbook of Mathematical Functions." Dover, New York.

ALVI, S. H. (1975). Flow characteristics of sharp-edged orifices, quadrant-edged orifices and nozzles. Ph. D. Thesis, Dept. Civ. Eng., Ind. Inst. Sci., Bangalore.

ANANDA MURTHY, K. R., and HAMMITT, A. G. (1958). "Investigation of the Interaction of a Turbulent Boundary Layer with Prandtl-Meyer Expansion Fan at M = 1.88," Rep. No. 434. Dept. Aero. Eng., Princeton University, Princeton, New Jersey.

ANTONIA, R. A., CHAMBERS, A. J., PHONG-ANANT, D., RAJAGOPALAN, S., and SREENIVASAN, K. R. (1979). Response of atmospheric surface layer turbulence to a partial solar eclipse. *J. Geophys. Res.* **84**, 1689.

BACK, L. H., and SEBAN, R. A. (1967). Flow and heat transfer in a turbulent boundary layer with large acceleration parameter. *Proc. Heat Transfer Fluid Mech. Inst.* **20**, 410.

BACK, L. H., MASSIER, P. F., and GIER, H. L. (1964). Convective heat transfer in a convergent-divergent nozzle. *Int. J. Heat Mass Transfer* **7**, 549.

BADRI NARAYANAN, M. A. (1968). An experimental study of reverse transition in two-dimensional channel flow. *J. Fluid Mech.* **31**, 609.

BADRI NARAYANAN, M. A., and RAMJEE, V. (1969). On the criteria for reverse transition in a two-dimensional boundary layer flow. *J. Fluid Mech.* **35**, 225.

BADRI NARAYANAN, M. A., RAJAGOPALAN, S., and NARASIMHA, R. (1974). "Some Experimental Investigations on the Fine Structure of Turbulence," Rep. No. 74 FM 15. Dept. Aero. Eng., Ind. Inst. Sci., Bangalore.

BADRI NARAYANAN, M. A., RAJAGOPALAN, S., and NARASIMHA, R. (1977). Experiments on the fine structure of turbulence. *J. Fluid Mech.* **80**, 237.

BANKSTON, C. A. (1970). The transition from turbulent to laminar gas flow in a heated pipe. *J. Heat Transfer* **92**, 569.

BATCHELOR, G. K. (1953). "The Theory of Homogeneous Turbulence." Cambridge Univ. Press, London and New York.

BATCHELOR, G. K., and PROUDMAN, I. (1954). The effect of rapid distortion of a fluid in turbulent motion. *Quart. J. Mech. Appl. Math.* **7**, 83.

BATCHELOR, G. K., and TOWNSEND, A. A. (1948). Decay of turbulence in the final period. *Proc. R. Soc. London, Ser. A* **194**, 527.

BEER, J. M., CHIGIER, N. A., DAVIES, T. W., and BASSINDALE, K. (1971). Laminarisation of turbulent flames in rotating environments. *Combust. Flame* **16**, 39.

BLACKWELDER, R. F., and KOVASZNAY, L. S. G. (1972). Large scale motion of a turbulent boundary layer during relaminarisation. *J. Fluid Mech.* **53**, 61.

BLOY, A. W. (1975). The expansion of a hypersonic turbulent boundary layer at a sharp corner. *J. Fluid Mech.* **67**, 647.

BRADSHAW, P. (1969a). A note on reverse transition. *J. Fluid Mech.* **35**, 387.

BRADSHAW, P. (1969b). The analogy between streamline curvature and buoyancy in turbulent shear flow. *J. Fluid Mech.* **36**, 177.

BRADSHAW, P. (1973). Effects of streamline curvature on turbulent flow. *AGARDograph* **169**.

BRANOVER, G. G., GELFGAT, YU. M., and TSINOBER, A. B. (1967). Turbulent mhd flows in prismatic and cylindrical pipes. *NASA Tech. Transl.* **F-482**.

BRANOVER, H., and GERSHON, P. (1976). "MHD Turbulence Study, Part 2," Rep. ME 6-76. Dept. Mech. Eng., Ben-Gurion University.

BRANOVER, H., GERSHON, P., and YAKHOT, A. (1977). "MHD Turbulence and Two-phase Flow," Rep. ME 5-77. Dept. Mech. Eng., Ben-Gurion University.

BROWN, G. L., and THOMAS, A. S. W. (1977). Large structure in a turbulent boundary layer. *Phys. Fluids* **20**, S243.

BUSINGER, J. A., and ARYA, S. P. S. (1974). Height of the mixed layer in the stably stratified planetary boundary layer. *Adv. Geophys.* **18A**, 73.

CHAPMAN, D. R., WIMBROW, W. R., and KESTER, R. H. (1952). Experimental investigation of base pressure on blunt trailing edge wings at supersonic velocities. *Natl. Advis. Comm. Aeronaut., Rep.* **1109**.

CHEN, C. P., and PUEBE, J. L. (1964). Sur l'écoulement radial divergent d'un fluide visqueux incompressible enter deux plans paralleles. *C. R. Hebd. Seances Acad. Sci.* **258**, 5353.

CHEN, K. K., and LIBBY, P. A. (1968). Boundary layers with small departures from the Falkner-Skan profiles. *J. Fluid Mech.* **33**, 273.

COLES, D. (1962). Interfaces and intermittency in turbulent shear flow. *In* "Mécanique de la turbulence," pp. 229–248. CNRS, Paris.

COLES, D. (1978). A model for flow in the viscous sublayer. *In* "Workshop on Coherent Structure of Turbulent Boundary Layers" (C. R. Smith and D. E. Abbott, eds.), p. 462. Lehigh University, Bethlehem, Pennsylvania.

COON, C. W., and PERKINS, H. C. (1970). Transition from the turbulent to the laminar regime for internal convective flow with large property variations. *J. Heat Transfer* **92**, 506.

CROW, S. C. (1969). Distortion of sonic bangs by atmospheric turbulence. *J. Fluid Mech.* **37**, 529.

DHAWAN, S., and NARASIMHA, R. (1958). Some properties of boundary layer flow during transition from laminar to turbulent motion. *J. Fluid Mech.* **3**, 418.

DRYDEN, H. L., and SCHUBAUER, G. B. (1947). The use of damping screens for the reduction of wind tunnel turbulence. *J. Aeron. Sci.* **14**, 221.

DUTTON, R. A. (1960). "The Effect of Distributed Suction on the Development of Turbulent Boundary Layers." ARC R & M No. 3155.

ECKERT, E. R. G., and RODI, W. (1968). Reverse transition turbulent-laminar for flow through a tube with fluid injection. *J. Appl. Mech.* **35**, 817.

ELLISON, T. H. (1957). Turbulent transport of heat and momentum from an infinite rough plane. *J. Fluid Mech.* **2**, 456.

ELLISON, T. H., and TURNER, J. S. (1960). Mixing of dense fluid in a turbulent pipe flow. *J. Fluid Mech.* **8**, 514.

FIEDLER, H., and HEAD, M. R. (1966). Intermittency measurements in the turbulent boundary layer. *J. Fluid Mech.* **25**, 719.

FRAIM, F. W., and HEISER, W. H. (1968). The effect of strong longitudinal magnetic field on the flow of mercury in a circular tube. *J. Fluid Mech.* **33**, 397.

FREYMUTH, P. (1975). Search for the final period of decay of the axisymmetric turbulent wake. *J. Fluid Mech.* **68**, 813.

FULLER, L., and REID, J. (1958). "Experiment in Two-dimensional Base Flow at M = 2.4." ARC R & M No. 3064.

GARDNER, R. A., and LYKOUDIS, P. S. (1971). Magneto-fluid-mechanic pipe flow in a transverse magnetic field. Part 1. Isothermal flow. *J. Fluid Mech.* **47**, 737.

GAVIGLIO, J., DUSSAUGE, J.-P., DEBIEVE, J. F., and FAVRE, A. (1977). Behavior of a turbulent flow, strongly out of equilibrium, at supersonic speeds. *Phys. Fluids* **20**, S179.

GLOBE, S. (1961). The effect of a longitudinal magnetic field on pipe flow of mercury. *Trans. ASME* **83**, 445.

HALL, A. A. (1938). "Measurements of the Intensity and Scale of Turbulence." ARC R & M No. 1852.

HALLEEN, R. M., and JOHNSTON, J. P. (1967). "The Influence of Ratation on Flow in a Long Rectangular Channel—An Experimental Study," Rep. MD 18. Thermosci. Div., Dept. Mech. Eng., Stanford University, Stanford, California.

HARTMANN, J., and LAZARUS, F. (1937). Hg-dynamics. II. *K. Dan. Vidensk. Selsk., Mat.-fys. Medd.* **15**, 7.

HUA, H. M., and LYKOUDIS, P. S. (1974). Turbulence measurements in a magnetofluid-mechanic channel. *Nucl. Sci. Eng.* **54**, 445.

JEFFREYS, H. (1928). Some cases of instability in fluid motion. *Proc. R. Soc. London. Ser. A* **118**, 195.

JOHNSTON, J. P., HALLEEN, R. M., and LEZIUS, D. K. (1972). Effects of spanwise rotation on the structure of two-dimensional fully developed turbulent channel flow. *J. Fluid Mech.* **56**, 533.

KIRKO, I. M. (1965). "Magnetohydrodynamics of Liquid Metals." Consultants Bureau, New York.

KLINE, S. J., REYNOLDS, W. C., SCHRAUB, F. A., and RUNSTADLER, P. W. (1967). The structure of turbulent boundary layers. *J. Fluid Mech.* **30**, 741.

KREITH, F. (1965). Reverse transition in radial source flow between two parallel planes. *Phys. Fluids* **8**, 1189.

LANDAU, L. D., and LIFSHITZ, E. M. (1959). "Fluid Mechanics." Pergamon, Oxford.

LAUFER, J. (1962). Decay of non-isotropic turbulent field. *In* "Miszellaneen de angewandte Mechanik, Festschrift Walter Tollmien," Akademie-Verlag, Berlin.

LAUFER, J. (1975). New trends in experimental turbulence research. *Ann. Rev. Fluid Mech.* **7**, 307.

LAUNDER, B. E. (1963). "The Turbulent Boundary Layer in a Strongly Negative Pressure Gradient," Rep. No. 71. Gas Turbine Lab., Massachusetts Institute of Technology, Cambridge.

LAUNDER, B. E. (1964). "Laminarisation of the Turbulent Boundary Layer by Acceleration," Rep. No. 77. Gas Turbine Lab., Massachusetts Institute of Technology, Cambridge.

LAUNDER, B. E., and STINCHCOMBE, H. S. (1967). "Non-normal Similar Turbulent Boundary Layers," Imp. Coll. Note TWF/TN 21. Dept. Mech. Eng.

LEHNERT, B. (1955). An instability of laminar flow of mercury caused by an external magnetic field. *Proc. R. Soc. London, Ser. A* **233**, 299.

LEUTHEUSSER, H. H., and LAM, K.-W. (1977). Laminar-to-turbulent transition in accelerated fluid motion. *Proc. Congr. IAHR, 17th,* 1977 p. 124.

LEZIUS, D. K., and JOHNSTON, J. P. (1971). "The Structure and Stability of Turbulent Wall Layers in Rotating Channel Flow," Rep. No. MD-29. Thermosci. Div., Dept. Mech. Eng., Stanford University, Stanford, California.

LIEPMANN, H. W. (1943). Investigations on laminar boundary layer stability and transition on curved boundaries. *Natl. Advis. Comm. Aeronaut., Wartime Rep.* **W107** (ACR 3H 30).

LIEPMANN, H. W. (1945). Investigations of boundary layer transition on concave walls. *Natl. Advis. Comm. Aeronaut., Wartime Rep.* **W87** (ACR 4J 28).

LIGHTHILL, M. J. (1970). Turbulence. In "Osborne Reynolds and Engineering Science Today" (D. M. McDowell and J. D. Jackson, eds.), pp. 83–146. Manchester Univ. Press.

LOCK, R. C. (1955). The stability of the flow of an electrically conducting fluid between parallel planes under a transverse magnetic field. *Proc. R. Soc. London, Ser. A* **233**, 105.

LYONS, R., PANOFSKY, H. A., and WOLLASTON, S. (1964). The critical Richardson number and its implications for forecast problems. *J. Appl. Meteorol.* **3**, 136.

McELIGOT, D. M., COON, C. W., and PERKINS, H. C. (1970). Relaminarization in tubes. *Int. J. Heat Mass Transfer* **13**, 431.

MACPHAIL, D. C. (1944). "Turbulence Changes in Contracting and Distorted Passages," RAE Rep. No. 1928.

MAGEE, P. M., and McELIGOT, D. M. (1968). Effect of property variation on the turbulent flow of gases in tubes: The thermal entry. *Nucl. Sci. Eng.* **31**, 337.

MILES, J. W. (1961). On the stability of heterogeneous shear flows. *J. Fluid Mech.* **10**, 496.

MOFFATT, H. K. (1967). On the suppression of turbulence by a uniform magnetic field. *J. Fluid Mech.* **28**, 571.

MOLLER, P. S. (1963). Radial flow without swirl between parallel plates. *Aeron. Quart.* **14**, 163.

MONIN, A. S., and YAGLOM, A. M. (1971). "Statistical Fluid Mechanics." MIT Press, Cambridge, Massachusetts.

MOREAU, R. (1969). On mhd turbulence. In "Turbulence of Fluids and Plasmas" (J. Fox. ed.), pp. 359–372. Polytechnic Press, Brooklyn.

MORETTI, P. H., and KAYS, W. M. (1965). Heat transfer in turbulent boundary layer with varying free stream velocity and varying surface temperature—an experimenting study. *Int. J. Heat Mass Transfer* **8**, 1187.

MORKOVIN, M. V. (1955). Effects of high acceleration on a turbulent supersonic shear layer. *Proc. Heat Transfer Fluid Mech. Inst.*

MURGATROYD, W. (1953). Experiments on mhd channel flow. *Philos Mag.* [7] **44**, 138–1354.

NARASIMHA, R. (1957). On the distribution of intermittency in the transition region of a boundary layer. *J. Aeron. Sci.* **24**, 711.

NARASIMHA, R. (1977). "The Three Archtypes of Relaminarisation," Rep. No. 77 FM 7. Dept. Aero. Eng., Ind. Inst. Sci., Bangalore.

NARASIMHA, R., and SREENIVASAN, K. R. (1973). Relaminarization in highly accelerated turbulent boundary layers, *J. Fluid Mech.* **61**, 417.

NARASIMHA, R., and VISWANATH, P. R. (1975). Reverse transition at an expansion corner in supersonic flow. *AIAA J.* **13**, 693.

NASH-WEBBER, J. L., and OATES, G. C. (1972). An engineering approach to the design of laminarizing nozzle flows. *Am. Soc. Mech. Eng. [Pap.]* **72-FE-19**.

NICHOLL, C. I. H. (1970). Some dynamical effects of heat on a boundary layer. *J. Fluid Mech.* **40**, 361.

OKAMOTO, T., and MISU, I. (1977). Reverse transition of turbulent boundary layer on plane wall of two-dimensional contraction. *Trans. Jpn. Soc. Aerosp. Sci.* **20**, 1.

OWEN, P. R. (1969). Turbulent flow and particle deposition in the trachea. *In* "Circulatory and Respiratory Mass Transport" (G. E. W. Wolstenholme and J. Knight, eds.), p. 236. Churchill, London.

PAO, Y. H. (1969). Undulance and turbulence in stably stratified media. *In* "Clear Air Turbulence and Its Detection," p. 73. Plenum, New York.

PATEL, V. C. (1965). Calibration of the Preston tube and limitations on its use in pressure gradients. *J. Fluid Mech.* **23**, 185.

PATEL, V. C., and HEAD, M. R. (1968). Reversion of turbulent to laminar flow. *J. Fluid Mech.* **34**, 371.

PENNELL, W. T., ECKERT, E. R. G., and SPARROW, E. M. (1972). Laminarization of turbulent pipe flow by fluid injection. *J. Fluid Mech.* **52**, 451.

PERKINS, H. C., and WORSOE-SCHMIDT, P. M. (1965). Turbulent heat and momentum transfer for gases in a circular tube at wall to bulk temperature ratios to seven. *Int. J.Heat Mass Transfer* **8**, 1011.

PERKINS, H. C., SCHADE, K. W., and McELIGOT, D. M. (1973). Heated laminarising gas flow in a square duct. *Int. J. Heat, Mass Transfer* **16**, 897.

POPPLETON, E. D. (1971). "Exploratory Measurements of the Flow in the Wing-tip Vortices of a Lockheed Hercules," Aero. Tech. Note 7104. University of Sydney, Sydney, Australia.

PRANDTL, L. (1929). Einfluss stabilisierender Krafte auf die Turbulenz. *Gesammelte Abh.* **2**, 778.

PRANDTL, L., and REICHARDT, H. (1934). Einflusse von Warmeschichtung auf die Eigenschaften einer turbulenten Strömung. *Dtsch. Forschungsges.* No. 21.

RAMAPRIAN, B. R., and SHIVAPRASAD, B. G. (1978). The structure of turbulent boundary layers along mildly curved surfaces. *J. Fluid Mech.* **85**, 273.

RAMJEE, V. (1968). Reverse transition in a two-dimensional boundary layer flow. Ph.D. Thesis, Dept. Aero. Eng., Ind. Inst. Sci., Bangalore.

RAMJEE, V., BADRI NARAYANAN, M. A., and NARASIMHA, R. (1972). Effect of contraction on turbulent channel flow. *Z. Angew. Math. Phys.* **23**, 105.

RAO, K. N., NARASIMHA, R., and BADRI NARAYANAN, M. A. (1971). The "bursting" phenomenon in a turbulent boundary layer. *J. Fluid Mech.* **48**, 339.

REYNOLDS, O. (1884). The two manners of motion of water. *Proc. R. Inst. G. B.* **11**, 44.

RICHARDSON, L. F. (1920). The supply of energy to and from atmospheric eddies. *Proc. R. Soc. London, Ser. A* **97**, 354.

ROSENHEAD, L., ed. (1963). "Laminar Boundary Layers." Oxford Univ. Press, London and New York.

ROTTA, J. C. (1967). Effect of streamwise wall curvature on compressible turbulent boundary layers. *Phys. Fluids* **10**, S174.

ROWE, M. (1970). Measurements and computations of flow in bends. *J. Fluid Mech.* **43**, 771.

SAJBEN, M., and FAY, J. A. (1967). Measurement of the growth of a turbulent mercury jet in a coaxial magnetic field. *J. Fluid Mech.* **27**, 81.

308   *R. Narasimha and K. R. Sreenivasan*

SCHRAUB, F. A., and KLINE, S. J. (1965). "A Study of the Structures of the Turbulent Boundary Layer with and without Longitudinal Pressure Gradients, "Rep. No. MD-12. Thermosci. Div., Stanford University, Stanford, California.

SCHUBAUER, G. B., and SKRAMSTAD, H. K. (1947). Laminar boundary layer oscillations and stability of laminar flow. *J. Aeron. Sci.* **14**, 69.

SCORER, R. S. (1958). "Natural Aerodynamics." Pergamon, Oxford.

SENOO, Y. (1957). The boundary layer on the end wall of a turbine nozzle cascade. *Am. Soc. Mech. Eng.* [*Pap.*] **A-172**.

SERGIENKO, A. A., and GRETSOV, K. V. (1959). Transition from turbulent to laminar boundary layer. *Dokl. Akad. Nauk SSSR* **125** (RAE Translation No. 827).

SHERCLIFF, J. A. (1956). The flow of conducting fluids in circular pipes under transverse magnetic fields. *J. Fluid Mech.* **1**, 644.

SHERCLIFF, J. A. (1965). "A Textbook of Magnetohydrodynamics." Pergamon, Oxford.

SHIVAPRASAD, B. G. (1976). An experimental study of the effect of "mild" longitudinal curvature on the turbulent boundary layer. Ph.D. Thesis, Dept. Aero. Eng., Ind. Inst. Sci., Bangalore.

SIBULKIN, M. (1962). Transition from turbulent to laminar flow. *Phys. Fluids* **5**, 280.

SIMPSON, R. L., and SHACKLETON, C. R. (1977). "Laminariscent Turbulent Boundary Layers: Experiments on Nozzle Flows," Proj. SQUID Tech. Rep. No. SMU-2-PU.

SIMPSON, R. L., and WALLACE, D. B. (1975). "Laminariscent Turbulent Boundary Layers: Experiments on Sink Flows," Proj. SQUID Tech. Rep. No. SMU-1-PU.

SINGH, P. I., and UBEROI, M. S. (1976). Experiments on vortex stability. *Phys. Fluids* **19**, 1858.

SO, R. M. G., and MELLOR, G. L. (1972). An experimental investigation of turbulent boundary layers along curved surfaces. *NASA Contract. Rep. NASA* **CR-1940**.

SO, R. M. G., and MELLOR, G. L. (1973). Experiment on convex curvature effects in turbulent boundary layer. *J. Fluid Mech.* **60**, 43.

SPENCE, D. (1956). The development of turbulent boundary layer. *J. Aeron. Sci.* **23**, 3.

SREENIVASAN, K. R. (1972). "Notes on the Experimental Data on Reverting Boundary Layers," Rep. No. 72 FM2. Dept. Aero. Eng., Ind. Inst. Sci., Bangalore.

SREENIVASAN, K. R. (1974). Mechanism of reversion in highly accelerated turbulent boundary layers. Ph.D. Thesis, Dept. Aero. Eng., Ind. Inst. Sci., Bangalore.

SREENIVASAN, K. R., and NARASIMHA, R. (1974). Rapid distortion of shear flows. *Aero. Soc. Indian Silver Jubilee Tech. Conf., 1974* Paper 2.3.

SREENIVASAN, K. R., and NARASIMHA, R. (1978). Rapid distortion of axisymmetric turbulence. *J. Fluid Mech.* **84**, 497.

STEINER, A. (1971). Reverse transition of a turbulent flow under the action of buoyancy forces. *J. Fluid Mech.* **47**, 503.

STERNBERG, J. (1954). "The Transition from a Turbulent to a Laminar Boundary Layer," Rep. No. 906. Ballistic Res. Lab, Aberdeen.

STEWART, R. W. (1959). The problem of diffusion in a stratified fluid. *Adv. Geophys.* **6**, 303.

STRICKLAND, J. H., and SIMPSON, R. L. (1975). Bursting frequencies obtained from wall shear stress fluctuations in a turbulent boundary layer. *Phys. Fluids* **18**, 306.

TAYLOR, G. I. (1929). The criterion for turbulence in curved pipes. *Proc. R. Soc. London, Ser. A* **124**, 243.

TETERVIN, N. (1967). "An analytical Investigation of the Flat Plate Turbulent Boundary Layer in Compressible Flow," NOL TR 67-39, Aerodyn. Res. Rep. No. 286. Nav. Ord. Lab., Silver Spring, Maryland.

THOMANN, H. (1968). Effect of streamwise wall curvature on heat transfer in a turbulent boundary layer. *J. Fluid Mech.* **33**, 283.

THOMPSON, B. G. J. (1973). "The Production of Boundary-Layer Behaviour and Profile Drag for Infinite Yawed Wings," RAE Tech. Rep. No. 73091.

TOWNSEND, A. A. (1954). The uniform distortion of homogeneous turbulence. *J. Mech. Appl. Math.* **7**, 104.

TOWNSEND, A. A. (1956). "The Structure of Turbulent Shear Flow." Cambridge Univ. Press, London and New York.

TOWNSEND, A. A. (1958). Turbulent flow in a stably stratified atmosphere. *J. Fluid Mech.* **3**, 361.

UBEROI, M. S. (1956). Effect of wind tunnel contraction on free stream turbulence. *J. Aerosp. Sci.* **23**, 754.

VISWANATH, P. R., and NARASIMHA, R. (1972). "Base Pressure on Sharply Boat-tailed Aft Bodies," Rep. No. 72 FMI. Dept. Aero. Eng. Ind. Inst. Sci., Bangalore.

VISWANATH, P. R., and NARASIMHA, R. (1975). "Supersonic Flow Past a Sharply Boat-tailed Step." Rep. No. 75 FM13. Dept. Aero. Eng. Ind. Inst. Sci., Bangalore.

VISWANATH, P. R., NARASIMHA, R., and PRABHU, A. (1978). Visualization of relaminarizing flows. *J. Indian Inst. Sci.* **60**, 159.

VIVEKANANDAN, R. (1963). A study of boundary layer expansion fan interactions near a sharp corner in supersonic flow. M.Sc. Thesis, Dept. Aero Eng., Ind. Inst. Sci., Bangalore.

WALLIS, R. A. (1950). "Some Characteristics of a Turbulent Boundary Layer in the Vicinity of a Suction Slot," Aerodyn. Note 87. Aero. Res. Lab., Dept. Supply, Melbourne, Australia.

WEBSTER, C. A. G. (1964). An experimental study of turbulence in a density-stratified shear flow. *J. Fluid Mech.* **19**, 221.

WILSON, D. G., and POPE, J. A. (1954). Convective heat transfer to gas turbine blades. *Proc., Inst. Mech. Eng., London* **168**, 861.

ZAKKAY, V., TOBA, K., and DUO, T.-J. (1964). Laminar, transitional and turbulent heat transfer after a sharp corner. *AIAA J.* **2**, 1389.

Author Index

Numbers in italics refer to the pages on which the complete references are listed.

Subject Index

A

Abelian group, 164
Accelerated flow
  burst parameters, 263
  relaminarization, 246–267
Airy function, 18, 89, 91, 93
Amplification, standing wave, 123
Amplitude
  enhancement, 96
  function, 71, 103
Asymptotic approximation, 63–64

B

Bead-spring model, 158–162
Bessel function, 92–93
BKZ fluid model, 155
Blasius law, 296
Body force, 240, 288
Boundary fiber, force, 29
Boundary layer
  convex, 269–274
  flow, 248
  regions, 261
  thickness, 255, 287–288
Boundary value
  displacement, 26–27, 35
  problems, 7–9
  traction problems, 9–11
Bursting, *see* Turbulent bursting

C

Cantilever
  deformation, 8–9
  edge layer, 33
  finite deflection, 45–46
Cauchy stress tensor, 147–148, 159, 174, 191
  simple fluid, 172–173

Cauchy–Green tensor, 171
Caustic barrier, 76–79
Caustic reflection, 79–81
Caustic refraction, 79–81
Caustics, 76–81
  amplitude enhancement, 96
  curve, 86–87
  cusped, 85–87, 91
  phase function, 81–82
  reflection condition, 95–102
Channel flow
  correlation coefficient, 294
  relaminarization, 229
  velocity fluctuation, 294
Characteristic equation, 119, 121, 125–126
  for deformation rate, 190
Coefficient function, 113
  roots, 116–118
  for round island, 114
Coiled pipe, relaminarization, 299
Coiled tube, relaminarization, 222–223
Commutator product, two tensors, 164
Commutative motion
  kinematics, 163–179
  schematic, 168
  stress, 163–179
Configurational convection–diffusion
  equation, 158–159
Convex boundary layer, 269–274
Corolis force, 275–276
Corotating channel, 277–278
Corotating side, 275
Corotational strain
  measure, 152
  rate, 190
  tensor, 149
Correlation coefficient, 231, 294
Couette flow
  base, 198–209
  circular, 193–198
  nonslip boundary condition, 199